Urban Regeneration: Challenges and Opportunities for the Landscape

Urban Regeneration: Challenges and Opportunities for the Landscape

Guest Editors

Celestina Fazia
Carmela Mariano
Francesca Moraci
Kh Md Nahiduzzaman
Laura Ricci
Francesca Perrone

MDPI

Basel • Beijing • Wuhan • Barcelona • Belgrade • Novi Sad • Cluj • Manchester

Guest Editors

Celestina Fazia
Department of Engineering and Architecture
"Kore" University of Enna
Enna
Italy

Carmela Mariano
Department of Planning, Design and Technology of Architecture
Sapienza University of Rome
Rome
Italy

Francesca Moraci
Department of Architecture and Design
University of Reggio Calabria "Mediterranea"
Reggio Calabria
Italy

Kh Md Nahiduzzaman
Citinnov for Integrated Territorial Planning and Smart Cities
Mohammed VI Polytechnic University (UM6P)
Rabat
Morocco

Laura Ricci
Department of Planning, Design and Technology of Architecture
Sapienza University of Rome
Rome
Italy

Francesca Perrone
Department of Planning, Design and Technology of Architecture
Sapienza University of Rome
Rome
Italy

Editorial Office
MDPI AG
Grosspeteranlage 5
4052 Basel, Switzerland

This is a reprint of the Special Issue, published open access by the journal *Land* (ISSN 2073-445X), freely accessible at: https://www.mdpi.com/journal/land/special_issues/2316R7TMFH.

For citation purposes, cite each article independently as indicated on the article page online and as indicated below:

Lastname, A.A.; Lastname, B.B. Article Title. *Journal Name* **Year**, *Volume Number*, Page Range.

ISBN 978-3-7258-4619-1 (Hbk)
ISBN 978-3-7258-4620-7 (PDF)
https://doi.org/10.3390/books978-3-7258-4620-7

Cover image courtesy of Alessandro Agostinelli
Torre del Fiscale area, South-east district of Rome.

Contents

About the Editors

Celestina Fazia

Celestina Fazia. Architect, PhD in Spatial Planning, MS in "Research Technicians specialized in new technologies for land defense". She is an associate professor of Urban Planning (Cear 12/B) at the Department of Engineering and Architecture, Kore University of Enna. She has decades of experience in the field of urban, land, and environmental planning in both research and professional settings. A former TDi Researcher at the DArTe of the Mediterranea University of Reggio Calabria and previously RTD at the Kore University of Enna, she has been a research fellow and adjunct professor at the two universities. She is a member of national and international research groups, including the establishment of the Strait Integrated Area; the Technical Table for the drafting of a Feasibility Study for the improvement of the connection systems in the Strait Area; and the Italian Contribution to the UN Resolution on the city crisis.

She is responsible for numerous research conventions, including international ones, such as "SS_Consumption Stop Soil consumption: between reuse and regeneration", signed between the departments of the Kore University of Enna and the University of Oradea (Romania); 'RaRe_Reuse and regeneration contexts: disused defense sites, infrastructure and brownfield sites' between the departments of the universities Oradea, UM6P Polytechnic University of Morocco, and Kore University of Enna; "AmbeS_1 Environment and Health, protection and safeguard"; MOD_RetEc, with the city of Modica; and A.Ma.Te Sponde, with the DArTe of the Mediterranea University of Reggio Calabria. As a consultant for public entities and companies, she coordinated and drafted the Masterplan for the Strait Area_Consiglio Regionale della Calabria; she was senior consultant for the International Cooperation sector and former member of the EIA-VAS-IPPC Nucleus of the Region of Calabria; and consultant for the Territorial Analysis of the stable crossing system of the Strait of Messina, Parsons T.G.

Carmela Mariano

Carmela Mariano, architect, PhD, is a full professor of Urban Planning at the Department of Planning, Design and Technology of Architecture, Sapienza University of Rome. She is the Director of the School of Specialisation in Natural and Territorial Heritage, Faculty of Architecture. From 2018 to 2024, she was the head of the master's degree program in Architecture—Urban Regeneration. From 2015 to 2021, she was the PDTA Teaching Project Commission Coordinator. She is also a member of the University Research Commission evaluation and the Steering Committee of the Laboratorio Progetto Roma, PDTA Department. She is a member of the Scientific Teaching Council of the Second-level Advanced Professional Course "Natural Capital and Protected Areas. Planning, Design and Management" of the PDTA Department. She is also a member of the Teachers' Board of the PhD program "Planning, Design, Architecture Technology" of the PDTA Department.

She is the principal investigator for many international and national research programs that focus on the following topics: polycentric dimension of the contemporary city and metropolitan territorial governance models; relationships between climate change and urban regeneration strategies as terms of comparison for the innovation and experimentation of urban planning; and the role of the urban project in urban regeneration strategies that can insert invariant elements endowed with urban meaning and strong legibility into the urban palimpsest as the foundation of the generative processes that contribute to the construction of the public city, housing policies, and quality of living.

The results of this research have been published in books, scientific journals, and specialized volumes. She has carried out design and experimentation activities in the field of town and country planning and technical/scientific consultancy on behalf of public bodies as a priority for the in-depth study of the main issues concerning this disciplinary debate.

Francesca Moraci

Francesca Moraci. Architect, PhD in Spatial Planning, MS in Economic Policy and Planning and Fullbright scholar in Economics and Public Policies (NU Boston). She is a full professor of Urban Planning at the Mediterranea University of Reggio Calabria, where she has held apex academic positions; she was a member of the Board of Directors of "Ferrovie dello Stato Italiane" and ANAS spa (2015/21) and a member of the MIMS Commission for the reform of the national urban planning law, urban standards, and the "Testo Unico dell'Edilizia". She has coordinated national research on competitive calls (PON, POR, and URBACT, including the goal related to the PNRR Innovation System pilot project TECH4YOU technologies for climate adaptation). She has served as a consultant for public bodies, ministries, and international engineering companies. She has participated in numerous studies, projects, and general and sector plans. She is a member of numerous Technical and Scientific Committees of national and international interest (Eurispes Infrastructure and Logistics Observatory, Abitacolo, QVQC, CRPPN Sicily Region, Parsons Transportation group Ltd; Sustainable Infrastructure Association; Technical Committee Permanent Interregional Conference Strait Area, Port Committee AP of Messina; etc.). She was among the 15 experts from MIT for the National Strategic Plan for Portuality and Logistics. She is part of the 100 women who are projected to change Italy. She has received various awards and recognitions for scientific, professional, and managerial activities. She is a full member of the National Institute of Urban Planning and vice president of the scientific society Accademia Urbana, which she co-founded. She is a member of the CS of the States General of Women and a member of the Belisario Foundation. She is curator of "La Biennale dello Stretto" and the author of numerous publications.

Kh Md Nahiduzzaman

Kh Md Nahiduzzaman is an Associate Professor at Citinnov for Integrated Territorial Planning and Smart Cities, Mohammed VI Polytechnic University (UM6P), Morocco. He has previously held academic positions at the University of British Columbia (UBC) Okanagan, Canada; King Fahd University of Petroleum & Minerals (KFUPM), Saudi Arabia; and the Royal Institute of Technology (KTH), Sweden.

His research focuses on urban and territorial planning, human-centric smart cities, and climate resilience. He has led projects totaling over USD 5 million in research funding and authored more than 100 publications, including journal articles, book chapters, and conference proceedings. His work on flood resilience, floodplain mapping, and land use planning has contributed significantly to policy development in British Columbia.

Prof. Nahiduzzaman collaborates globally with governments, industries, and communities on applied research projects. He serves as the Editor-in-Chief of *City Development: Issues and Best Practices* (ICCCASU), "Executive" and "Book Review" Editor of the *Journal of Urban Management* (Elsevier), and holds editorial roles in frontier journals, including *Journal of Urban Planning and Development* (ASCE), *Discover Cities* (Springer), etc.

His recent books—*Making Sense of Planning and Development for Post-Pandemic Cities* (Springer, 2024) and *The Forms of Resilience in Urban Contexts* (La Penseur, 2025)—critically explore transformative planning and urban resilience in evolving global contexts.

Laura Ricci

Laura Ricci, Architect, PhD in Territorial and Urban Planning, is a full professor of Urban Planning at Sapienza University of Rome. She was the Director of the Department of Planning, Design and Technology of Architecture (PDTA) (2015-2021) and the Director of the School of Specialisation in Natural and Territorial Heritage at Sapienza University of Rome (2016-2023). Since 2022, she has been coordinator of the PDTA doctoral program, and since 2017, she has been the director of the master's program "Natural Capital and Protected Areas: Planning, Design and Management". She is the president of the Scientific Society Accademia Urbana. She has been active in design and experimentation in the field of urban and territorial planning. She served as the general consultant to the City of Rome for the new PRG of Rome (1994-2012). She was a member of numerous expert commission and coordinated numerous university, CNR, and third-party research projects. She is principal investigator of Thematic Line 4 "*Actualization_Patrimonio culturale e rigenerazione urbana. Per una dimensione multidisciplinare e sperimentale*" of the Extended Partnership 5 - CHANGES - Spoke 8, Sustainability and Resilience of Tangible Cultural Heritage, and of the Thematic Line 4 "*Nuove Regole, strumenti, procedure, riferimenti operativi del piano urbanistico per un approccio ecosostenibile alla rigenerazione urbana*"—Next Generation EU, Rome Technopole, as well as of several executive protocols of cooperation agreements with foreign universities. She has written books, articles, and essays, particularly focusing on topical issues and disciplinary innovation, including local planning innovation, governance of metropolization processes, the urban project, urban regeneration and urban welfare, and the PRG of Rome. She carries out intensive dissemination activities through the organization and scientific coordination of numerous national and international conferences and seminars.

Francesca Perrone

Francesca Perrone, landscape architect and PhD, is a researcher in urban and landscape planning (Cear 12/B) at the Faculty of Architecture, Department of Planning, Design and Technology of Architecture (PDTA), Sapienza University of Rome, Italy. She earned her Bachelor of Science in Landscape Architecture with honors and recommended publication from Sapienza University of Rome. She then obtained a postgraduate degree in "GEOinformation and Geographic Information Systems for systemic analysis of territory and geographic data" (GEO-GST) at the Department of Civil Engineering and Computer Science Engineering, University of Rome Tor Vergata, and is a holder of the international Esri ArcGIS User certificate. She completed her PhD in Planning, Design and Technology of Architecture, Sapienza University of Rome, where she subsequently conducted postdoctoral research.

She has engaged in issues ranging from land control to soil ecosystem services. She was a professor on contract at the Faculty of Architecture and the master's degree program "Urban planning for public administration" (URBAM), Sapienza University of Rome. She has conducted projects focusing on green infrastructures, urban regeneration, and public spaces. She contributed to the VAS PN Metro Plus e Città Medie Sud 2021–2027 that is currently under development. She has participated in national and international research groups. Her research interests center on sustainable urban, territorial, and landscape planning; territory management; land use; land take; soil ecosystem services; and cultural heritage as a strategic lever for the implementing urban regeneration strategies and projects.

Preface

Numerous aspects come into play when addressing the issues of urban regeneration and various interventions are necessary to make cities resilient and ready to respond to environmental, economic, and social challenges. Resilience is understood as a "transversal condition" in the reorganisation of a city, as the ability to deal with adverse events and externalities is induced by programmatic choices that will significantly change the reference structures and scenarios. The complexity of new urban scenarios is reflected in the increased difficulty in both understanding and describing the connotations of the contemporary city and thinking of new policies capable of capturing the signals of change and managing various aspects.

The city was almost always saturated and unable to update the range of services and the offer of public spaces in the past, which is something that has changed in recent times. Above all, cities that show a historical delay in the adaptation of public spaces, services, and collective equipment are unprepared at present because they are unable to keep up with the change dynamics. It is essential to recognise the changing demand for public spaces, which corresponds to the evolution of living models, to analyse the potential of the growing diffusion of digital technology and the availability of local spaces in the social transition towards the future city. Consequently, this will enable an investigation into how the development of new digital media spaces influences design.

For these reasons, it will be necessary to do the following:

- reduce land consumption and promote de-impermeabilization;
- encourage re-naturalisation, focusing on enhancing ecosystem services in land use activities;
- restore landscape and green infrastructure by emphasising their contribution to climate mitigation;
- implement multi-level regulations and behavioural pathways for combating climate change and strategies for the ecological transition of territories;
- incentivize the reuse of existing urban regeneration interventions as a strategic alternative to new land consumption. This Special Issue focuses on welfare tools, the supply of services and mobility, the landscape standard, the environment, the restoration of the landscape, historic architecture, and cultural heritage. The enhancement of the landscape is an opportunity to regenerate cities.
    - How important is space in the age of advanced technology and what are the expected requirements of its users in terms of quality, shape, size, and function?
    - Is one function and one type of user enough? Is the traditional organisation of space still feasible, and what are its uses and symbols? Above all, is the space residual (what remains of the physical commitment of roads, public buildings, and private spaces), "public", or is it an urban connective that both welds and cuts through the city and places demands on it?
    - If we shift the focus away from the undifferentiated demand for basic services and towards the increasingly diversified almost individual requests (the hyper-technologies), the issue remains complex and prompts numerous reflections.

Measures and policies to activate city modernization processes can be implemented through the urban regeneration interventions envisaged by the new European Programming. In the transition towards the era of prevailing technology—the era of the smart city—urban public spaces and services must renew themselves by looking at the challenges that the contemporary world poses and that the future demands by developing proactive performing, and intelligent cities. At the same time, the city must be resilient.

Therefore, the cities in which we live today must renew themselves and face the complexity of the issues that arise by the fact that they are "contemporary" cities, cities of change. However, there exists a structural gap: the city is unable to follow the "times" and the speed of the social and physical transformations of the city itself.

In the absence of a punctual and adequate design response, the city undergoes planning inertia, a lack of adequate redevelopment interventions, and a rapid extension of the residential, commercial, industrial, and touristic fabric. Cities and territories suffer and experience dilemmas and profound transformations: segregation, immigration, integration, the density and heterogeneity of the urban settlement, the transformations of the use of time and space in communities and cities, urbanisation and sprawl, mobility and new populations in transit, being perpetually in crisis, the environmental paradigm, the new information and communication technologies, and the effect of not being a city.

The complexity of the new challenges to be faced and how to transform the city has opened a discussion on the current issues that animate the scientific debate and the urban/territorial policy agenda.

The idea of developing a Special Issue arose from a variety of issues addressed in the context of the debate on urban regeneration and its relationship to territory, landscape, and current and future transformations.

The topic has in fact solicited points of reflection and in-depth analysis in numerous contexts, stimulating interesting multidisciplinary contributions from faculty members and doctoral students that have been accepted and published because of their scientific contributions to the topics. The goal of this Special Issue is to yield scholarly knowledge that provides insights into the advances in the models of urban regeneration and innovative approaches for resilience for the enhancement of the landscape as an opportunity.

Celestina Fazia, Carmela Mariano, Francesca Moraci, Kh Md Nahiduzzaman, Laura Ricci, and Francesca Perrone
Guest Editors

land

Article

Cultural Landscape as a Resource for Urban Regeneration in Rupea (Romania)

Georgeta Gabriela Rățulea, Codrina Csesznek, Mariana Borcoman and Daniela Sorea *

Department of Social and Communication Sciences, Transilvania University of Brașov, 25, Eroilor Bd, 500030 Brașov, Romania; gratulea@unitbv.ro (G.G.R.); codrina.csesznek@unitbv.ro (C.C.); m.borcoman@unitbv.ro (M.B.)

* Correspondence: sorea.daniela@unitbv.ro

Abstract: Cultural heritage plays a key role in communities' sustainable development. The culture-led development highlights the local cultural resources and specifics while being assisted by contemporary tourist interest in niche offers. At the same time, culture-led development could reinforce a process of urban regeneration. The purpose of this paper is to highlight the urban regeneration potential of culture-led development in the case of a small town from Transylvania (Romania), Rupea, by identifying local characteristics that define this town and its surroundings as a cultural landscape and also by suggesting methods for capitalizing on this cultural landscape in heritage tourism. Data collected from six interviews with cultural stakeholders, bibliographic research on archaeological discoveries, and local tourism potential, as well as through ethnographic methods, support the approach of the Rupea area as a cultural landscape. The main dimensions of this cultural landscape are the interethnic character of the area and the multitude of archaeological discoveries that indicate its habitation in the Paleolithic. Tourist capitalization could support the urban regeneration of Rupea in a culture-led development approach by arranging routes that highlight the specifics of the Romanian, Saxon, Hungarian, and mixed villages in the Rupea area and/or the points of archaeological interest.

Keywords: cultural landscape; culture-led development; niche tourism; multiethnic space

1. Introduction

Cultural heritage is an identity resource and a source of community cohesion [1]. It is an instrument of legitimization [2] and support for local cultural diversity [3], strengthening people's connection with their places of origin [4]. An effective approach to cultural heritage, which is dynamic, requires the combination of conservation expertise with cultural management expertise [1]. Conservation should be for the benefit of the local people, and their involvement is a necessary condition for keeping alive and enriching the heritage [5] as a resource for sustainable development and well-being [6]. In this context, participatory practices come naturally [2]. Heritage, which has a strategic role in local socio-economic development [7], is linked to cultural institutions, local people, and communities alike, and the management of heritage resources should ensure their intergenerational transmission [8].

Nowadays, there is a clear trend toward using culture as a tool for development [9,10]. Cultural resources are considered a guarantee of sustainable development [11], and culture-led development discourse largely takes into account the specificity of local communities and cultures while placing urban and rural development strategies under the umbrella of culture-led global policies [12–14]. The space of possible human cultures is vast, but some cultural configurations are more consistent with cognitive and social constraints than others. This leads to a "landscape of possibilities that our species has explored over millennia of cultural evolution" [15] (p. 1). Cultural management thus becomes the support of sustainable regional development [16]. This development is boosted by contemporary tourist interest in differentiated and personalized, even niche products [17].

Niche tourism is more sustainable and ethical than mass tourism [18]; controlled-size groups [17] are smaller and more experienced in niche tourism [18]. Niche tourists have similar needs and special interests related to authenticity [19]; they seek authentic experiences [17,18]. Niche tourism, as an expression of special cultural interests, could even represent a solution for preserving authenticity [18]. For example, speaking of cultural niche tourism, culinary traditions could be harnessed, given their potentially crucial role in tackling issues linked to food and landscape [20].

Unlike mass tourism, which is based on the generally recognized reputation of the destinations to be visited and which, through the huge flows of tourists, has imposed an industrial approach to related services, in niche tourism, the attractiveness is linked to the possibility of differentiation and personalization of the offer [17]. From this perspective, niche tourism, which represents an alternative to mass tourism, has great potential to generate emerging forms. It is less aggressive than mass tourism in terms of its impact on the lives of communities in the vicinity of tourist attractions. Furthermore, niche tourism, such as that centered on cultural heritage, can contribute to the imposition of a sustainable approach to tourism [19].

Heritage tourism is a niche within cultural tourism [17], which "relies on living and built elements of culture and folkways of today, for they too are inheritances from the past; other immaterial heritage elements, such as music, dance, language, religion, foodways and cuisine, artistic traditions, and festivals; and material vestiges of the built and cultural environment, including monuments, historic public buildings and homes, farms, castles and cathedrals, museums, and archaeological ruins and relics" [21] (pp. 3–4). Ethnographic tourism is part of heritage tourism. It is centered on traditional settlements, ethnic groups well defined in the territory, and traditional cultural practices [22].

In this study, we highlight the urban regeneration potential of culture-led development in the case of a small town in Transylvania (Romania), Rupea. For this purpose, we aim to achieve two objectives: to identify the local characteristics that define the city of Rupea and its surroundings as a cultural landscape, and to design methods for capitalizing on this cultural landscape in heritage tourism.

2. Literature Review

Tourism represents a strategic way to revitalize areas in economic decline [18] and could develop low-income communities [23]. The first initiatives focused on the idea of using cultural resources for economic and social development emerged in large cities that were concerned with becoming more attractive in the increasingly competitive global economic market [14,24]. However, in recent decades, culture-led development strategies that were based on cultural heritage, tourism, and creativity have been frequently implemented in rural communities as well [25]. According to Lysgård (2019), "Most research on culture-led policies has examined them in urban contexts, especially in large cities, but studies of the construction of such policies in small towns and rural communities are expanding" (p. 11). Referring to culture-led policies guiding development, the same author points out that these are parts of a global policy discourse reconstructed through local traditions and practices, materialization, and institutionalization [13].

Other authors [14,24,26,27] highlight the link between the concept of "culture-led development" and newer concepts such as "creative city", "cultural vibrancy", "creative economy", "creative districts", "creative tourism", or "sustainable culture", which have recently emerged and generated innovative approaches to developmental processes amid increasing competition between large cities on the international market as well as concerns about sustainable development.

Tourism plays an important role in this approach to development based on the valorization of cultural resources, which is the case for both cities and peri-urban areas, the latter being "characterized by emerging consumption spaces with leisure and tourism facilities and residential communities" [24] (p. 1). Yet, beyond their economic function, led-culture strategies and actions can contribute to community development in terms of increasing community participation, strengthening local identity, consolidating social networks, and enhancing social inclusion [24,28]. We consider that culture-led development, as previously described, fits very well with "asset-based community development" [29], an approach that starts with identifying local resources and consists of bottom–up actions and decisions for community development, having community members at its core and thus encouraging participation as well as the development of local networks; this approach differs from what the authors mentioned above have called the "needs-based approach" (p. 99), which relies more on external interventions for change and is based on top–down decisions and actions, with little community participation.

When studying the landscape, the first association that usually comes to mind is with geography, but the landscape is a major object of interest for community development as well as for tourism, taking into account that "cultural landscapes provide abundant and diverse ecosystem services for human well-being" [30] (p. 1). In the past decades, we have witnessed the development of traditional tourism that is influenced by tourists' interests in authenticity [31] (p. 254).

The concept of cultural landscape was introduced to academia by the German geographer Friedrich Ratzel (1895–1896) and became central to the Berkeley School of Geography [32]. Its classical definition was given by Carl O. Sauer in 1925, cited in [33] (p. 1147): "The cultural landscape is fashioned from a natural landscape by a cultural group. Culture is the agent; the naturals are the medium; the cultural landscape is the result".

According to the World Heritage Convention, cultural landscapes represent the "combined works of nature and of man", and "they are illustrative of the evolution of human society and settlement over time, under the influence of the physical constraints and/or opportunities presented by their natural environment and of successive social, economic, and cultural forces, both external and internal [...]. Cultural landscapes often reflect specific techniques of sustainable land use, considering the characteristics and limits of the natural environment they are established in and a specific spiritual relation to nature" [34].

According to the Cultural Landscape Foundation [35], "cultural landscapes are landscapes that have been affected, influenced, or shaped by human involvement" (p. 1). They may be of four types: Designed Landscapes, Ethnographic Landscapes, Historic Sites, and Vernacular Landscapes. The same given landscape may have characteristics of several types of cultural landscapes. Defining landscape as an area means that we are talking about a territory that is well-defined and organized. Also, speaking of the perception of people, that means that must be taken into consideration their aesthetic judgment. Different landscapes result from the interaction between natural processes and human activities. When we talk about landscapes, we cannot accomplish this without considering history, economy, or ecology. A landscape can be improved with cultural elements, and its design results from the treatment of the external form of traditional cultural elements in order to achieve the aesthetic effect of highlighting and amplifying the traditional cultural elements [36]. In traditional cultural landscapes, which are nowadays objects of interest for tourism and implicitly lead to community development, an important aspect is connecting landscapes with their broader heritage surroundings. This focuses on historical trails and their adjacent cultural and natural heritage as a holistic cultural landscape [37].

A cultural landscape that could be built around a small city, as is the case in our study, can also be approached in association with the concept of "urban regeneration". Culture-led development is a fundamental dimension of current strategies for urban regeneration as well as for increasing the competitiveness of cities in the economic market; culture is seen as "the crucial ingredient" in the processes of reinventing the image of cities, attracting tourists, and increasing the quality of life of their inhabitants [10].

As a former socialist country and as part of the CEE (Central and East Europe) region, Romania has followed a similar path of urban development as many other cities in this part of the world [10]; thus, unlike Western European cities where culture has become an important factor of socio-economic regeneration, some countries of Eastern Europe have turned into a "Wild East" after the fall of communism, a change characterized by "incoherence, deregulations (...), and clientelism" [10] (p. 73) in the field of urban development. In such a context, culture has played a marginal role in urban regeneration and local development strategies. In Eastern Europe, some traditional settlements were subject to uncontrolled structural and spatial transformations, which caused the deformation of centuries-old spatial systems of high cultural and natural value. This is why local initiatives and actions in the field of industry policy are supposed to play a significant role in consolidating the image of a harmonious landscape in public awareness [38]. However, matters have started to change lately as Eastern European countries have become increasingly attracted to recognizing and exploiting their cultural capital [10]. But we must also take into consideration another challenge encountered by the conservation of cultural landscapes, namely urbanization, and bear in mind that many traditional cultural landscapes worldwide are currently undergoing rapid urbanization [30]. Therefore, the need to better care for the urban landscape as a cultural, material, and regenerative resource is urgent and inevitable [39].

A solution cultivated across Europe in the last decades for protecting the cultural landscape is citizen participation, which has become a formal requirement in landscape and heritage planning all over Europe. The concept of landscape is to be elaborated with regard to the ways in which embodied practices manifest themselves and create place and the intertwinement of the cultural and the natural [40]. The European Landscape Convention (2000) [41] encourages public participation in defining landscape heritage values and identifying strategies for its protection, management, and planning [42]. Furthermore, the traditional knowledge and the ancestors' ways of life recognized as cultural heritage have become a driving force towards achieving the 2030 Agenda for Sustainable Development [43]. According to Torralba et al. (2023) [44], landscape approaches are gaining momentum on both scientific and policy agendas. The concept of landscape is to be elaborated with regard to the politics of the community [45], since the protection of cultural landscapes is a challenge for local management. For management to succeed, management approaches must recognize and respect the core values and experiences that are at the heart of culturally significant landscapes [46]. The challenge is to assess the values and keep the balance between tourism development and cultural heritage protection [40].

As far as Romania is concerned, investments in both culture and culture-led development are important leads for local developmental strategies. At a national level, there are strategic objectives aimed at "protecting and enhancing cultural heritage, contemporary creation, and cultural diversity" [47] (p. 6), as well as "implementing reforms and investments in culture" [47] (p. 13).

As Sepe (2013) argued, urban regeneration should be closely linked to the history and culture of the place in order to shape or strengthen the identity of the place, and future design strategies should focus on both physical constructions and the built environment, as well as on connections between users and intangible elements such as history, cultural identity, memories, and emotional experiences [48].

3. Materials and Methods

The concept of "asset-based development" [29] has been applied as a starting point in the methodological approach in order to identify the characteristics of Rupea's area that could be considered local resources able to contribute to the definition and promotion of this zone as a cultural landscape. As previously mentioned, this concept denotes a form of community development that is focused on identifying and using local resources while emphasizing community participation as well as bottom–up development strategies. According to the authors mentioned above, the asset-based approach means that stakeholders and community residents are involved in the process of identifying resources and determining how these assets can best be used for community development. Pioneered by Kretzmann and McKnight in the early 1990s, asset-based community development is anchored in local assets and capacity building [49]. Jaye et al. (2022) [50] consider that the asset-based approach is very suitable for rural areas and remote environments because it is oriented towards the assets that can be found in such communities and areas. Jakes et al. (2015), cited in [50], show that "the advantages of asset-based approaches include the recognition of local knowledge, resources, and leaders, and the identification of historical narratives and their enduring influence" (p. 285). Thus, this approach is "citizen-led, relationship-driven, and capacity-building" [29] (p. 100), which makes it very suitable as a starting point for our socio-ethnographic study that focuses on the idea of culture-based development as well as the potential of an area as a cultural landscape.

The data we consider relevant for the purpose of this study come from three sources:

(1) Interviews with cultural stakeholders.

Between July 2022 and April 2023, we conducted non-directive interviews with three members of the staff of cultural organizations currently working in Rupea (logged I1–I3): the Romanian local resident acknowledged as an authority in the field of ethnographic traditions and author of a recent paper (2023) [51] on houses in Rupea (I4), and the retired technician who had prepared the construction documentation, rehabilitation, and commercial transactions for houses in the Rupea area for over 50 years (I5). We also conducted an in-depth interview with an archaeologist directly involved in the archeological excavations of recent years in the perimeter of Rupea and in its immediate vicinity, co-author of many articles on the results of these excavations (I6) [52–62]. All these stakeholders are well-known in the Rupea area for their involvement in the cultural field. We have analyzed the interviews thematically.

(2) Document analysis. We have covered the bibliography related to the archaeological discoveries in the area, from basic documentation [63,64] to recent articles [52–62]. We have also used the results of older archive research, capitalized by one of the authors of the study in two medieval history papers about Rupea and Sighișoara [65,66], and more recent monographic articles regarding the tourist potential of the Rupea area [67–71].

(3) The ethnographic method. We have made numerous observations on-site and had plenty of informal discussions with the locals. More specifically, in 2019–2023, we conducted several working visits to Rupea, where we had informal meetings and conducted short interviews with citizens interested in the cultural life and development of the town and the area. We thus collected qualitative data in the form of opinions and social representations about cultural heritage resources from more than 25 citizens, using the snowball method for their selection. During these conversations, the six stakeholders mentioned above were recommended for more in-depth research. We have also capitalized on our own knowledge regarding the Rupea area, given that two of the authors come from this area. They know it very well and are concerned with studying its historical and sociological aspects.

Data analysis consisted of two stages. First, we identified those pieces of information regarding cultural resources that could be used in future culture-led development processes, particularly information that carries the purpose of strengthening the idea of cultural landscape in the Rupea area. We combined data from the three sources mentioned above into a unitary and coherent presentation. We indicated the source for each accessed document. Likewise, in the case of the information collected from the processing of the interviews, we indicated from which of the interviews the opinions expressed come.. Explicitly unassigned information has been obtained through the ethnographic method. The second stage involved a creative approach aimed at building potential cultural tourism routes that are easy to implement and carry a participatory dimension in the sense of involving the locals in the processes of sustainable local development.

4. Rupea—General Description of the Area

Rupea is located in Brașov County, in southeastern Transylvania. It is a town that represents the economic, legal (through the Court of Rupea), and cultural center of the area under the same name. This status is due to Rupea's location on the northwest side of the county, at a relatively long distance from the nearest cities in the counties of Brașov and Mureș (Brașov, Făgăraș, and Sighișoara). Brasov is the center of the ethnographic area of Țara Bârsei and the county seat of Brașov. It has, like Rupea, a history of multi-ethnic settlement but is much larger and more socio-economically developed than Rupea. Large industrial enterprises have been operating in Brașov since the first half of the last century, and it is currently one of the most important tourist destinations in Romania. Făgăraș is the center of another well-known ethnographic area in Romania, namely Olt Land; the town of Făgăraș is of medium size, and the area has a majority Romanian population. Sighișoara is a medium-sized town in Mures County. Being an important tourist destination for its significance as a former Saxon chair and for the medieval architecture preserved in its historical center, the medieval fortress is still inhabited today; in the whole of Mures County, Hungarians represent a significant part of the population [72].

At the 2022 population census in Rupea, 4907 inhabitants were registered, a decrease of 6.87% compared to the previous census, the one in 2011. Among the inhabitants, 3137 (63.9%) declared themselves Romanian, 706 (14.4%) Hungarians, 484 (9.9%) Roma, and 52 (1.1%) Saxons [73]. The Rupea area represents 22% of Brașov County and includes, in addition to the city of the same name, 12 townships (TAU, Territorial Administrative Unit) [69] (Figure 1).

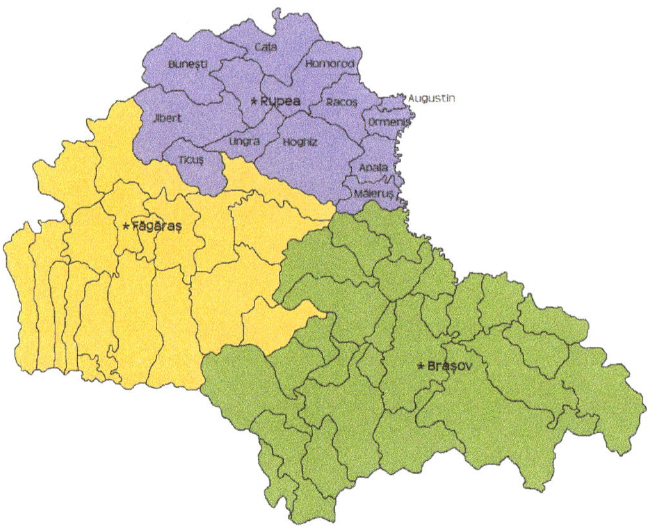

Figure 1. Rupea, with its TAU, in the north of Brașov County. Own adaptation according to the information in [69], on the support https://ro.wikipedia.org/wiki/List%C4%83_de_comune_din_jude%C8%9Bul_Bra%C8%99ov (accessded on 25 October 2023).

To the east and northeast of the Rupea area, towards the Eastern Carpathians, there is an area mainly inhabited by ethnic Hungarians. To the west lies the Hârtibaciului Plateau, and to the southeast, Țara Bârsei, areas with a high density of Saxon villages. To the southwest, the Rupea area borders Țara Făgărașului, a Romanian residential area that has been well-established throughout the history of Transylvania [74–78]. To the northwest and north lies the Târnave Plateau, also a predominantly Romanian living area.

The ethnic distribution of the population in the Rupea area is slightly different from that in the city: 41.7% Romanians, 23.1% Hungarians, 34.0% Roma, 1.2% Germans, and 0.1% other minorities [70]. Therefore, the Rupea area is multi-ethnic. There are Romanian villages, predominantly Hungarian villages, and villages that were predominantly Saxon, currently inhabited by Romanians and Roma. In fact, in almost all villages in the area, Roma people live as a minority group.

Romanians represent the local population. The Romanian people have come into being in the same territory they live in today. According to the dominant historical theory, the process of ethnogenesis took place in the first half of the first millennium AD, after the conquest of Dacia by the Romans, led by Emperor Trajan [79,80]. Szeklers have been attested in the area since the 12th century as defenders of the intra-Carpathian Hungarian kingdom's borders. In this context, they received privileges and the right to manage the territory on the eastern border of Transylvania [80]. After 1900, the Szeklers referred to themselves as Hungarians. The Saxons were colonized in the area by Hungarian royalty in the 13th century. *Andreanum Diploma* (1224) indicates the locality of Drăușeni, located east of Rupea, as the border of Pământul Crăiesc. The direction of the Saxon colonization and its stages can be followed with the help of the antiquity of the Romanesque-style constructions in the territory [81]. Following this trail helps us draw the conclusion that the settlement of the Saxons in the Rupea area is subsequent to their settlement in Sibiu—the most important Saxon fortress in Transylvania—and Cincu—the easternmost chair of the Saxon region of Sibiu. Until the enforcement of the *Edict of Concivility* issued by Emperor Joseph II in Transylvania, the Saxons lived in the central area of Rupea, and the conterminous areas were occupied by the local Romanians and Szekler craftsmen who came later in the 18th century from the neighboring villages.

After World War II, the Saxons in the Rupea area began emigrating to Germany. Although the process developed slowly in communist Romania, it intensified after 1989. As a result, the share of ethnic Saxons in the Rupea area has significantly decreased. This decrease is visible in the results of successive censuses of the Romanian population in the period 1956–2021 (Table 1), more precisely comparing the 1977 census with that of 1992. Detailed information regarding the Roma in the area is missing. Their presence in the city of Rupea is recorded around 1800 [65], where they are described as aids of the Saxons and Romanians in various household chores.

Table 1. Population of Rupea by ethnicity. Sources: [72,82].

Census Year	Stable Population	Romanians	Hungarians, Szeklers	Roma	Saxons	Others
2021	4907	3137	706	484	52	-
2011	5269	3591	975	360	82	-
2002	5759	4063	1245	338	110	-
1992	6326	4331	1428	371	193	-
1977	6640	3617	1502	242	1269	3
1966	6273	3414	1455	69	1327	2
1956	4691	2532	602	162	1385	3

According to the data available at the National Institute of Statistics, in 2022, the Rupea area had the accommodation capacity shown in Table 2 below. The table also shows the number of tourists who benefited in 2022 from the available accommodation. There are no available data for four of the TUAs in the area (Apața, Augustin, Cața and Ormeniș) nor for the number of tourists staying at motels in Rupea or in guesthouses in Homorod and Jibert.

Table 2. Accommodation capacity and the number of tourists staying in the Rupea area in 2022. Source: [82].

Locality/TUA	Total Places	Hotels: Number of Places/Number of Tourists Accommodated in 2022	Motels: Number of Places/Number of Tourists Accommodated in 2022	Tourist and Agritourism Guesthouses: Number of Places/Number of Tourists Accommodated in 2022	Apartments and Rooms for Rent: Number of Places/Number of Tourists Accommodated in 2022
Rupea	172	90/879	74/-	8/1096	-
Bunești	109	-	-	91/5107	18/241
Hoghiz	68	-	-	68/632	-
Homorod	16	-	-	16/-	-
Jibert	20	-	-	20/-	-
Măieruș	52	-	36/490	16/542	-
Racoș	12	-	-	-	12/352
Ticuș	8	-	-	-	8/585

5. Results

5.1. The Cultural Heritage Resources of the Rupea Area

In the center of Rupea, there is a fortified evangelical church dating from the beginning of the 14th century [67]. Like other churches belonging to the Saxons in Transylvania, this one was a Catholic church until the parishioners joined the Reformation. There is also a Roman Catholic church in the city, dating from the beginning of the 19th century, built on the land of a former Franciscan monastery from the 18th century that had been destroyed by a fire. It is a baroque church located near the pedestrian access road to the Rupea citadel from the city. Its parishioners are Hungarian. So are the parishioners of the Unitarian and Reformed (Calvin) churches in the city, currently housed in newer buildings.

The Orthodox from Rupea have two churches. They are parish churches built in the 18th and 19th centuries, respectively. The oldest of these, dedicated to Saint Nicholas, was built as an Orthodox church but functioned as a Greek Catholic church from 1819 to 1948 [51]. This change in confession was the result of the establishment of the Romanian Church, united with Rome in Transylvania, as part of the Catholic Church. At the end of the 17th century, the Orthodox Metropolitan of Transylvania converted to Catholicism together with part of the subordinate clergy, accepting the proposal of Catholic Vienna to grant privileges to the Romanians who followed their Metropolitan and to guarantee the preservation of the specific Greek (Byzantine) practices of worship in the newly established church, the Greek Catholic church [83].

The process of the Romanians' conversion to Greek Catholicism continued until the 19th century and was accompanied by martyr resistance and reprisals from the Austrian army's side. The Romanians from Rupea, who refused to convert to Greek Catholicism (I4), built a new Orthodox church at the end of the century [51]. They first built a small wooden church, and then in 1892, they built a stone and brick church around it (I4). The communist authorities installed in Romania after the Second World War banned the Greek Catholic cult and forced the priests and believers to convert to Orthodoxy. The Greek Catholic places of worship turned into Orthodox places. After 1989, the Greek Catholic Church was recognized by the Romanian state, and in many areas of the country, its old parishes were re-established. This did not happen in Rupea. Both Romanian churches are currently Orthodox.

The stakeholders interviewed and all the citizens we questioned agree that the current main local tourist attraction is the Rupea Fortress. It is a medieval fortress transformed into a fortress with three enclosures by the Saxons. The fortress had dwellings that could be used as needed in the 10th century. The citadel was rehabilitated with European funds between 2009 and 2013 and is currently visited by a multitude of tourists [84]. It is visible from both directions from the European road E60 (Figure 2).

Figure 2. Rupea Fortress. Source: authors' collection.

A section of the Brașov Ethnography Museum [85] operates in the city of Rupea under the name of the "Gheorghe Cernea" Ethnography Museum. Separate from it, a private collection of Romanian clothes, fabrics, furniture, ceramics, tools, and icons of local provenance from the 18th–20th centuries could be exhibited. This collection was gathered by Ana and Vasile Borcoman, a couple from Rupea who work in Rupea as teachers and are housed on their property in the city. As most of the locals claim, the two were important actors in shaping the local residents' interest in the preservation of cultural heritage.

Two Romanian cultural associations are active in the city: "Junii Cetății" ("The Striplings of the Fortress"), which derives from the Rupea House of Culture old men's choir (founded in 1963), and "Ramidava XXI", a more recent association. The activities of the first-mentioned association are focused on musical performances and presentations of popular customs and traditions [86]. Association members consider themselves responsible for keeping and perpetuating the tradition. Among them are elderly people, acknowledged and respected as connoisseurs of popular customs, folk costumes, and practices related to significant events in the life of the community. They have participated in numerous shows in Romania and other European countries, "Poland, Belgium, Germany, beautiful countries" (I1). Many of their costumes are "over 100 years old and still retain their beauty", carefully preserved in the old dowry boxes of local brides (I1). Others were recently sewn by "women who still have patience and skill" (I1). The women in the association are very attentive to the significance of the garments; they know what piece can be worn by whom and in what context. Worn by ignorant, alienated clothes lose their "true value", they become simple festive clothes (I1). Part of the public success of the association's artistic performances is due to the authenticity of the attire worn.

The second cultural association from Rupea, "Ramidava XXI", organizes folk dance courses for children, folk craft workshops, various festivals, and other cultural events [71,87]. Their dancers, who learn from a young age (among them preschoolers, I2) the complicated steps of a wide variety of popular dances from the area, impress audiences in the tours and festivals they participate in. The organization of local cultural events can become a financial challenge for the association, which often covers part of the expenses from its own funds (I2), but the efforts are rewarded by the appreciation and joy of the community. More importantly, "someone must remind the elders and tell the young people who they are and what it means to be from Cohalm" (I3). Cohalm is the alternative name of Hungarian origin (Kőhalom) of the locality of Rupea. Each generation is obliged to carry on the tradition and honor its ancestors. The assumed purpose of the efforts of the members of Ramidava XXI is to convey the significance of traditional customs (I3), "which should not be seen as simple performances". Together, the two associations strengthen the prestige of the Romanian cultural heritage of the area. From discussions with local residents, it appears that the activities of the two associations benefit from broad community support.

The traditional houses of the Romanians and Saxons in Rupea were structured similarly: a large room on the first floor facing the street, with access by an external staircase and through another smaller room, and a cellar on the ground floor. The merchants' houses in the center had a commercial area and several rooms upstairs (I4). After the First World War, Romanians began to build houses with three rooms on the first floor and to transform the side of the cellar facing the street into living space, but these were houses located on the streets further from the borders of the settlement. The Saxons lived downtown, and the areas of the city inhabited by Romanians, Saxons, and Roma were compact. On some streets, the central Saxon residential area continues with the Romanian one until the periphery of the settlement (I4). The Romanians were successful in buying land for houses in the central area from the Saxons, but this was conducted relatively late and with difficulty (the first purchase contract was concluded in 1926, I4). Some of the Saxon properties on the main street were expropriated during the communist period, and the Town House of Culture and the hospital were built in their place. Some of the Saxon houses inhabited by the Romanians after the Saxons left for Germany have been modified. But in the city center, there is a core of buildings that still preserve the old Saxon burgh architecture.

Most of the Saxon or former Saxon villages (e.g., Jimbor, currently having a majority Hungarian population) in the vicinity of Rupea have fortified evangelical churches. In the churches' surrounding areas, i.e., in the center of the village, there were open terraces where the young people of the village danced to the music of the local brass band on Sundays and other holidays after the service. The girls decorated these dance spaces with flowers and fabrics. "It was always a joy to pass by" (I5). The church of Viscri, a village where the current king of the UK owns the property, is registered in the UNESCO heritage. These churches are points of interest on various tourist routes [88]. In August, the "Haferland Week" Festival is organized in the area between Rupea and Sighișoara (Haferland means "Land of Oats" in German), with traditional crafts, arts, and gastronomy [89]. In the olden days, the Saxons in the area grew oats and were famous for them. August is the month of return tourism, when traditionally Transylvanian Saxon residents in Germany return to their native places for vacations. Most of the Saxon households in the area were, at the time when the Saxons had not yet left, well organized. "From the gate to the edge of the garden, everything was in its place. And in the house, on the farm, and in the stable, everything was arranged functionally, almost as if in a mechanism. So that the householders do not waste time, which is precious" (I5). In the Saxon residential areas, the streets are wide, and there is generous space for maneuvers between the houses and the road as an interface between the private space delimited by the facade of the house and the high gate, with a masonry vault, and the public space. In order to border the pedestrian circulation, there are narrow layers of flowers in this space.

But the most beautiful flowers are in the Hungarian villages in the area (I5). The houses are decorated with attention to color details, and some courtyards have carved or openwork wooden gates. In front of every yard, there are flowers; as if in a housewives' competition, "you enter the village and you cheer up; it is as if the whole village is laughing; you can see the joy of living" (I5). The admiration for the organization of the living space of the Saxons and Hungarians also appears as an important topic in the discussions with the Romanian locals during our working visits to the Rupea area.

Near the town of Rupea is the village of Hoghiz, the capital city of the TAU under the same name. In the village, there are three noble castles from the 16th century located in vast parks: Bethlen–Haller Castle (sometimes called "The Fortress"), Kalnoky Castle, and Guthman–Valenta Castle. The history of their residence and the repeated changes of owners until communist and then post-communist Romania strengthened the cultural heritage value of these castles.

The village of Racoșul de Jos (TAU Racoș) is also located in the Rupea area. In the center of the village is Sükösd–Bethlen Castle. Attested from the 17th century, it successively belonged to several Hungarian noble families, as well as the village. The castle is currently open to the public, and the area available for touring has been expanding as the rehabilitation process continues [90]. The village is famous for the nearby geological reserve, consisting of a volcanic crater, a formation of basalt columns, and "Lacul de smarald" ("Emerald Lake"), a lake formed in the crater of an old quarry [91]. At Racoșul de Jos, the "Ziua Pietrei" ("Stone's Day") Festival takes place in August. On this particular occasion, there are organized guided tours to geological sites, concerts, dance performances, art exhibitions, craft workshops and fairs, and culinary fairs [92].

The Rupea area is one where signs of habitation (tools, ceramics, and the remains of houses) have been discovered since the Paleolithic. In the perimeter of the city of Rupea and in its vicinity, recent archaeological research has revealed several traces of human presence from this era [52], from the Neolithic [53–55], Eneolithic [53,56–61], the bronze age [53], the iron age, the period of the Dacian kingdom, and the period after the conquest of Dacia by the Romans [62]. Some traces were identified through the extensive analysis of the landscape's configuration, while others were anticipated in connection with the sources of salt in the area. In other cases, the information has been collected from elderly locals who discovered shards during yearly spring plowing.

In Figure 3 below, the points of archaeological interest are indicated. As they are on the land of the city or the surrounding villages, the points are marked R1-11 (from Rupea), H1 (from Homorod), H1 Hoghiz (from Hoghiz), M1 (from Mercheașa), and U1-2 (from Ungra). The Rupea fortress and the Blumenthal Valley are distinctly marked; they are old sites known by these names in specialized literature. In Table 3, the historical periods related to each of the archaeological sites in Figure 3 are indicated.

Figure 3. Archaeological discoveries were made near the city of Rupea. Location according to S. Gridan (I6), on support https://wiki.openstreetmap.org/ (accessed on 16 September 2023).

Table 3. Traces of human presence in the city of Rupea and its surroundings from prehistory and Antiquity to the Middle Ages.

No.	Archaeological Site	Historical Periods
1	Blumenthal	Hallstatt, Roman
2	Rupea Fortress	Neolithic, Copper, Bronze, Medieval
3	R1	Bronze
4	R2	Bronze?, Roman, Medieval
5	R3	Hallstatt, Bronze, Medieval
6	R4	Palaeolithic, Neolithic, Copper, Post Roman
7	R5	Palaeolithic, Copper, Hallstatt, Roman, Post Roman
8	R6	Hallstatt, Roman, Post Roman
9	R7	Palaeolithic, Neolithic, Copper, Bronze
10	R8	Neolithic, Hallstatt, Roman
11	R9	Copper
12	R10	Hallstatt
13	R11	Bronze?, Hallstatt?, Medieval
14	H1	Copper, Bronze, Medieval
15	H1 Hoghiz	Copper, La Tène
16	M1	Bronze
17	U1	Neolithic, Copper, Hallstatt
18	U2	Neolithic, Copper, Bronze, Roman

Archaeological sites are not marked (indicated) in the field at present. The sporadic excavations carried out there cover small areas that can be fully investigated (i.e., down to the oldest/deepest habitation level) in an intensive field campaign. This fact is mainly due to the general underfunding of archaeological research in Romania and, secondarily, to the increased interest of treasure hunters in new archaeological sites. These are prospectors for whom the value of a site is given exclusively by the possibility of finding gold, silver, or precious stones there.

5.2. *Rupea Area as Cultural Landscape*

The geographical location makes the Rupea area one of the cultural confluences and stimulates interethnic coexistence. As a space of cultural interference, the Rupea area could be valued as a cultural landscape.

A second dimension of this potential cultural landscape is represented by the multitude of local archaeological discoveries, discoveries that indicate- as highlighted above, the human presence starting from the Paleolithic in the area. Archaeologists link this abundance of sites to the positioning of Rupea at the intersection of travel routes in the territory on the surrounding hilly peaks, in correlation with the existence of exploitable salt resources. Where there is salt, there is a good chance that there is also a human presence [54].

Perceptible inter-ethnic interferences in architecture, garb, traditions, and gastronomy, on the one hand, together with the high density of points of archaeological interest related to different periods, are defining the Rupea area as a cultural landscape. In addition to these and supporting them is the openness to the organization of cultural events of various kinds by the "Junii Cetății" and "Ramidava XXI" associations.

The two defining dimensions of the cultural landscape can be enhanced by configuring distinct local tourist routes. In the case of ethnic interference, the routes should cross both Romanian, (former) Saxon, Hungarian, and mixed villages, highlighting the specific local architecture. Churches, as architectural expressions of the believers' relationship with divinity [93], are the main objectives of these routes. In many villages, there are small ethnographic museums near and/or under the administration of the churches. Visiting them as objectives on the tourist routes in the Rupea area supports their development and would also encourage the establishment of new ones. Eastern European Christianity, especially rural Christianity, allows associations of pre-Christian practices and beliefs [94,95], and village museums could highlight these associations, which would be spectacular.

The area targeted by the routes to highlight neighborhoods and ethnic interference is easily accessible by car and is covered by national and county roads in good working condition. A spring–summer–autumn alternative to personal-car tourism could be the bicycle. For the Saxon part of the area, there is already a low-difficulty mountain bike route that connects the fortified Saxon churches [96]. The route could be extended through the Hungarian villages to the east and northeast of small health resorts of local interest. From the discussions with the locals, it emerged that the small health resorts located east of the Rupea area represent pleasant destinations for short trips of a few days, which led us to propose routes to them on the multicultural dimension of the cultural landscape.

These tourist routes could enhance the value of the multi-ethnic space of the Rupea area by including a gastronomic tourism component. In this area, ethnic similarities and differences in gastronomy can be easily noticeable regarding soups and stews [97]. Easy to prepare in large quantities and spectacularly diverse, they could represent the core of the area's gastronomic brand. The use of brands is generally effective in promoting intangible cultural heritage [98]. Local emblematic foods and cultural-gastronomic routes with well-defined branding destinations are attractive to "authenticity seekers" [99]. Furthermore, the importance of soups and stews in local gastronomy is a more widespread characteristic feature in Transylvania [100]. The plateau in front of the main access gate to the Rupea Fortress, the park area in the center of the city, and the former market area of the medieval settlement are suitable places for the sale of local agri-food products. In addition to soups and brews, the locals could offer dairy and meat products prepared according to traditional recipes, bread and traditional pastries, even seasonal vegetables and fruits from their own production, or honey, jam, and syrups.

In order to highlight the traces of human presence in the area dating back to the Paleolithic, we consider it effective to organize guided pedestrian tours of the archaeological sites, with tours ending in an exhibition space dedicated to the artifacts discovered at these sites. The distances between most points of archaeological interest in the perimeter of the city of Rupea and the vicinity allow such an approach in one-day tours. Currently, most of the objects discovered in the area's sites can be found at the warehouses of the Brașov County Museum of History. For these and future discoveries, an exhibition center could be set up right in Rupea Fortress. Many of its visitable rooms are currently empty. The locals had been repurposing the rooms throughout time, and in accordance with their history of use, some that had been intended as private living spaces could also be set up for exhibition, displaying furniture and objects that are specific to both the inhabitants' activities in the periods of temporary refuge in the fortress and the settlement(s) it served. The proximity of the exhibition spaces intended for living in prehistory and in the Middle Ages strengthens the cultural landscape dimension of the continuity of living in the area.

The funds obtained from the guided tours would be of great use for the current regime of sporadic excavations at the sites in Rupea. When the raised funds equal the expected expenses for the next site to be opened and researched, that one will undergo full research and then be arranged for tourist visits. As one site after another is surveyed and developed for visitation, guided tours to the unmarked sites of sporadic discoveries will become rigorously configured historical tourism routes.

6. Discussion

Although this is ethical niche tourism, sustainable and oriented towards authenticity through the specific interests of the beneficiaries [18], we believe that the order in which the research of the sites is conducted should take into account their attractiveness for treasure hunters. The sites referred to as most attractive to the treasure hunters should be the first ones to be investigated by archaeologists—as soon as possible and as quickly as possible—in order to avoid or limit the hunters' destructive interventions. The sites' location should be roughly indicated in the guided tours at the beginning of this self-sustainable program of excavations (conducted with funds from tourist visits to points of archeological interest). To the same effect, it would be useful to specify in the guided tours the type of artifacts that archaeologists expect to find at the other sites, namely tools and ceramics, perhaps bone ornaments, and certainly not objects made of precious metals.

The involvement of the local community/communities in the logistical support of the archaeological research through accommodation and food would reduce the financial pressure. The women from the area and surroundings are used to and happy to cook together [74,97,100] and to strengthen their community prestige through hospitality. Tourism social entrepreneurship (TSE) represents an effective way of managing this type of logistical support for archaeological excavations in the area. TSEs, which strengthen local social networks as a support for regional economic interactions [101], catalyze the sustainable development of host communities [23,102]. It boosts the local economy, educates local tourism actors, creates sustainable livelihoods [23], provides innovative solutions at the community level [19], and effectively mobilizes local resources [103]. Romanians' representations of the country's economic situation and development prospects are rather lacking in optimism [104], and government predictions should be carefully corroborated with local social particularities [104,105]. In this context, TSE should be taken into account as a way of managing several activities related to Rupea as a cultural landscape.

The valorization of the Rupea area as a cultural landscape that we propose thus recognizes the role attributed by Baltà Portolés [1] to cultural heritage in strengthening community identity and cohesion. Through TSE, the condition of the involvement of local people in the preservation and enrichment of heritage, considered necessary by Gray and Kuokkanen [5], is met. Moreover, community participation, social inclusion, and the strengthening of social networks are features associated with culture-led development [24]. The valorization that we propose as cultural landscape can be part of a development strategy for the Rupea area—with the city of Rupea in the center and the neighboring rural localities—which credits the specifics of local communities and cultures, a feature highlighted by Montaldo et al. [14] as defining culture-led development. Our study is part of the trend of expanding culture-led development research towards small towns and rural communities, a trend signaled by Lysgård [13]. As Bajec and Kranjc [43] show, the ways of life of ancestors as cultural heritage support sustainable development within the horizon of the 2030 Agenda. The valorization of the Rupea area as a cultural landscape that we propose is also one oriented towards sustainable local development.

Newsome, Moore, and Dowling [106] identify three macro markets in niche tourism (cultural, event, natural area) and seven secondary markets (adventure tourism, ecotourism, festival, heritage, nature-based, religious, and sports tourism). The tourist itineraries that we propose in the cultural landscape of Rupea combine offers from several of these markets. We are talking about adventure, festivals, heritage, and nature-based tourism. The itineraries approach can be personalized; they are comfortable for small groups and have a flexible schedule. The itineraries can be valued differently, as the tourists' interests in relation to the local offer are configured differently. Even with regard to festival tourism, like "Stone's Day" from Racoș, the interests of the participants are well anchored in the local cultural peculiarities, and the number of participants allows differentiated satisfaction of these interests. In all forms of niche tourism with potential for development in the Rupea cultural landscape, the discovery component is significant. Tourists interested in the cultural landscape are looking for their own experiences and representations of the local heritage. This approach represents a viable alternative to mass tourism, an alternative that we recommend for the Rupea area.

Tourism related to Rupea's cultural landscape dimensions is a niche one, with groups of tourists of the kind described by Dujmović [18], i.e., small in size, with experience and well-defined interests. At the same time, the combination of the routes to the archaeological sites, those centered on traditional dwellings and ethnic neighborhoods situated between the hills on the edge of the Hârtibaciu plateau and the extensions of the volcanic Eastern Carpathians, allows the Rupea area to be approached as a holistic cultural landscape through the local touristic offer, bestowing a landscape resembling those defined by Otero et al. [37].

Taking over the Rupea area as a cultural landscape would not only bring benefits to the locals. It involves the effective management of relationships between community members involved in TSE as well as between the members and the community as a whole. It also assumes control over potential conflicts in inter-ethnic relations, a subject of historical depth and still sensitive in Transylvania. In this context, the protection of the cultural landscape actually represents a challenge for local management, as considered by Tuvikene [45]. Maintaining the balance between tourism development and cultural heritage protection [40] is an important dimension of this challenge. The danger of the aggressive, treasure-hunting investigation of archaeological sites included in tourist routes, which we have mentioned above, is a local connotation of a wider challenge related to the sustainable management of heritage tourism. The fairly common temptation to maximize profit by lowering the quality of tourism products and services offered even under the TSE regime is also subject to this challenge. But it is a challenge worth taking on, like any challenge aimed at the sustainable development of an area.

As mentioned above, the cultural landscape of the Rupea area is defined by three dimensions, two of which are primary (multicultural settlement and archaeological finds) and secondary (activities of cultural associations). From discussions with local people, it appears that they attach much more importance to the secondary dimension than to the main dimensions. However, this result can be considered a specific asset [29] in this area and used in cultural management and community development projects.

7. Conclusions

The transformation of the city of Rupea into the central node for tourist routes centered on local archaeological sites or on the ethnic specifics of traditional dwellings in the area is an approach to urban regeneration. It is a non-invasive approach that can be managed as a source of benefits for local people and involves the protection of local cultural heritage. The assumption of the Rupea area as a cultural landscape with the dimensions that have been highlighted represents a way of valorizing the past by orienting it towards the future: the effective current management of the identified resources, implicitly their protection and conservation, outlines the tourist prestige of the area and ensures its preservation. Thus, assuming the Rupea area as a cultural landscape can be an important step in the process of urban regeneration as well as the sustainable development of both the city and the area.

In the presentation of the area's heritage resources, we only referred to those most well-known. This could be a limitation of the present study. Each of the villages in the Rupea area has at least one church, a village museum, a monument dedicated to heroes, a site, or a celebration that can be used on niche tourist routes. We considered their exhaustive presentation out of the economy of our argument, but they convey consistency to the cultural landscape. In the future, we aim to explore the cultural heritage assets of these communities in more detail and also broaden the scope of research both geographically and in terms of collecting data from the local community members, with the aim of integrating their vision and voices into local culture-led development strategies.

Through this research, we aimed to highlight the urban regeneration potential of the Rupea area through specific mechanisms for culture-led development. Our approach is intended to stimulate the interest of potential tourists or visitors to the area who can discover new values in this cultural landscape. At the same time, the results of our research favor the knowledge of more isolated places by highlighting local cultural resources and their potential to be capitalized on by the tourist market. The entry of these isolated places into the tourist circuit can also favor return tourism. All these aspects can contribute to community development and increase cohesion and a sense of local belonging. The results of our research can be a source of solutions for local operators in the tourism field, as well as for projects by local administrations and cultural management institutions.

Author Contributions: Conceptualization, G.G.R., C.C., M.B. and D.S.; methodology, C.C. and D.S.; validation, G.G.R., C.C., M.B. and D.S.; investigation, G.G.R., C.C., M.B. and D.S.; writing—original draft preparation, G.G.R., C.C. and D.S.; writing—review and editing, G.G.R., C.C. and D.S.; visualization, M.B.; supervision, G.G.R., C.C., M.B. and D.S. All authors have read and agreed to the published version of the manuscript.

Funding: This research received no external funding.

Data Availability Statement: No new data were created or analyzed in this study. Data sharing is not applicable to this article.

Conflicts of Interest: The authors declare no conflict of interest.

References

1. Baltà Portolés, J. Cultural Heritage and Sustainable Cities. Key Themes and Examples in European Cities. *UCLG Comm. Cult. Rep.* **2018**, *7*. Available online: https://www.agenda21culture.net/sites/default/files/report_7_-_cultural_heritage_sustainable_development_-_eng.pdf (accessed on 14 July 2022).
2. Mäkinen, K. Cultural heritage: Connecting people? *Monit. Glob. Intell. Racism* **2020**, *6*, 1–4. Available online: http://monitoracism.eu/cultural-heritage-connecting-people/ (accessed on 14 July 2022).
3. Calligaro, O. From "European cultural heritage" to "cultural diversity"? *Polit. Eur.* **2014**, *3*, 60–85. [CrossRef]
4. Wellbeing and the Historic Environment by Historic England. 2018. Available online: https://historicengland.org.uk/images-books/publications/wellbeing-and-thehistoric-environment/wellbeing-and-historic-environment (accessed on 14 July 2022).
5. Grey, S.; Kuokkanen, R. Indigenous governance of cultural heritage: Searching for alternatives to co-management. *Int. J. Herit. Stud.* **2020**, *26*, 919–941. [CrossRef]
6. Heritage, A.; Tissot, A.; Banerjee, B. Heritage and Wellbeing: What Constitutes a Good Life? International Centre for the Study of the Preservation and Restoration of Cultural Property (ICCROM). 2019. Available online: https://www.iccrom.org/projects/heritage-and-wellbeing-what-constitutesgood-life (accessed on 14 July 2022).
7. Barca, F. L'Anno europeo del patrimonio culturale e la visione europea della cultura. *Digit. Sci. J. Digit. Cult.* **2017**, *2*, 75–93.
8. EENC Paper. *Challenges and Priorities for Cultural Heritage in Europe: Results of an Expert Consultation*. 2013. Available online: https://eenc.eu/uploads/eenc-eu/2021/04/21/07d4959fa85d06c29a0e262d18398e75.pdf (accessed on 14 July 2022).
9. Lysgård, H.K. The 'actually existing' cultural policy and culture-led strategies of rural places and small towns. *J. Rural. Stud.* **2016**, *44*, 1–11. [CrossRef]
10. Nedučin, D.; Krklješ, M.; Zlatomir Gajić, Z. Post-socialist context of culture-led urban regeneration—Case study of a street in Novi Sad, Serbia. *Cities* **2019**, *85*, 72–82. [CrossRef]
11. Jeanotte, M.S. Story-telling about place: Engaging citizens in cultural mapping. *City Cult. Soc.* **2016**, *7*, 35–41. [CrossRef]
12. Loulanski, T. Cultural Heritage in Socio-Economic Development: Local and Global Perspectives. *Environ. J.* **2006**, *34*, 51–69.
13. Lysgård, H.K. The assemblage of culture-led policies in small towns and rural communities. *Geoforum* **2019**, *101*, 10–17. [CrossRef]
14. Montalto, V.; Alberti, V.; Panella, F.; Sacco, P.L. Are cultural cities always creative? An empirical analysis of culture-led development in 190 European cities. *Habitat Int.* **2023**, *132*, 102739. [CrossRef]
15. Poulsen, V.M.; DeDeo, S. Inferring Cultural Landscapes with the Inverse Ising Model. *Entropy* **2023**, *25*, 264. [CrossRef] [PubMed]
16. Borin, E.; Donato, F. What is the legacy of the European Year of Cultural Heritage? A long way from cultural policies towards innovative cultural management models. *Eur. J. Cult. Manag. Policy* **2020**, *10*, 1–17. [CrossRef]
17. Lotter, M.; Geldenhuys, S.; Potgieter, M. A conceptual framework for segmenting niche tourism markets. In Proceedings of the 6th International Adventure Conference (IAC), Segovia, Spain, 30 January–2 February 2018.
18. Dujmović, M. Komercijalizacija kulturne baštine u turizmu. *Soc. Ekol.* **2019**, *28*, 145–161. [CrossRef]
19. Boukas, N.; Chourides, P. Niche tourism in Cyprus: Conceptualising the importance of social entrepreneurship for the sustainable development of islands. *Int. J. Leis. Tour. Mark.* **2016**, *5*, 26–43. [CrossRef]
20. García-Martín, M.; Huntsinger, L.; Ibarrola-Rivas, M.J.; Penker, M.; D'Ambrosio, U.; Dimopoulos, T.; Fernández-Giménez, M.E.; Kizos, T.; Muñoz-Rojas, J.; Saito, O.; et al. Landscape products for sustainable agricultural landscapes. *Nat. Food* **2022**, *3*, 814–821. [CrossRef] [PubMed]
21. Timothy, D.J.; Nyaupane, G.P. Introduction: Heritage tourism and the less-developed world. In *Cultural Heritage and Tourism in the Developing World—A Regional Perspective*; Timothy, D.J., Nyaupane, G.P., Eds.; Routledge: London, UK, 2009; pp. 3–19.
22. Busuioc, M.-F.; Simion, T.; Niculescu, A.C.; Trifănescu, R. New opportunities for niche tourism in Romania: Ethnographic tourism. *Rom. Econ. Bus. Rev.* **2016**, *11*, 35–43.

23. Aquino, R.S.; Lück, M.; Schänzel, H.A. A conceptual framework of tourism social entrepreneurship for sustainable community development. *J. Hosp. Tour. Manag.* **2018**, *37*, 23–32. [CrossRef]
24. Li, J. Culture and tourism-led peri-urban transformation in China—The case of Shanghai. *Cities* **2020**, *99*, 102628. [CrossRef]
25. Balfour, B.; Fortunato, M.W.P.; Alter, T.R. The creative fire: An interactional framework for rural arts-based development. *J. Rural. Stud.* **2018**, *63*, 229–239. [CrossRef]
26. Marques, L.; Borba, C. Co-creating the city: Digital technology and creative tourism. *Tour. Manag. Perspect.* **2017**, *24*, 86–93. [CrossRef]
27. Tsang, K.K.M.; Siu, K.W.M. The 3Cs model of sustainable cultural and creative cluster: The case of Hong Kong. *City Cult. Soc.* **2016**, *7*, 209–219. [CrossRef]
28. Flint, W.R. *Practice of Sustainable Community Development. A Participatory Framework for Change*; Springer: New York, NY, USA, 2013.
29. Matarrita-Cascante, D.; Lee, L.H.; Nam, J.W. What elements should be present in any community development initiative? Distinguishing community development from local development. *Local Dev. Soc.* **2020**, *1*, 95–115. [CrossRef]
30. Liu, J.; Fang, Y.; Yan, Q.; Chen, C. Modern Zoning Plans versus Traditional Landscape Structures: Ecosystem Service Dynamics and Interactions in Rapidly Urbanizing Cultural Landscapes. *J. Environ. Manag.* **2023**, *331*, 117315. [CrossRef] [PubMed]
31. Li, X.; Wang, C. Understanding the relationship between tourists' perceptions of the authenticity of traditional village cultural landscapes and behavioural intentions, mediated by memorable tourism experiences and place attachment. *Asia Pac. J. Tour. Res.* **2023**, *28*, 254–273. [CrossRef]
32. Jones, M. The Concept of Cultural Landscape: Discourse and Narratives. In *Landscape Interfaces*; Palang, H., Fry, G., Eds.; Landscape Series; Springer: Dordrecht, The Netherlands, 2003; Volume 1, pp. 22–51. [CrossRef]
33. Wu, J. Landscape of culture and culture of landscape: Does landscape ecology need culture? *Landsc. Ecol.* **2010**, *25*, 1147–1150. [CrossRef]
34. UNESCO. World Heritage Convention. Cultural Landscapes. Available online: https://whc.unesco.org/en/culturallandscape/#1 (accessed on 14 July 2023).
35. The Cultural Landscape Foundation. About Cultural Landscapes. Available online: https://www.tclf.org/places/about-cultural-landscapes (accessed on 2 March 2022).
36. Liu, T. Application of Traditional Cultural Elements in Landscape Design. *Appl. Math. Nonlinear Sci.* **2023**. [CrossRef]
37. Otero, A.; Timothy, D.J.; Galí, N.; Vidal-Casellas, D. Historical Pathways as Promotion and Protection of the Cultural Landscape: Tourism and the Camí de Ronda on the Costa Brava. *PASOS Rev. De Tur. Y Patrim. Cult.* **2023**, *21*, 255–270. [CrossRef]
38. Gajdek, A.; Krupa, B.; Nowak, A. What is attractive rural landscape? Differences in the social and expert assessment of the changes in the rural landscape of the Carpathian region in Poland with regard to the need of its protection. *J. Mt. Sci.* **2023**, *20*, 501–515. [CrossRef] [PubMed]
39. Stilling, S.; Braae, E. Relational heritage: 'Relational character' in national cultural heritage characterisation tools. *Landsc. Res.* **2023**, *48*, 917–934. [CrossRef]
40. Yu, J.; Safarov, B.; Yi, L.; Buzrukova, M.; Janzakov, B. The Adaptive Evolution of Cultural Ecosystems along the Silk Road and Cultural Tourism Heritage: A Case Study of 22 Cultural Sites on the Chinese Section of the Silk Road World Heritage. *Sustainability* **2023**, *15*, 2465. [CrossRef]
41. European Environment Agency. European Landscape Convention. Available online: https://www.eea.europa.eu/policy-documents/european-landscape-convention (accessed on 2 September 2023).
42. Ducci, M.; Janssen, R.; Burgers, G.J.; Rotondo, F. Co-design Workshops for Cultural Landscape Planning. *Landsc. Res.* **2023**, *48*, 900–916. [CrossRef]
43. Bajec, J.F.; Kranjc, D. Significance of cultural heritage practices in karst landscape management: 2030 Agenda for Sustainable Development. *Carbonates Evaporites* **2023**, *38*, 10. [CrossRef]
44. Torralba, M.; Nishi, M.; Cebrián-Piqueras, M.A.; Quintas-Soriano, C.; García-Martín, M.; Plieninger, T. Disentangling the practice of landscape approaches: A Q-method analysis on experiences in socio-ecological production landscapes and seascapes. *Sustain. Sci.* **2023**, *18*, 1893–1906. [CrossRef]
45. Tuvikene, T. Absence, Presence, and Mobility. A Landscape Approach to an Unfinished Tram Project. *Transfers-Interdisciplinary J. Mobil. Stud.* **2022**, *12*, 47–59. [CrossRef]
46. Schlanger, S. More Than Meets the Eye: Integrating the Management of Landscape Character and Archaeological Resources for Culturally Significant Landscapes. *Adv. Archaeol. Pract.* **2023**, *11*, 42–51. [CrossRef]
47. Ministerul Culturii. Planul Strategic Instituțional al Ministerului Culturii 2024–2027 (Institutional Strategic Plan of the Ministry of Culture 2024–2027). Available online: http://www.cultura.ro/sites/default/files/inline-files/FINAL_PSI%20Ministerul%20Culturii%202024%202027.pdf (accessed on 17 June 2023).
48. Sepe, M. *Planning and Place in the City: Mapping Place Identity*; Routledge: London, UK, 2013.
49. Webber, S.; Butteris, S.M.; Houser, L.; Coller, K.; Coller, R.J. Asset-Based Community Development as a Strategy for Developing Local Global Health Curricula. *Acad. Pediatr.* **2018**, *18*, 496–501. [CrossRef] [PubMed]
50. Jaye, C.; McHugh, J.; Doolan-Noble, F.; Wood, L. Wellbeing and health in a small New Zealand rural community: Assets, capabilities and being rural-fit. *J. Rural. Stud.* **2022**, *92*, 284–293. [CrossRef]
51. Danciu, V. *Memoria Caselor Orașului Rupea*; Petru Maior: Reghin, Romania, 2023.

52. Doboș, A.; Gridan, S. Recente descoperiri paleolitice în sud-estul Transilvaniei: Microzona Rupea—Homorod—Ungra, judeţul Braşov. *Mater. Cercet. Arheol.* **2018**, *14*, 5–12.
53. Gridan, S.; Florian, C. Contribuţii la cunoaşterea neoliticului şi eneoliticului din sud-estul Transilvaniei, comuna Homorod (judeţul Braşov). In *ArheoVest III. Interdisciplinaritate în Arheologie și Istorie*; Forțiu, S., Stavilă, A., Eds.; JATE Press Kiadó: Szeged, Hungary, 2015; pp. 119–133.
54. Lazarovici, G.; Lazarovici, C.M.; Gridan, S. Surse de sare și procesul de neolitizare din S-E Transilvaniei. In *Anuarul Muzeului Etnografic al Transilvaniei*; Argonaut: Cluj-Napoca, Romania, 2018; pp. 303–319.
55. Oancă, M.; Lazarovici, G.; Gridan, S. Cercetări în stațiuni Starčevo-Criș din zona Rupea. In *Marea Unire de la Marea Neagră*; Colesniuc, S., Ed.; Celebris: Constanța, Romania, 2018; pp. 21–30.
56. Lazarovici, C.M.; Lazarovici, G.; Gridan, S.; Florian, C.; Aparaschivei, C.; Roman, C. Cercetări privind cultura Petreşti la Rupea. In *Marea Unire de la Marea Neagră*; Colesniuc, S., Ed.; Celebris: Constanța, Romania, 2018; pp. 31–44.
57. Lazarovici, G.; Lazarovici, C.M.; Gridan, S.; Pirău, H.; Apapraschivei, C.; Oancă, M.; Florian, C.; Roman, C. Cercetări arheologice la Rupea. Campania din 2018. *Acta Terrae Fogorasiensis* **2018**, *7*, 21–37.
58. Gridan, S.; Lazarovici, G.; Lazarovici, C.M.; Aparaschivei, C.; Gridan, O. Contribuții la cunoașterea culturii Petrești din sud-estul Transilvaniei- zona Rupea (jud. Brașov). In *ArheoVest VII1, Interdisciplinaritate în Arheologie și Istorie*; Forțiu, S., Ed.; JATE Press Kiadó: Szeged, Hungary, 2019; pp. 155–174.
59. Mârza, I.; Lazarovici, G.; Gridan, S. Analize mineralogice și determinări petrografice ale unor unelte și așchii descoperite în așezarea de la Rupea 7- Părăul Mălinilor (Comuna Homorod, jud. Brașov). *Acta Terrae Fogorasiensis* **2019**, *8*, 13–21.
60. Gridan, S. An eneolithic innovation from south-eastern Transylvania. The Ariușd-Cucuteni culture sickle blade from Rupea 4- La Movilă settlement (Rupea, Brașov county). *Acta Musei Tutovensis. Istor. Veche Și Arheol.* **2021**, *17*, 23–41.
61. Beldiman, C.; Gridan, S.; Beldiman, D.-M. Un obiect de podoabă preistoric din os descoperit în sud-estul Transilvaniei (com. Homorod, jud. Brașov). In *ArheoVest X1. Interdisciplinaritate în Arheologie și Istorie*; Forțiu, S., Ed.; Istros: Brăila, Romania, 2022; pp. 243–274.
62. Gridan, S.; Urdea, P.; Hegyi, A. Castrul de la Ungra, jud. Brașov. Cercetări multidisciplinare. In *ArheoVest V2. Interdisciplinaritate în Arheologie și Istorie*; Forțiu, S., Ed.; JATE Press Kiadó: Szeged, Hungary, 2017; pp. 851–883.
63. Costea, F. *Repertoriul Arheologic al Județului Brașov*; C2design: Brașov, Romania, 2004.
64. Repertoriul Arheologic Național (RAN). Available online: http://ran-tmp.cimec.ro/sel.asp (accessed on 2 September 2023).
65. Borcoman, M. *Așezări Transilvane: Rupea*; Editura Academiei: Cluj-Napoca, Romania, 2010.
66. Borcoman, M. *Sighișoara și Așezările Scaunului Omonim*; Argonaut: București, Romania, 2020.
67. Stancescu, A.M.; Neciob, R. Monografia turistică Orașul Rupea. 2021. Available online: https://www.primariarupea.ro/wp-content/uploads/MONOGRAFIE-_REDACTATA.pdf (accessed on 2 September 2023).
68. CULTEMA. Rupea-Cohalm. *Patrimoniu Cultural Pentru Dezvoltare Locală.* Available online: http://cultema.umpcultura.ro/files/promotionale_cultema/Brosura%20Rupea/brosura%20RUPEA%20RO%20web%20spread.pdf (accessed on 2 September 2023).
69. Consiliul Județean Brașov. Strategia de Dezvoltare Durabilă a Județului Brașov 2021–2030. In *Analiza Situației Actuale a Județului Brașov.*. Available online: https://addjb.ro/uploads/user_upload/nSDDJB__II__Analiza_situatiei_actuale_a_judetului_Brasov.pdf (accessed on 3 September 2023).
70. Asociaţia Transilvană Braşov Nord. Plan de Dezvoltare Locală Zona de Nord a Judeţului BRAŞOV. Available online: http://galatbn.ro/wp-content/uploads/2013/07/PDL-FINAL_varianta-initiala.pdf (accessed on 14 July 2022).
71. Primăria Rupea. Strategia de Dezvoltare Durabilă a Orașului Rupea. Available online: http://www.primariarupea.ro/img/upload/strategie-dezvoltare-durabila-rupea.pdf (accessed on 3 September 2023).
72. Institutul Național de Statistică. Tempo Online. Available online: http://statistici.insse.ro:8077/tempo-online/#/pages/tables/insse-table (accessed on 20 October 2023).
73. Recensământul Populației și Locuințelor. Rezultate Definitive: Caracteristici Etno-Culturale Demografice. Available online: https://www.recensamantromania.ro/rezultate-rpl-2021/rezultate-definitive-caracteristici-etno-culturale-demografice/ (accessed on 3 September 2023).
74. Băjenaru, C. *Ţara Făgăraşului în Timpul Stăpânirii Austriece (1691–1867)*; ALTIP: Alba Iulia, Romania, 2013.
75. Sorea, D.; Csesznek, C. The Groups of Caroling Lads from Făgăras, Land (Romania) as Niche Tourism Resource. *Sustainability* **2020**, *12*, 4577. [CrossRef]
76. Sorea, D.; Bolborici, A.M. The Anti-Communist Resistance in the Făgăraș Mountains (Romania) as a Challenge for Social Memory and an Exercise of Critical Thinking. *Sociológia Slovak Sociol. Rev.* **2021**, *53*, 266–286. [CrossRef]
77. Sorea, D.; Roșculeț, G.; Rățulea, G.G. The Compossessorates in the Olt Land (Romania) as Sustaina-ble Commons. *Land* **2022**, *11*, 292. [CrossRef]
78. Sorea, D.; Csesznek, C.; Rățulea, G.G. The Culture-Centered Development Potential of Communities in Făgăras Land (Romania). *Land* **2022**, *11*, 837. [CrossRef]
79. Giurescu, C.C. *Istoria Românilor*; ALL: București, Romania, 2015; Volume 1.
80. Pop, I.A. *Scurtă Istorie a Românilor*; Litera: București, Romania, 2023.
81. Nägler, T. *Asezarea Sașilor în Transivania*; Kriterion: București, Romania, 1992.

82. Institutul Național de Statistică. Direcția Județeană de Statistică Brașov. In *Recensământul Populației și Locuințelor 2021- Județul Brașov.*. Available online: https://brasov.insse.ro/wp-content/uploads/2023/06/RECENSAMANTUL-POPULATIEI-SI-AL-LOCUINTELOR-2021.pdf (accessed on 3 September 2023).
83. Giurescu, C.C. *Istoria Românilor. Vol.III: De la Moartea lui Mihai Viteazul Până la Sfârșitul Epocii Fanariote (1601–1821)*; ALL: București, Romania, 2015.
84. Ministerul Investițiilor și Proiectelor Europene. Proiecte de Succes. Available online: https://mfe.gov.ro/vechea-fortareata-rupea-reabilitata-din-fonduri-europene/ (accessed on 3 September 2023).
85. Muzeul de Etnografie "Gheorghe Cernea". Available online: http://ghidulmuzeelor.cimec.ro/id.asp?k=154&-Muzeul-de-Etnografie-Gheorghe-Cernea-RUPEA-Brasov (accessed on 14 November 2022).
86. Junii Cetății Rupea. Available online: https://www.juniicetatiirupea.ro/ (accessed on 2 September 2023).
87. Festival Aniversar Fanfara Țărănescă Rupea—100 de ani de Activitate. Available online: https://brasovtourism.app/ro/events/festival-aniversar-fanfara{-}{-}-r-nesc{-}{-}rupea{-}{-}-100-de-ani-de-activitate (accessed on 3 September 2023).
88. Bisericile Fortificate Săsești din Zona Rupea. Available online: https://www.traseeromania.ro/bisericile-fortificate-din-zona-rupea/ (accessed on 14 November 2022).
89. Haferland. Available online: https://haferland.ro/ (accessed on 20 August 2023).
90. Castelul Sükösd-Bethlen. Available online: https://www.castelulsukosdbethlen.ro/istoric/ (accessed on 14 November 2022).
91. Rezervația Naturală Complexul Vulcanic Racoș, Ghid Turistic și Geologic. Available online: https://amfostacolo.ro/impresii9.php?iid=86000&d=descopera-racos-si-comorile-ascunse-ale-zonei{-}{-}viscri-bunesti-bv (accessed on 14 November 2022).
92. Festivalul "Ziua Pietrei" la Racoș, Ediția a V-a. Available online: https://www.monitorfg.ro/2022/08/19/festivalul-ziua-pietrei-la-racos-editia-a-v-a-la-ce-actiuni-puteti-participa-in-cele-2-zile-de-festival/ (accessed on 14 November 2022).
93. Blaga, L. *Trilogia Culturii*; Humanitas: București, Romania, 2018.
94. Sorea, D. Two Particular Expressions of Neo-Paganism. *Bull. Transilv. Univ. Braşov Ser. VII* **2013**, *6*, 29–40.
95. Sorea, D.; Scârneci-Domnișoru, F. Unorthodox depictions of divinity. Romanian children's drawings of Him, Her or It. *Int. J. Child. Spiritual.* **2018**, *23*, 380–400. [CrossRef]
96. Amazingromania. Colinele Transilvaniei—MTB Printre Sate Săsești și Biserici Fortificate. Available online: https://amazingromania.net/mtb-in-colinele-transilvaniei/ (accessed on 3 August 2023).
97. Borcoman, M.; Sorea, D. Ethnic Soups from Rupea Area (Romania) as Resources for Sustainable Local Development. *Sustainability* **2023**, *15*, 943. [CrossRef]
98. Hammou, I.; Aboudou, S.; Makloul, Y. Social Media and Intangible Cultural Heritage for Digital Marketing Communication: Case of Marrakech Crafts. *Mark. Manag. Innov.* **2020**, *1*, 121–127. [CrossRef]
99. Gupta, V.; Sajnani, M. A study on the influence of street food authenticity and degree of their variations on the tourists' overall destination experiences. *Br. Food J.* **2019**, *122*, 779–797. [CrossRef]
100. Bolborici, A.-M.; Lupu, M.I.; Sorea, D.; Atudorei, I.A. Gastronomic Heritage of Făgăras, Land: A Worthwhile Sustainable Resource. *Sustainability* **2022**, *14*, 1199. [CrossRef]
101. Neumeier, S.; Pollermann, K. Rural Tourism As Promoter Of Rural Development—Prospects And Limitations: Case Study Findings From A Pilot Projectpromoting Village Tourism. *Eur. Countrys.* **2014**, *6*, 270–296. [CrossRef]
102. Sheldon, P.; Pollock, A.; Daniele, R. Social entrepreneurship and tourism: Setting the stage. In *Social Entrepreneurship and Tourism: Philosophy and Practice*; Sheldon, P.J., Daniele, R., Eds.; Springer International Publishing: Cham, Switzerland, 2017; pp. 1–18.
103. Altinay, L.; Sigala, M.; Waligo, V. Social value creation through tourism enterprise. *Tour. Manag.* **2016**, *54*, 404–417. [CrossRef]
104. Stroe, F. Arguments of a different type of Romanian social project. *Bull. Thetransilvaniauniversity Braşov Ser. VII* **2018**, *11*, 113–126.
105. Stroe, F. The lingering Hellenism. *Bull. Thetransilvania Univ. Braşov Ser. VII* **2022**, *15*, 175–182. [CrossRef]
106. Newsome, D.; Moore, S.A.; Dowling, R.K. *Natural Area Tourism: Ecology, Impacts and Management*; Channel View Publications: Bristol, UK, 2013.

land

MDPI

Article

From Urban Challenges to "ClimaEquitable" Opportunities: Enhancing Resilience with Urban Welfare

Marsia Marino

PDTA Department, Sapienza University of Rome, 00196 Rome, Italy; marsia.marino@uniroma1.it

Abstract: From the perspective of the scientific-disciplinary debate within urban planning, this research addresses the theme of the "new urban question" resulting from environmental concerns related to the climate crisis and socioeconomic issues that have now become structural. It then delves into the connection between urban environment quality and quality of life, ultimately questioning the role that territorial governance tools play in positively influencing the perception of well-being in cities. The overall objective of this contribution is to define an interpretative framework for experimental approaches in territorial governance. This overarching objective is articulated in the definition of two specific outcomes, pursued through an inductive methodology. The first one involves establishing an initial set of urban welfare indicators; the second entails defining strategies for planning, designing, and regenerating the public components of the city that could influence the indicators. Both outcomes are designed to be exportable to different territorial contexts.

Keywords: ClimaEquitable planning; local urban plan; new urban question; urban welfare; climate crisis; socioeconomic crisis; urban resilience

1. The New Urban Question: Between Socioeconomic and Environmental Challenges

The discussion surrounding the "new urban question" began in the 1970s when Manuel Castells introduced the concept in his book *The Urban Question: A Marxist Approach* (1979) [1]. Castells identified a so-called "city of well-being" as the central core of this new issue, emphasizing the importance of analyzing the deep-seated causes of urban development rooted in a complex network of economy, politics, and social well-being.

Later, Jacques Donzelot, in his book *Quand le ville se défait. Quelle politique face à la crise des banlieues?* (2008) [2] focused on the "social question," highlighting how it had long been associated solely with defending wage conditions, excluding the issue of social exclusion within cities from the debate. His investigation traces the "new urban question" back to what he calls the "logic of separation in the city," which has progressively compromised its capacity to "create society," leading to three concurrent trends of urban transformation within the same territorial reality: "relegation," "suburbanization," and "gentrification."

Danzelot argues that to ensure genuine social diversity, it is necessary to structurally intervene in inter-neighborhood and inter-municipal mobility to overcome the infrastructural barriers that separate relegated, peri-urban, and gentrified areas. This concept has recently been echoed by Mimar et al. (2022) in the article "Connecting intercity mobility with urban welfare" [3].

Danzelot's discussion of mobility also connects with what Giuseppe Campos Venuti refers to as "genetic anomalies", which have characterized the development of Italian cities since the 20th century. Unlike Western European cities that grew in close relation to the railway network, Italian cities developed in a completely subordinate manner in relation to road infrastructure, facilitating a "spread like an oil stain" expansion. Many urban dynamics scholars have attempted over the years to provide a definition for this condition: Francesco Indovina talks about "dispersed city" (1990), "metropolitan archipelago" (2009), and "territorial metropolis" (2010), Manuel Castells about "networked city" (2004), Augé about "non-place city" (1992) [4–8].

What differentiates the current urban crisis from the one that originated in the second half of the last century is the emergence of environmental concerns linked to the climate crisis, which add to the socioeconomic issues mentioned so far.

Indeed, «there are not two separate crises, social and environmental, but one complex socio-environmental crisis. To respond to this, we need a comprehensive approach to combat poverty while simultaneously caring for nature» [9], an approach that considers both socioeconomic and environmental demands at the same time.

Almost always, ecological and social issues are addressed at all levels of governance in a distinct manner, even though they are recognized as the two main challenges of contemporary society.

International organizations highlight how the processes of metropolization have, in recent decades, significantly influenced the form and structure of contemporary cities, with well-known consequences in terms of pollution, land consumption, lack of infrastructure, and a general sense of insecurity [10].

In this regard, Khan, Hildingson, and Garting (2020) [11] point out that there is an increasingly concrete risk that efforts to address ecological challenges may have a negative impact on equality and social well-being.

One of the most recent examples of how these closely related issues struggle to find common ground is the limited outcome of COP 27, held in Sharm el-Sheikh from 7–8 November 2022, regarding the establishment of a specific fund to compensate for the impacts of climate change on the territories of the most vulnerable countries.

The need to integrate socioeconomic and environmental demands is also strongly emphasized by the United Nations in the document *Strategy for sustainability management in the United Nations system, 2020–2030 Phase II: Towards leadership in environmental and social sustainability* [12], which builds on two previous reports: *A framework for advancing environmental and social sustainability in the United Nations system* from 2012 and *Advancing the Environmental and Social Sustainability Framework in the United Nations System* from 2014 [13,14].

1.1. Programmatic Measures Introduced by the EU

In this frame of reference, the European Union has implemented various policy-programmatic measures aimed at creating sustainable and inclusive urban communities. These initiatives stem from the thematic priorities outlined in the *European Urban Agenda* [15], addressing critical issues such as migrant inclusion, air quality, urban poverty, and adaptation to climate change.

Heading in the same direction is the *European Green Deal*, which aims to make the EU a zero-emissions society by 2050, balancing environmental and social dimensions, along with the pact *A Strong Social Europe for Just Transitions* and the *Just Transition Fund 2021–2027*, which emphasize themes of employment, equal opportunities, and social protection [16–18].

Furthermore, the COVID-19 pandemic has underscored the importance of "space"—both public and private—for people's quality of life, a concept emphasized by the *New European Bauhaus* initiative [19]. This initiative seeks to rethink cities and living spaces, making them more aesthetically pleasing, sustainable, and inclusive.

At the national level, Italy first developed its own *Urban Agenda for Sustainable Development* [20], and subsequently, in response to the COVID-19 pandemic, the *National Recovery and Resilience Plan* [21] was introduced to enhance digitization, ecological transition, and social inclusion.

This brief overview provides insight into how the EU and Italy are implementing measures and strategies to address environmental and social challenges, promoting sustainability and social inclusion within urban communities.

1.2. "Right to the Public City" and Urban Welfare

The convergence of socioeconomic and environmental concerns in the "new urban question" refers to what Henri Lefebvre calls the "right to the city." This encompasses the right to participate in decisions that affect individuals and the city they inhabit, regardless of the level from which the decision originates (central government, local administration, company, international organization, etc.).

The second aspect pertains to the right of city residents to physically access, occupy, and use urban space. Lefebvre's concept not only applies to existing urban space but also to the right to produce new urban space that meets citizens' needs. In a way, Lefebvre's thinking anticipates both the concept of urban welfare and the attribution of social value to urban regeneration. The latter is understood as a strategy for urban planning, social inclusion, economic development, and ecological transformation of territories. It considers the spontaneous nature of metropolization phenomena and is aimed at pursuing a unified and integrated public governance strategy capable of envisioning a new decentralized arrangement of urban structure. This new structure should be polycentric, sustainable, and accessible, aimed at achieving integration between informal and planned cities, as well as the introduction of functions and residences [22].

In this sense, the city becomes the ideal projection of a new urban welfare for communities, aiming to combine quality of life and urban environmental quality.

For these reasons, its definition requires high levels of integration, interdisciplinarity, interscalarism, and iterativity for a recomposition of the physical and socioeconomic components of change [23].

In this frame of reference, the concept of "urban welfare" refers to the ability of an urban system to provide settled inhabitants with an adequate level of well-being through the creation of facilities and spaces of collective interest, integrating the concepts of "urban standards" and "urban facilities" [24].

Considering the outlined conceptual framework, the primary purpose of this article is to investigate relations and interdependencies between urban life quality and potential innovative strategies for the planning, design, and regeneration of public components within the city (conceived by the author as the most effective reference parameter for assessing the overall health of cities), aimed at constructing a new urban welfare for settled communities. In this context, following a literature review outlined in the subsequent paragraph, the intention is to examine two cities characterized by the highest quality of life in the year 2022, one at an international level and another at a national Italian level, with a dual objective:

- Analyzing the indicators used to determine the ranking, distinguishing between socioeconomic and environmental indicators (considering the outlined thematic framework).
- Verifying the existence, within the local urban plans of the analyzed cities, of specific strategies that may have positively influenced the aforementioned indicators.

The first objective is preliminary to extract primitive urban welfare indicators exportable to other contexts, while the second aims to define strategies for the planning, design, and regeneration of the public city, from which to develop specific criteria and quantifiable parameters exportable in subsequent phases of the research.

The expected outcome of this article is, therefore, to establish a useful theoretical, methodological, and operational reference for applying, in future phases of the research, primitive urban welfare indicators to a sample city in order to assess the quality of life in cities (as perceived by their inhabitants in terms of their relationship with public components) at the time of assessment and after the potential implementation of corrective strategies, also broadly outlined in this paper. This aims to understand how indicators might be affected by the application of specific strategies.

2. Literature Review

The close relationship between quality of life and urban environmental quality is strongly emphasized by the *European Urban Agenda*, launched in Amsterdam in 2016 [15].

The underlying concept is that of a city as a "well-being factory" [25] that places people at the center of urban development processes.

A noteworthy example demonstrating the multidisciplinary nature of the connection between quality of life and urban environmental quality is provided by a study conducted by James F. Sallis, published in 2016 in the international medical journal The Lancet. This study offers an objective assessment of the relationship between features of the built environment and the inclination of city inhabitants towards physical activity. It demonstrates that in dense cities with a good presence of green areas (considering their accessibility), the population is encouraged to walk to reach services and activities [26].

Another interesting perspective for understanding the relationship between quality of life and urban environmental quality in a multidisciplinary context is the map of health for human habitat at the local scale [27]. This map provides some intriguing insights for integrating urban planning and health, identifying five areas of consideration:

- Land use and transport planning;
- Social services;
- Economic regeneration;
- Integrated transportation;
- Integrated planning of resources for energy, water, food, waste, etc.

It is necessary to adopt new indicators of vulnerability and sustainability to better comprehend these connections and interdependencies and to define and guide a new model of urban welfare capable of incorporating both socioeconomic and environmental dimensions.

In this regard, the research within which this article is placed has preliminarily analyzed various datasets of socioeconomic and environmental vulnerability and sustainability indicators (Table 1) [28–32].

Table 1. Within the scope of the project "Urban welfare, public city, and rights: Strategies, tools, mechanisms for innovation in the local plan within a climate-resilient perspective" (within which this contribution is inserted), the author analyzed various datasets containing indicators of socioeconomic and environmental vulnerability and sustainability.

Characteristics of the Analyzed Indicator Sets	**Sets of Analyzed Indicators**					
	ISTAT (2017)	**"Territori civili" (Legambiente e Caritas Italiana, 2020)**	**SDGs Measures on a Regional Scale (Istat, 2022)**	**BES Report 2021**	**Global Liveability Index 2022 (EIU)**	**33rd Survey on Quality of Life (Sole 24 Ore, 2022)**
Scope of Application	Italian national, urban scale	Regional	Regional	National, urban scale	International, urban scale	National, urban scale
Structure	Eight reference domains with 27 specific indicators (mainly socioeconomic)	40 social indicators and 30 environmental indicators, divided into various dimensions	Over 200 indicators defined by the Inter-Agency Expert Group on SDG (Sustainable Development Goals) Indicators	Integrated analysis of economic, social, and environmental phenomena divided into 12 domains	Five main categories with various specific indicators (both socioeconomic and environmental)	Six synthetic indicators are divided into sub-indicators (including both socioeconomic and environmental aspects)
Main focus	Exposure to socioeconomic vulnerabilities of the analyzed cities, with specific reference to peripheral areas.	Relationship between fragility and socioeconomic and environmental resources of Italian regions.	Monitoring progress towards Sustainable Development Goals	Evaluation of the well-being, defined as fair and sustainable, of Italian cities	Evaluation of the city with the best quality of life on a global scale.	Evaluation of the city with the best quality of life on a national scale in Italy.

Based on the analysis conducted and for the purposes of this research, which aims to connect the quality of life in cities with the quality of the urban environment, it was deemed appropriate to consider the two datasets explicitly intended to assess the quality of life in major international and Italian cities. Indeed, thanks to the *Global Liveability Index* (by The Economist Group–Economist Intelligence Unit) [33] and the *33rd Survey on Quality of Life* by Il Sole 24 Ore [34] it has been possible to identify Vienna (AT) and Bologna (IT) as the cities with the best quality of life for the year 2022, the former at an international level and the latter at a national Italian level. These cities were therefore selected for more specific analyses to investigate the connections and interdependencies between innovative strategies for planning, designing, and regenerating urban components (primarily public ones)–envisaged by the respective local planning instruments–and the quality of life.

2.1. The International Context: The Global Liveability Index (2022)

Every year, the Economist Intelligence Unit (EIU) compiles the *Global Liveability Index* to determine the most liveable city in the world out of 172 cities assessed. This assessment is based on various indicators aimed at investigating political stability, culture, environment, education, and infrastructure.

The set of indicators proposed by the EIU is divided into five main categories, each of which carries a different weight (expressed in percentage) in the total evaluation. Each category is defined by various specific indicators that provide both quantitative and qualitative insights [33].

From the data analysis, it is interesting to note that the impact of restrictions implemented in response to the COVID-19 pandemic has significantly influenced the considered values. The results for 2022 closely approach those from pre-pandemic times (2019) and see the city of Vienna reclaiming the top position, which it held in 2018 and 2019 but lost in 2020 and 2021 because of the health crisis.

Generally, the greatest progress is observed in cities in Western Europe, while the worst rankings are held by cities experiencing armed conflicts. In fact, many African cities occupy the lowest positions, and Damascus ranks last out of all cities, at 172nd place, due to the decade-long war in Syria.

The top ten cities in the rankings, in order, are: Vienna (Austria); Copenhagen (Denmark); Zurich (Switzerland); Calgary (Canada); Vancouver (Canada); Geneva (Switzerland); Frankfurt (Germany); Toronto (Canada); Amsterdam (Netherlands); Osaka (Japan); Melbourne (Australia).

2.2. The Italian National Context: 33rd Survey on Quality of Life (2022)

The 33rd Survey on Quality of Life by Il Sole 24 Ore in 2022 extensively addresses the repercussions of major shocks on the territory in recent years: the pandemic, the war in Ukraine, high energy costs, and inflation. Once again, a categorization is proposed, encompassing both synthetic indicators and sub-indicators.

Overall, the analysis for 2022 indicates an increasing gap between the regions of Southern and Northern Italy. Bologna emerges as the city with the highest quality of life; this marks the fifth occurrence, following 2000, 2004, 2011, and 2020. Bolzano and Florence secure the second and third positions, respectively.

Like the *Global Liveability Index*, the survey by Il Sole 24 Ore notes a decline in some metropolitan cities, notably Milan, which ranked second in 2021 but has dropped to eighth place this year. This is primarily attributed to the "Wealth and Consumption" indicator, which records higher values for mid-sized cities in 2022. Despite this decline, Milan still leads in the "Business and Employment" indicator but is penalized by the high percentage (over 60% in the city) of rental fees in relation to average income.

Rome experiences a drop of 18 positions, landing at 31st place, just below Genoa (27th). Turin follows at 40th place (down 9 from 2021), affected by poor air quality and a high incidence of reported crimes. Palermo ranks at 88th, while Naples is at 98th due to higher population density and perceived street-level insecurity [34].

From the analysis of the two datasets, it emerges that Vienna and Bologna are the cities with the highest quality of life for the year 2022, internationally and nationally, respectively. In examining these cities, it was initially interesting to scrutinize the socioeconomic and environmental vulnerability and sustainability indicators used to determine their top positions in the rankings. Subsequently, the focus shifted to investigating whether the territorial governance tools of these cities provide indications, guidelines, and strategies for planning, designing, and regenerating urban components (primarily public ones) that positively influence the well-being perception of their inhabitants.

It is important to highlight the diversity among the analyzed datasets, one at an international level and the other at a national level within Italy, which leads to the utilization of distinct indicators. Indeed, indicators and categories are different, and the results on the liveability of cities are the result of the indicators applied. This is also due to the different application scales of the datasets. The author, through the direct comparison of these indicators (Section 3.1), has chosen to harness this differing scope between the analyzed datasets (on an international and national scale) as an opportunity to integrate two distinct perspectives for a more comprehensive understanding of urban dynamics in their entirety.

3. Methodology

As previously mentioned, the overarching aim of this research is to ascertain whether there are connections between innovative territorial governance methods achieved through the establishment of guidelines and specific strategies for planning, designing, and regenerating the urban components (primarily public ones) outlined in local planning instruments and the excellent results attained by the two cities under examination (Vienna and Bologna) in international rankings assessing quality of life. The goal is to outline theoretical-methodological and operational references that can be applied to different urban contexts and to experiment with a "Clima-Equitable" innovation in the local urban plan as future developments of the research.

This objective materializes in the articulation of two distinct outcomes (target):

- Definition of an initial set of urban welfare indicators—exportable to different territorial contexts—to assess the quality of components of the city (primarily public ones) and how they impact the well-being of settled inhabitants.
- Definition of general strategies for the planning, designing, and regenerating of these components, also exportable to different territorial contexts.

To achieve these two targets, the research adopts an inductive methodology structured into two main phases (Figure 1):

- Phase 1 | Synthesis and categorization of only the indicators referring to urban structure and components, adapted from the two datasets referenced in Sections 2.1 and 2.2.
- Phase 2 | Analysis of the structure of local plans of the analyzed cities to bring out elements useful for defining strategies for the planning, designing, and regenerating urban components—primarily public ones.

3.1. *Phase 1 | Synthesis and Categorization of Indicators Related to Urban Structure and Components*

This paragraph provides a comparative analysis (Table 2) of only the indicators showing explicit relationships with urban components. The indicators have been synthesized and categorized based on those used in the two datasets mentioned in Section 2.1, *The Global Liveability* Index (column "EIU indicators" in Table 1), and Section 2.2, the *33rd Survey on Quality of Life* (column "S24O indicators" in Table 1).

Once the suitable indicators were selected (highlighted in green in Table 1)—through an inductive approach—each of them was classified into reference macro-categories (conceptualized by the author to be exportable to different territorial contexts) to obtain useful references for defining new urban welfare indicators.

Specifically, each indicator was classified into a general category (in the "Gen. Cat." column of Table 2), to which, in turn, an explicit impact on the urban structure was associated (in the "Impact U.S." column of Table 2). Subsequently, the type of indicator was also indicated, i.e., whether it was an indicator of socioeconomic vulnerability/sustainability (indicated in Table 2 with the acronym "SE") or an indicator of environmental vulnerability/sustainability (indicated in Table 2 with the acronym "EN"). Additionally, the reference system most influenced by each of them was indicated as an environmental system (indicated in Table 2 with the acronym "E.S"), settlement system (indicated in Table 2 with the letter "S.S"), or mobility and services system (indicated in Table 2 with the acronym "MS.S").

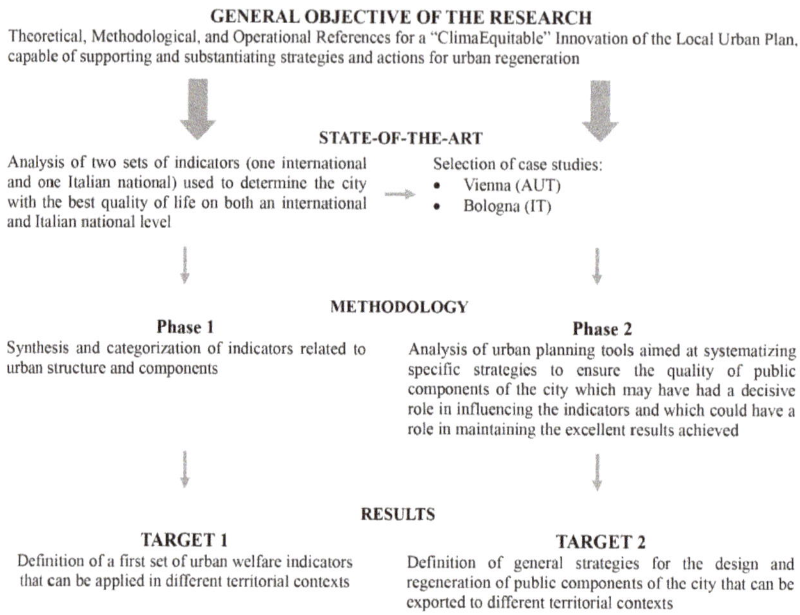

Figure 1. Schematic representation of the adopted methodology.

Table 2. Selection and comparative analysis of indicators of socioeconomic and environmental vulnerability/sustainability. The indicators (EIU, 2022; Sole 24 Ore, 2022) have been categorized into reference macro-categories (column "Gen.Cat."), and for each of them, the impact on urban components (column "Impact U.S.") in socioeconomic terms (column "SE") and environmental terms (column: "EN") has been evaluated, as well as the reference system: environmental, settlement, mobility, and services ("E.S," "S.S," "MS.S"). Conceptualization by Marsia Marino (2023).

Gen. Cat.	Impact U.S.	EIU Indicators	Indicator Typologies		System			S24O Indicators	Indicators		System		
			SE	EN	E.S	S.S	MS.S		SE	EN	E.S	S.S	MS.S
Social stability	Perception of public space	Presence of petty crime	✔			✔		Crime index—total number of crimes reported	✔			✔	
		Presence of violent petty crime	✔			✔		Robberies on public streets	✔			✔	
								Sustainable public lighting	✔	✔		✔	
		Perception of fear	✔			✔		Perception of fear	✔	✔		✔	
		Presence of disorders	✔			✔							

Table 2. *Cont.*

Gen. Cat.	Impact U.S.	EIU Indicators	Indicator Typologies		System			S24O Indicators	Indicators		System		
			SE	EN	E.S	S.S	MS.S		SE	EN	E.S	S.S	MS.S
Health care	Coverage and quality of healthcare facilities	Availability of private healthcare facilities	✔				✔						
		Evaluation of private healthcare facilities	✔				✔						
		Availability of public health facilities	✔				✔						
		Evaluation of public health facilities	✔				✔						
Instruction	Coverage and quality of school facilities	Availability of private school facilities	✔				✔						
		Evaluation of private school facilities	✔				✔						
		Availability of public school facilities	✔				✔						
		Evaluation of public school facilities	✔				✔						
Culture and free time	Coverage and quality of cultural and leisure venues	Availability of cultural facilities/places	✔				✔	Availability of restaurants (including mobile catering)	✔				✔
								Availability of museum heritage	✔				✔
								Availability of agritourism companies	✔				✔
								Availability of libraries	✔				✔
								Bar availability	✔				✔
								Cultural offer (shows per thousand inhabitants)	✔				✔
Sport	Coverage and quality of facilities equipped for sports	Availability of sports equipment/facilities	✔				✔	Availability of gyms, swimming pools, wellness centers and spas	✔		✔		✔
Environment	Effects of climate change on urban climate	Humidity/temperature		✔	✔			Consecutive days without rain		✔	✔		
		Perception of discomfort about the climate		✔	✔			Energy consumption	✔	✔	✔		
								Air quality		✔	✔		
	Green mobility							Motorization rate (cars in circulation per 100 inhabitants)		✔	✔		
								Pedestrian areas		✔	✔		
								Presence of cycle paths		✔	✔		
Infrastructural accessibility	Mobility infrastructure	Road quality	✔				✔						
		Quality of public transport	✔	✔			✔						
		Quality of connections to and from the city	✔				✔						

Table 2. *Cont.*

Gen. Cat.	Impact U.S.	EIU Indicators	Indicator Typologies		System			S24O Indicators	Indicators		System		
			SE	EN	E.S	S.S	MS.S		SE	EN	E.S	S.S	MS.S
Building heritage	Public and private residences	Availability of good quality residential accommodation	✔			✔		Average rental rates	✔			✔	
								Average home sales price	✔			✔	
								Living space (average surface area calculated on the basis of the average family members)	✔			✔	
								Burglaries at home	✔			✔	✔
								Population density (residents per km2)	✔	✔		✔	✔
								Legally resident immigrants	✔			✔	

The comparative analysis presented in Table 2 reveals that the *Global Liveability Index* dataset, used to determine the city with the best quality of life, relies almost exclusively on socioeconomic vulnerability/sustainability indicators. On the other hand, the dataset from the *33rd Survey on Quality of Life* includes several environmental vulnerability/sustainability indicators. However, the latter has a significant gap in evaluating healthcare and educational facilities compared to the former, while it provides a more detailed breakdown of indicators related to real estate assets [33,34].

3.2. Phase 2 | Analysis of the Local Urban Plans of Vienna and Bologna

3.2.1. Vienna–SPTEP 2025

The approach of Vienna's local plan is encapsulated in its description: «The task for the future which Vienna is facing now is, to put it in simple terms, to create adequate living space, jobs as well as infrastructure to ensure local supplies, education, and recreation. "Adequate" does not only mean appropriate in quantitative terms but also adjusted to the needs of a city that has emerged as the most liveable city—or, in the 'worst case' as one of the most liveable cities—in the world in all international rankings for many years. Hence, Vienna is also challenged in qualitative terms» [35].

The Urban Development Plan of Vienna 2025 (from here on abbreviated as STEP 2025) is strongly oriented towards the public dimension of the city, especially in relation to three key points, closely interconnected with each other, the efficiency of which—as emphasized in the Plan—is directly proportional to the quality of life of the inhabitants:

1. Housing: great attention is paid to the provision of subsidized housing and social mix.
2. Green and open spaces: significant focus on the "undeveloped" space of the city, understood as public space (including the road network) and green areas.
3. Mobility: public mobility is understood as the backbone of the city—great attention is paid to green and cycle-pedestrian mobility.

The Plan is structured into four thematic areas:

1. Vienna: setting the stage, which defines the vision of the Plan.
2. Vienna: building the future, which provides general guidelines for the quality of the urban structure.
3. Vienna: reaching beyond its borders, in which the terms of urban development are defined from the perspective of a regional metropolis.
4. Vienna: networking the city, which outlines the principles related to mobility, social infrastructure, public spaces, and green areas.

For the purposes of this research, thematic areas 2, "Vienna: building the future," and 4, "Vienna: networking the city," are particularly interesting. These are the areas in which guidelines and strategies for the planning, designing, and regeneration of city components (housing, mobility, public spaces, and green areas) are explicitly detailed, determining urban welfare and the high quality of life in Vienna.

It should be noted that the STEP 2025 Plan delegates the definition of these strategies (and also some specific criteria/parameters) to detailed prescriptive documents, one for the "historic and consolidated city," another for the "urban expansion areas," and another for the "transformation or underutilized areas."

For the sake of brevity, in Appendix A, some interesting strategies and quali-quantitative references for the planning and regeneration of city components (predominantly public ones) of the historic city are provided, synthesized from the detailed document attached to the STEP 2025 Plan, Gründerzeit Action Plan (2018) [36].

3.2.2. Bologna–PUG 2021

Bologna is one of the first municipalities in the Emilia-Romagna region to approve the General Urban Plan (from here on, abbreviated as PUG) in 2021, as required by the new regional law 24 of 2017.

The new PUG of the Municipality of Bologna was finally approved on 26 July 2021, after being adopted on 7 December, 2021, and identifies three thematic priorities:

1. Quality of the environment.
2. Quality of life.
3. Quality of infrastructure.

The Plan is structured around three general objectives ("Resilience and Environment," "Habitability and Inclusion," "Attractiveness and Employment"), each of which is articulated into urban strategies (a total of 12) and specific actions that define priorities and establish guidelines for urban policies and regulations for urban and building interventions in the city's transformation. The first objective, "Resilience and Environment," focuses on the recovery and redevelopment of existing areas rather than expansion beyond urban space.

The second, "Habitability and Inclusion," aims to improve the quality of life both in the city center and in the suburbs to build a liveable and inclusive city.

The third, "Attractiveness and Employment," aims to implement and strengthen the most important infrastructure as a driving force for urban regeneration.

Furthermore, the Plan identifies 24 territorial frameworks within the municipal territory, corresponding to parts of the city defined as "urban areas" by regional law (art.34 new regional law 24 of 2017). Each part of the city is characterized by recognizable spatial, functional, and environmental relationships and constitutes a shared reference for those who live in that area.

Through local strategies, the Plan pursues specific objectives of urban and ecological-environmental quality and identifies specific actions, interventions, or policies to be implemented in subsequent phases of planning and implementation.

For each urban area, the following is specified:

- The location of major public interventions, either underway or already included in planning instruments.
- Opportunities and issues.
- Functional and meaningful connections.

Unlike the Vienna STEP 2025 Plan, the Bologna PUG defines some guidelines—useful for defining strategies for the planning, design, and regeneration of city components—already in the definition of "specific actions" while leaving detailed elements to establish rules for the application of these guidelines to specific "fields of application" [37–48].

In Appendix B, a summary of the study of the PUG is proposed to extract connections and interdependencies between the excellent perception of the inhabitants' quality of life and innovative methods of territorial governance.

4. Results

4.1. Target 1 | Urban Welfare Indicators

Based on the comparative analysis presented in Table 2 (Section 3.1), it was possible to synthesize an initial set of urban welfare indicators that can be applied to different territorial contexts. This allows for the assessment of the quality of urban components and how they impact the well-being of settled inhabitants (Target 1).

The indicators from the two analyzed datasets were grouped together. Also, in this case, each of them was linked to a "Reference System" and a "Reference Category."

For each indicator, the "Impact on the urban structure" was highlighted, and it was indicated whether it was more appropriately considered a socioeconomic and/or environmental indicator (Table 3).

Table 3. Initial categorization of urban welfare indicators exportable to different territorial contexts.

Urban Welfare Indicators					
Reference System	**Reference Category**	**Impact on the Urban Structure**	**Indicators**	**Indicator Typologies**	
				SE	**EN**
Environmental system	Climate change	Effects of climate change on urban climate	Humidity/Temperature		✔
			Perception of discomfort about the climate		✔
			Consecutive days without rain		✔
			Energy consumption	✔	✔
			Air quality		✔
	Sports and health	Coverage and quality of sports facilities	Availability of sports equipment/facilities	✔	
			Availability of gyms, swimming pools, wellness centers and spas	✔	
Settlement-morphological system	Social stability	Perception of public space	Presence of petty crime	✔	
			Presence of violent petty crime	✔	
			Crime index-total number of crimes reported	✔	
			Robberies on public streets	✔	
			Perception of fear	✔	
			Sustainable public lighting	✔	
			Presence of disorders	✔	
	Building heritage	Public and private residences	Availability of good quality residential accommodation	✔	
			Population density (Residents per km^2)	✔	✔
			Home burglaries	✔	
			Average rental rates	✔	
			Average home sales price	✔	
			Living space (average surface area based on average family members)	✔	
			Legal resident immigrants	✔	

Table 3. *Cont.*

Urban Welfare Indicators					
Reference System	**Reference Category**	**Impact on the Urban Structure**	**Indicators**	**Indicator Typologies**	
				SE	**EN**
Mobility and service system	Health care	Coverage of healthcare facilities	Availability of private healthcare facilities	✔	
			Availability of public health facilities	✔	
		Quality of healthcare facilities	Evaluation of private healthcare facilities	✔	
			Evaluation of public health structures	✔	
	Instruction	Coverage of school facilities	Availability of private school facilities	✔	
			Availability of public school facilities	✔	
		Quality of school facilities	Evaluation of private school facilities	✔	
			Evaluation of public school facilities	✔	
	Culture and free time	Coverage and quality of cultural and leisure venues	Availability of cultural facilities/places	✔	
			Cultural offer (shows per thousand inhabitants)	✔	
			Availability of sports equipment/facilities	✔	
			Availability of restaurants (including mobile catering)	✔	
			Availability of museum heritage	✔	
			Availability of agritourism companies	✔	
			Availability of libraries	✔	
			Bar availability	✔	
			Availability of gyms, swimming pools, wellness centers and spas	✔	
	Infrastructural accessibility	Mobility infrastructure	Road quality	✔	
			Quality of public transport	✔	✔
			Quality of connections to and from the city	✔	
	Green mobility	Coverage, quality and use of green mobility infrastructure	Pedestrian areas		✔
			Presence of cycle paths		✔
			Motorization rate (Cars in circulation per 100 inhabitants)		✔

4.2. Target 2 | Strategies for the Planning, Design, and Regeneration of the Urban Components

Based on the analysis of the local urban planning tools of the two analyzed cities, which highlight guidelines and strategies for the planning, design, and regeneration of the city components (Sections A and B), it was possible to synthesize and categorize some general and exportable strategies, as required by Target 2 (Table 4).

As in the case of Table 3, the strategies derived from the two local plans were grouped together. Since the aim is to allow a comparison between urban welfare indicators and city components to enable their monitoring and evaluation, for each criterion, the most relevant "Reference System" and "Reference Category" have also been specified. The expected impact of each action on the urban structure has also been explicitly stated. In this case, the nature of the strategies (socioeconomic and/or environmental) has also been explicitly mentioned.

Table 4. Initial categorization of strategies for the planning, design, and regeneration of city components—mostly public ones—exportable to different territorial contexts.

Strategies for Planning, Designing, and Regenerating Urban Components (Predominantly Public)					
Reference System	**Reference Category**	**Impact on the Urban Structure**	**Strategies**	**Criteria/Parameter Typologies**	
				SE	**EN**
Environmental system	Climate change	Effects of climate change on urban climate	Promote the recovery and upgrading of existing building heritage		✔
			Complete the parts of the city where transformation is not yet complete	✔	✔
			Promote reuse and urban regeneration of built areas and anthropized land	✔	✔
			Envision interventions for the unsealing and depavement of soil		✔
			Safeguard biodiversity and the main ecosystem services of hills and plains		✔
			Improve the quality of surface waters		✔
			Maintain natural watercourse flows and reduce withdrawals from groundwater		✔
			Enhance the quality of surface waters		✔
			Ensure the regular flow of water in the mouths of streams and culverts		✔
			Mitigate the urban heat island effect and introduce measures for building climate adaptation	✔	✔
			Reduce the population's exposure to pollution and anthropogenic risks	✔	✔
			Promote and incentivize various forms of energy efficiency and equitable access to low-impact energy services	✔	✔
			Plan the deployment of energy production plants from renewable sources by creating local distribution networks	✔	✔
	Connection between city and countryside		Promote innovative practices in peri-urban agriculture	✔	✔
			Enhance peri-urban parks, improving their accessibility for tourism	✔	✔
			Develop networks of safe paths and trails connected to national and European tourist routes	✔	✔

Table 4. *Cont.*

Strategies for Planning, Designing, and Regenerating Urban Components (Predominantly Public)					
Reference System	**Reference Category**	**Impact on the Urban Structure**	**Strategies**	**Criteria/Parameter Typologies**	
				SE	**EN**
Settlement-morphological system	Social stability	Perception of public space	In case of elevation, ensure it is proportional to the street width	✔	
			Promote practices for regenerating public space along the street (creating parklets)	✔	
			Construct open spaces and public buildings with high architectural and environmental quality	✔	
			Renew the street space in terms of formal and environmental quality, accessibility, and safety	✔	✔
			Preserve the habitability and characteristics of the historic city	✔	
			Enhance the specialized fabrics of the historic city	✔	
			Guarantee the conservation of architectural and cultural heritage of historical interest	✔	
	Building heritage	Public and private residences	Promote the refunctionalization of the ground floor of buildings	✔	
			Ensure social diversity through the integration of rent-controlled housing units in buildings designated for private residences and the juxtaposition of new social housing structures with buildings designated for private residences	✔	
			In case of elevation, ensure it is proportional to the street width	✔	
			Promote the shared use of semi-public spaces in residential buildings (understood as a "common room")	✔	
			Promote the recovery and improvement of existing building heritage	✔	✔
			Complete the parts of the city where transformation is not yet complete	✔	✔
			Promote the increase and innovation of rental housing supply	✔	
			Promote the increase of social housing supply	✔	
			Experiment with new forms of housing	✔	
			Introduce functional and typological mixes in specialized areas near residential fabrics	✔	

Table 4. *Cont.*

Strategies for Planning, Designing, and Regenerating Urban Components (Predominantly Public)					
Reference System	**Reference Category**	**Impact on the Urban Structure**	**Strategies**	**Criteria/Parameter Typologies**	
				SE	**EN**
Mobility and service system	Culture and free time	Coverage and quality of cultural and leisure venues	Support a balanced spread of spaces for culture	✔	
			Promote local services and commercial activities	✔	
	Infrastructural accessibility	Mobility infrastructure	Implement, where possible, a hierarchical mobility system (Woonerf)	✔	
			Strengthen the rail network	✔	✔
	Green mobility	Coverage, quality, and use of green mobility infrastructure	Integrate alternative mobility models in public spaces (e.g., charging stations for electric cars)		✔
			Enhance urban green infrastructure	✔	✔
			Create urban blue infrastructures	✔	✔
			Extend and integrate the main framework of the urban and extra-urban cycling network	✔	✔

5. Discussion

The results presented in Tables 3 and 4 correspond to the two explicit outputs, namely:

- Definition of a preliminary set of urban welfare indicators—exportable to different territorial contexts—to assess the quality of city components (mostly public ones) and how they impact the well-being of settled inhabitants.
- Defining some strategies for the planning, designing, and regenerating of these components, which are also exportable to different territorial contexts.

To demonstrate the relationship between the summarized results in Tables 3 and 4, this paragraph presents a reflection on some urban regeneration strategies envisaged by the local plans of the two analyzed cities. These strategies have contributed to attaining the top position in international quality of life rankings or are anticipated to strengthen this standing in the near future.

For the sake of brevity, only a few indicators from Table 3 (at least one from each of the three identified reference systems: environmental, settlement-morphological, and mobility systems) have been selected. These are then correlated with some strategies (belonging to the same reference systems) chosen from Table 4. This relationship is clearly depicted in Table 5.

Regarding the Environmental System, "Perception of discomfort about the climate" and "Air quality" are two indicators referenced from the two analyzed datasets (the first in the EIU dataset, the second in the S24O dataset) and included in Table 3, "Initial categorization of urban welfare indicators." These indicators fall under the "Climate Change" category and measure the impact of climate change on urban structures. Among the strategies identified in the local plans of Vienna and Bologna and summarized in Table 4, "Mitigate the urban heat island effect" and "Envision interventions for the unsealing and depavement of soil" were the most effective in improving these two indicators [49,50].

Regarding the Settlement-morphological System, "Perception of fear" is an indicator found in both the EIU and S24O datasets, also included in Table 3, "Initial categorization of urban welfare indicators." This indicator belongs to the "Social stability" category and measures the population's "Perception of public space." Among the strategies outlined in the local plans of Vienna and Bologna and synthesized in Table 4, "Renew the street space in terms of formal and environmental quality, accessibility, and safety" appears to be the most impactful in improving this indicator.

Table 5. Relations and interdependencies between Indicators and Strategies.

Relations and Interdependencies between Indicators and Criteria/Parameters						
Indicators from Table 3	**Strategies from Table 4**	**Reference System**	**Reference Category**	**Impact on the Urban Structure**	**Typologies**	
					SE	**EN**
Perception of discomfort about the climate (EIU) and Air Quality (S24O)	Mitigate the urban heat island effect and envision interventions for the unsealing and depavement of soil	Environmental	Climate change	Effects of climate change on urban climate		✔
Perception of fear (EIU and S24O)	Renew the street space in terms of formal and environmental quality, accessibility, and safety	Settlement-morphological	Social stability	Perception of public space	✔	
Road quality (EIU) and Presence of cycle paths (S24O)	Implement, where possible, a hierarchical mobility system (Woonerf)	Mobility	Infrastructural accessibility	Mobility infrastructure	✔	

Concerning the Mobility System, "Road quality" and "Presence of cycle paths" are two indicators referenced from the two analyzed datasets (the first in the EIU dataset, the second in the S24O dataset) and included in Table 3 "Initial categorization of urban welfare indicators." These indicators belong to the "Infrastructural accessibility" category and measure the quality of "Mobility Infrastructure." Among the strategies outlined in the local plans of Vienna and Bologna and synthesized in Table 4, "Implement, where possible, a hierarchical mobility system (Woonerf)," "Integrate alternative mobility models in public spaces (e.g., charging stations for electric cars)," and "Enhance urban green infrastructure" appear to be the most effective in improving these indicators.

6. Conclusions

It should be specified that both results are preliminary considerations that represent a reference from a methodological perspective. The overall goal of the contribution was to provide theoretical, methodological, and operational references for innovating the local urban plan with a climate-equitable approach, i.e., oriented toward sustainable urban development from both an environmental and socioeconomic standpoint.

In this sense, the aim was to demonstrate, starting from virtuous best practices, an explicit relevance between quality of life and quality of the urban environment, made evident by the fact that both the city of Vienna and Bologna (the two cities with the best quality of life for the year 2022) contain explicit references, strategies, criteria, and parameters within their local planning instruments aimed at ensuring the quality of urban components.

Furthermore, it was intended to show how it is possible to deduce, from these best practices, both exportable urban welfare indicators and strategies for a climate-equitable innovation of local planning instruments through the definition of guidelines for the planning, design, and regeneration of urban components.

Certainly, both the indicators presented in Table 3 and the strategies in Table 4 should be seen as initial results, subject to improvement and integration. Indeed, there are numerous datasets on socioeconomic and environmental vulnerability and sustainability from which further references can be deduced to investigate connections between quality of life and quality of the urban environment (both at the international and Italian national levels). An interesting development in research could involve integrating specific indicators selected from the datasets mentioned in Table 1:

- ISTAT Dataset (2017) prepared for the Commission of Inquiry on the Degradation of Cities and Suburbs [45].
- Dataset prepared by Legambiente and Caritas Italiana for the report "Territori civili. Indicators, maps, and best practices towards integral ecology" (2020) [46].
- Statistical measures for monitoring the SDGs at the regional level [47].
- BES Report (2021) [48].

Regarding the result presented in Table 4, it might be interesting, in future developments of the research, to differentiate the strategies based on the field of application. This would involve defining specific and differentiated guidelines for urban regeneration interventions within the historic and consolidated city, planning and design for urban expansion and transformation areas.

In conclusion, this research aimed to emphasize the importance of considering the quality of urban components and the urban environment in general as a determining factor in citizens' perception of well-being. Additionally, it laid the groundwork for future developments in two directions:

- Implementation of both urban welfare indicators and strategies through the definition of specific criteria/parameters for the planning, design, and regeneration of urban components with a climate-equitable approach;
- Validation and verification of urban welfare indicators in disadvantaged contexts and the application of criteria to delineate site-specific and climate-equitable urban regeneration interventions.

Funding: This research was funded by the Sapienza University of Rome, Research Project "Urban welfare, public city, and rights. Strategies, tools, mechanisms for an innovation of the local plan in a climate-proof perspective" (P.I. Marsia Marino). Protocol No. AR22218168348034.

Data Availability Statement: The data presented in this study are available in article.

Acknowledgments: The research project is part of a broader research project titled "Urban welfare, public city, and rights: A public governance strategy for the city and contemporary territories," which is funded by a scientific agreement between the Department PDTA at Sapienza University of Rome, and the Lazio Region–Regional Directorate of Housing Policies and Territorial, Landscape and Urban Planning (P.I. Laura Ricci).

Conflicts of Interest: The author declares no conflict of interest.

Appendix A

Table A1. Quali-quantitative parameters derived from the local urban plan of Vienna (STEP 2025–Annex: Gründerzeit Action Plan).

Quali-quantitative parameters: The "parterre street" and the "common room"
The Gründerzeit Action Plan (2018) highlights two fundamental characteristics that the public and semi-public spaces of Vienna's historic and consolidated city should incorporate in order to ensure urban quality: • The street is understood as a "parterre." • Public and semi-public spaces as a "common room." From the study of the Gründerzeit Action Plan (2018) [31], it is evident that both of these characteristics are closely connected to the morphology and typological-functional division of the buildings.Below are some qualitative-quantitative parameters identified by the Plan for achieving the concepts of "parterre street" and "common room."

Table A1. *Cont.*

<table>
<tr><th colspan="2">The "parterre street"</th></tr>
<tr><td colspan="2">It has been verified that the heights of the buildings comprising a "Superblock" (morphological configuration of the districts in Vienna's historic and consolidated city) are all equal and not proportionate to the width of the street each one faces. For this reason, the plan provides precise indications regarding the possibility of elevation (Parameter 1) in relation to the width of the streets on which the buildings face, differentiating, in this way, the street fronts proportionally, with particular emphasis on the corners (only those facing squares or public spaces that are intended to be emphasized) (Figure A and B). In the elevation process, the social mix is taken into account, providing for certain housing units and rent control prices (the private sector is encouraged in this practice through volumetric bonuses or through the possibility of changing land use). There is also the possibility of integrating new social housing buildings into the existing fabric (Figure C).Furthermore, in the characterization of high-quality public space, mobility plays a fundamental role (Parameter 2); therefore, STEP 2025, in general, and the Gründerzeit Action Plan in particular, place great emphasis not only on concepts of alternative mobility—such as, for example, car sharing (Figure D)—but also on a hierarchy of flows in urban regeneration interventions, envisioning green and slow mobility within the superblock (woonerf model) and a drivable one between one superblock and another (Figure E).</td></tr>
<tr><td colspan="2">Parameter 1: Building Upward</td></tr>
<tr><td>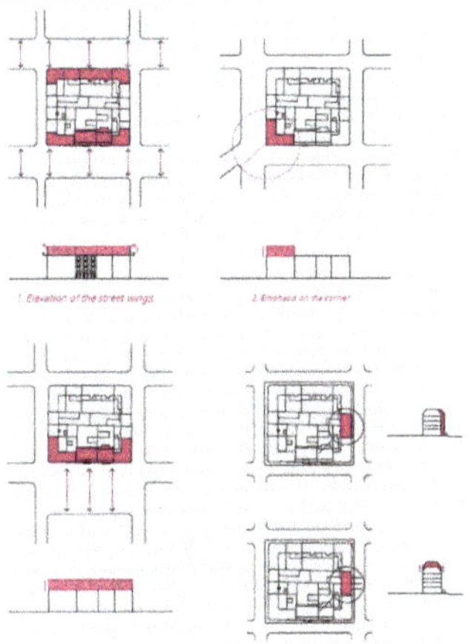</td><td>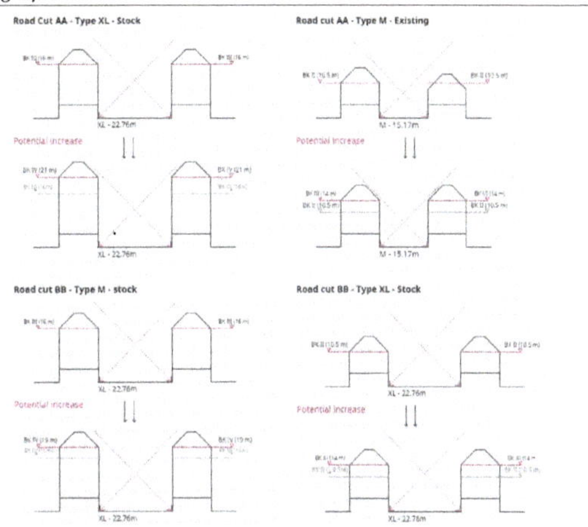
</td></tr>
<tr><td>Figure A. Possible building upward in relation to the road width. Source: Gründerzeit Action Plan (2018).</td><td>Figure B. Indications relating to possible building upward. Source: Gründerzeit Action Plan (2018).</td></tr>
<tr><td>Description of Figure A:
This action allows for the possibility of adding extra floors to the buildings that make up the perimeter of the superblock. The aim is to meet the increasing demand for housing while ensuring urban quality (and thus managing the transformation) and preserving social diversity within the same residential building. The action provides four possibilities for adding extra floors based on the building/public space relationship:
• Elevation of street wings: There is the possibility of adding extra floors to the parts of the building facing wider streets. However, any extra floors must take into account shading to avoid negatively affecting the apartments on the first floors of the buildings in front;
• Emphasis on corners: There is the possibility of adding extra floors to the corners of some superblocks, but only in cases where it is necessary to enhance or emphasize the public space in front;
• Defining the space: In some cases, adding extra floors is possible to better define the public space, with the intention of giving it greater spatial recognition. Again, special attention will be paid to shading to avoid negatively impacting the public space. A predominantly shaded public space throughout the day is less attractive;
• Adapting the existing city: In cases where a building protrudes beyond the perimeter of the superblock and thus narrows the sidewalk, an increase in building volume is not desirable. In this case, compensation may be provided to discourage this practice, such as increasing the building's classification.</td><td>Description of Figure B:
The image provides a detailed description of the possibility of adding extra floors, quantifying it in relation to the width of the street on which the building faces:
• Buildings facing the street, each with a width of 16 m on a street of approximately 22 m | possibility of adding extra floors up to 21 m for XL type superblocks, up to 19 m for M type superblocks;
• Buildings facing the street, each with a width of 10 m on a street of approximately 15 m | possibility of adding extra floors up to 14 m for M and XL type superblocks.</td></tr>
</table>

Table A1. *Cont.*

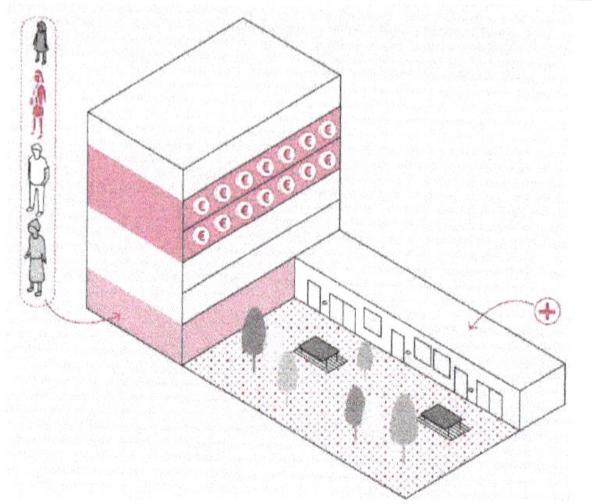	*Description of Figure C:* Due to the increasing demand for housing, there is a growing trend of buying and constructing apartments for investment purposes. This trend negatively impacts the social mix within residential buildings. Therefore, if a building undergoes actions aimed at improving the quality and value of its apartments (such as through increased volume or changes in use), it is suggested that a portion of the apartments be offered at controlled rental prices. The possibility of establishing agreements in this regard with social associations or organizations is also envisaged. Additionally, where possible, there should be provisions for integrating new constructions with dedicated units for socioeconomically vulnerable groups or those in need. Furthermore, for new constructions, a minimum quota of social housing should be considered in relation to the size of the project.
Figure C. Social mixitè through the integration of housing at controlled prices in buildings used for private residences and the combination of new social housing structures with buildings used for private residences. Source: Gründerzeit Action Plan (2018).	
Parameter 2: Mobility	
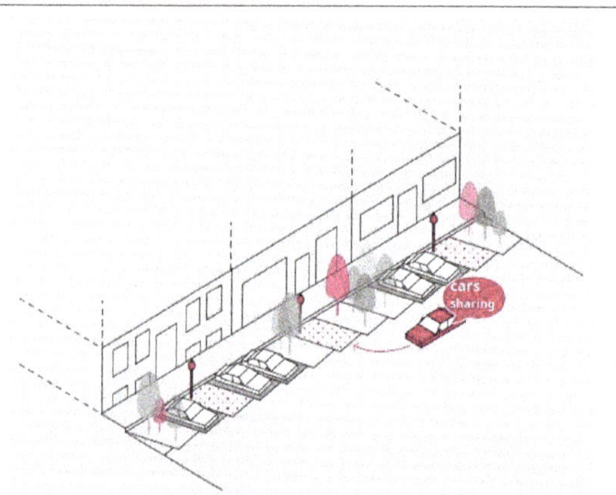	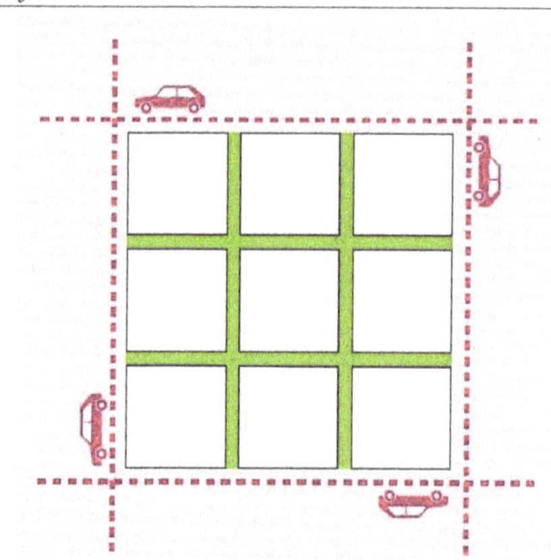
Figure D. Integration of alternative mobility models in public space. Source: Gründerzeit Action Plan (2018).	**Figure E.** Woonerf. Source: Gründerzeit Action Plan (2018).
Description of Figure D: Alternative mobility measures (such as car sharing) have a dual benefit: on the one hand, they have a lower impact in terms of pollution, and on the other, they lead to a reduction in parking spaces, allowing for increased space for pedestrians and bike lanes. Encouraging a form of condominium or neighborhood car sharing is also recommended. Furthermore, a reduction in parking spaces is planned, with the number of spaces being reduced based on the number of units per individual building. Exceptions may be made as a form of compensation for the creation of additional quality.	*Description of Figure E:* The extension of the restricted traffic zone is planned for secondary streets (inside the superblock), where the concept of a "woonerf" can be implemented.
Public and semi-public space as a "common room"	
Another aspect to consider in order to ensure a quality public space is the function of the ground floor of buildings (Figure F). This is considered particularly important because it influences both the space in front (public space) and the inner courtyard of the building block (semi-public space). In this sense, the ground floor of buildings in the Gründerzeit area should accommodate functions related to the building, such as bicycle parking (to free up the backyard) and commercial functions (Parameter 1). Furthermore, the emphasis that the Vienna city administration puts on the role of private individuals in managing the public space in front of their property is particularly relevant. This should include the possibility of a social and shared use of the sidewalk (Figure G). As for the semi-public space (backyard), great importance is attributed to its potential role in social aggregation aimed at strengthening neighborly relations. To this end, Parameter 2 is aimed at avoiding privatization initiatives and encouraging the shared use of spaces (Figure H).	

Table A1. *Cont.*

Parameter 1: Social and Shared Use of the Sidewalk	
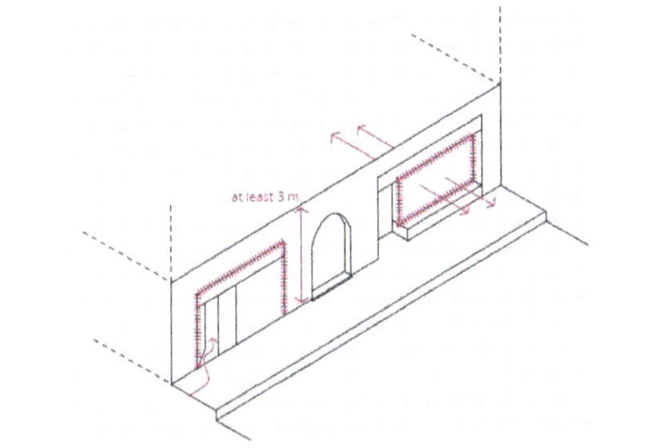	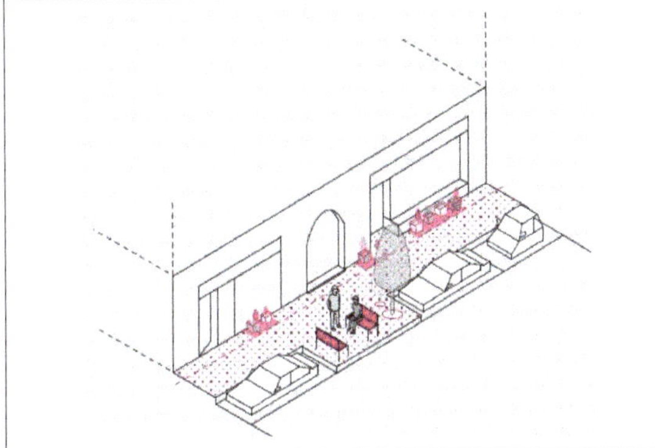
Figure F. Guidelines for the design and refunctionalization of the ground floor of buildings. Source: Gründerzeit Action Plan (2018).	**Figure G.** Suggestions for the regeneration of public space facing the street (parklet). Source: Gründerzeit Action Plan (2018).
Description of Figure F: In addition to contributing to the social use of the sidewalk as a common and shared space, the function of the ground floor also plays a fundamental role in the criterion of the "street parterre." To this end, for new constructions, it will be necessary for the spaces located on the ground floor to have a minimum height of 3 m in order to promote the social use of the spaces, such as playrooms, playgrounds, and commercial functions. Currently, the low heights of these spaces favor secondary uses, such as storage and garages, which do not encourage positive interaction with the public space in front. It is also desirable to provide for a flexible use of the ground floor premises, including through the design of an open structure.	*Description of Figure G:* The use of the ground floor of buildings is also connected to the management of the public space in front. The Gründerzeit Action Plan (2018) encourages private individuals to take responsibility for the care, maintenance, and animation of the public space, also through initiatives like "Street Life Wien," which aims to involve and encourage citizens to use the streets and public spaces as a "common room" [51].
Parameter 2: Backyard Function as a Social Gathering Space	
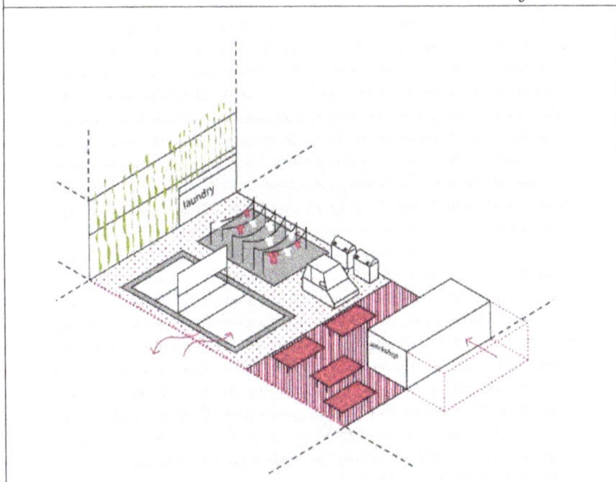	*Description of Figure H:* In order to promote the use of semi-public spaces as a "common room," the regeneration of these spaces is encouraged by clearing them of bicycles and recycling bins (for example, by allocating some ground floor areas for this purpose) and by promoting common and shared activities such as laundry, workshops, and playgrounds. For new constructions, it is preferable to limit vertical partitions in order to favor spacious and shared areas.
Figure H. Suggestions for the regeneration of semi-public space within the building blocks (backyard). Source: Gründerzeit Action Plan (2018).	

Appendix B

Table A2. Guidelines and quali-quantitative parameters derived from the local urban plan (PUG) of Bologna: Explanation of "Objectives," "Urban Strategies," "Specific Actions," and "Fields of Application".

Objectives	Urban Strategies	Actions	Field of Application
Resilience and Environment	Promote the regeneration of anthropized soils and counteract soil consumption	Promoting the recovery and improvement of existing building heritage	Urbanized territory
			Rural territory
		Completing unfinished parts of the city where transformation is not complete	Incomplete city parts
		Promoting interventions for the reuse and urban regeneration of built-up areas and anthropized soils	Historic city
			Planned parts with implementing urban planning tool
			City parts under construction
		Including measures for the de-sealing and de-pavement of soils	Municipal territory
			River areas
	Develop the urban eco network	Protect biodiversity and the main ecosystem services of the hills and plains	Hill rural territory
			Plain rural territory
		Strengthen urban green infrastructure	Urbanized territory perimeter
		Building a blue urban infrastructure	Municipal territory
			Bodies of water in major basins
			Active riverbeds and bodies of water in basins
		Maintain natural flows in the riverbed and reduce withdrawals from aquifers	Municipal territory
			Primary non-potable water networks
		Improve the quality of surface waters	Channels to be restored-areas 20 m away
			Minor hydrographic network areas 50 m away
			Covered network areas 100 m away
			Municipal territory
	Prevent and mitigate environmental risks	Contain natural risks	Areas in distress
			Areas of possible evolution and influence of distress
			Areas with an inclination for territorial transformation
		Ensure regular drainage of water in the entrances of streams and covered ditches	Inlets of hillside streams and culverted hillside ditches/upstream area
			Inlets of hillside streams and culverted hillside ditches/first 150 m from the upstream area
		Mitigate the urban heat island effect	Areas of microclimatic fragility
		Reduce the population's exposure to pollution and anthropogenic risks	Municipal territory
			Areas with high noise pollution/areas facing the main infrastructures
			Areas with high noise pollution/areas underlying the nominal routes
	Support the energy transition and circular economy processes	Promote and incentivize various forms of energy efficiency and ensure equitable access to low environmental impact energy services.	Municipal territory
		Plan the deployment of energy production facilities from renewable sources by creating local distribution networks.	Municipal territory
		Promote the circular economy of construction and excavation materials.	Municipal territory
		Increase recycling and reduce waste production	Municipal territory
			Collection and reuse centers for urban waste-first 100 m

Table A2. *Cont.*

Objectives	Urban Strategies	Actions	Field of Application
Habitability and inclusion	Extend access to the house	Promote the increase and innovation of rental housing supply	Urbanized territory
		Promote the increase of social housing supply.	Areas where to increase the supply of ERS
		Experiment with new forms of housing.	Municipal territory
		Introduce functional and typological mixes in specialized areas near residential fabrics.	Specialized areas near residential fabrics
	Ensure the creation of a balanced network of quality equipment and services	Promote the redevelopment and establishment of territorial amenities	Areas at risk of social marginality
			Municipal territory
		Support a balanced distribution of spaces for culture	Perimeter of the urbanized territory
		Foster local services and commercial activities	Municipal territory
			Perimeter of the urbanized territory
		Promote sustainable urban logistics	Municipal territory
			Perimeter of the urbanized territory
	Redesign spaces and equipment	Make the city universally accessible	Municipal territory
		Create open spaces and public buildings of high architectural and environmental quality	Areas at risk of social marginality
			Perimeter of the urbanized territory
		Renew the street space in terms of formal and environmental quality, accessibility, and safety	Municipal territory
			Accessibility to the backbone network of local public transport
	Preserve the characteristics of the historic urban landscape by renewing its role	Preserve the habitability and characteristics of the historic city	Fabrics of the historic city-nucleus of ancient formation
			Fabrics of the historic city-garden neighborhoods
			Fabrics of the historic city-compact fabric
			Buildings without particular interest in the fabrics of the historic city (ES)
			Buildings facing Via dell'Indipendenza, Via Ugo Bassi, and Via Rizzoli
		Enhance the specialized fabrics of the historic city	Fabrics of the historic city-specialized
		Ensure the conservation of the architectural and cultural heritage of historical significance	Point elements of interest
			Arcades
			Parks of historical interest
			Historical and urbanistic relevance
			Buildings of cultural and testimonial interest
			Buildings of historical and architectural interest
			Buildings of historical and architectural interest of the Modern era
		Enhance the architecture and cultural and testimonial agglomerates of the Second Half of the Twentieth Century	Agglomerations of cultural and testimonial interest of the Second Twentieth Century
			Buildings of interest and pertinence—buildings of cultural and testimonial interest of the Second Twentieth Century

Table A2. *Cont.*

Objectives	Urban Strategies	Actions	Field of Application
Attractiveness and work	Support overall urban reinfrastructure	Reconstruct the unified map of infrastructure networks, nodes, intersections, and managers	Urbanized territory
		Ensure the improvement of urban infrastructure with urban and building transformation interventions	Perimeter of the urbanized territory
		Promote the distribution and coordination of digital infrastructure	Perimeter of the urbanized territory
		Qualify the role and visibility of the city's access gates and create a system of mobility centers	Mobility centers and priority areas of metropolitan urban regeneration
			City access gates
		Improve the functionality of the highway-ring road system, mitigating impacts and redeveloping contact areas with the city	Areas affected by the enhancement project within the highway-ring road system
			Highway-ring road system areas 100 m away
		Build the urban tram network	Urbanized territory
		Extend and integrate the backbone of the urban and extra-urban cycling network	Municipal territory
	Promote the widespread establishment of economic activities in conditions of environmental compatibility	Ensure existing businesses have regulatory and procedural flexibility	Urbanized territory
		Identify new production needs, directing them toward the reuse and regeneration of urbanized areas	Planned production areas
			Perimeter of urbanized territory
		Promote innovation in planned production areas through the diversification of uses	Planned production areas
		Foster the establishment of innovative companies and the promotion of innovation centers	Areas near innovation centers
			Technopole
			Perimeter of urbanized territory
	Support the qualification of metropolitan hubs integrated into living places inserted in the context	Bologna Guglielmo Marconi Airport: support a development that is mindful of its relationship with the city	Bologna Guglielmo Marconi Airport
		Bologna Centrale Railway Station and Bologna Bus Station: integrate access, transit, and parking areas with quality urban functions	Bologna Central Railway Station and Bus Station
		University of Bologna-Alma Mater Studiorum: enhance and connect the campus facilities	Campus of Alma Mater Studiorum University of Bologna
		Healthcare Centers of Excellence: support the process of adapting facilities to social and environmental changes, improving accessibility conditions	Centers of excellence in healthcare
		Bologna Trade Fair: develop the multifunctionality of the hub, improving access methods at different scales	Bologna Fair
		Renato Dall'Ara Stadium: regenerate the facility and its relationships with the city	Renato Dall'Ara Stadium
		North-East District (CAAB, FICo Eataly World, Meraville, Business Park, University): integrate components and implement new infrastructure for access	Northeast District
	Qualify the relationship between urban territory and extra-urban territory	Promote innovative practices of peri-urban agriculture	Hillside rural territory
			Plain rural territory
		Enhance peri-urban parks, improving their usability for tourism	Periurban parks
		Develop networks of safe routes and paths connected to national and European tourist itineraries	Cycling and pedestrian tourist routes
			Hiking trails
			Hillside rural territory
			Plain rural territory

References

1. Castells, M. *The Urban Question a Marxist Approach*; MIT Press: Cambridge, MA, USA, 1979.
2. Donzelot, J. *Quand la Ville se Défait. Quelle Politique Face à la Crise des Banlieues?* Points: Paris, France, 2008.
3. Mimar, S.; Soriano-Paños, D.; Kirkley, A.; Barbosa, H.; Sadilek, A.; Arenas, A.; Gómez-Gardeñes, J.; Ghoshal, G. Connecting intercity mobility with urban welfare. *PNAS Nexus* **2022**, *1*, pgac178. [CrossRef] [PubMed]
4. Indovina, F. Metropoli territoriale e sviluppo economico-sociale. In *Economia e Società Regionale*; FrancoAngeli: Milano, Italy, 2010.
5. Indovina, F. *Dalla Città Diffusa All'arcipelago Metropolitano*; FrancoAngeli: Milano, Italy, 2009.

6. Indovina, F. La Città diffusa. In *La Città Diffusa*; Indovina, F., Matassoni, F., Savino, M., Sernini, M., Torres, M., Vettoretto, L., Eds.; Daest-IUAV: Venezia, Italy, 1990; pp. 21–43.
7. Castells, M.; Rizzo, C. *Città Delle Reti*; Marsilio: Padova, Italy, 2004.
8. Augè, M. *Non-Lieux: Introduction à une Anthropologie de la Surmodernité*; Seuil: Paris, France, 1992.
9. Francesco, P. *Laudato Si': Enciclica sulla cura della casa Comune*; Libreria editrice Vaticana: Città del Vaticano, 2020.
10. Ricci, L. Nuova questione urbana e nuovo welfare. Una rete di reti per la costruzione della città pubblica. In *Città Pubblica e Nuovo Welfare. Una Rete di Reti Per la Rigenerazione Urbana*; Ricci, L., Crupi, F., Iacomoni, A., Mariano, C., Eds.; Urbanistica Dossier: Roma, Italy, 2021.
11. Khan, J.; Hildingsson, R.; Garting, L. Sustainable Welfare in Swedish Cities: Challenges of Eco-Social Integration in Urban Sustainability Governance. *Sustainability* **2020**, *12*, 383. [CrossRef]
12. United Nation. Strategy for Sustainability Management in the United Nations System, 2020–2030 Phase II: Towards Leadership in Environmental and Social Sustainability. 2021. Available online: https://unsceb.org/sites/default/files/2022-03/CEB.2021.2.Add_.1-Strategy%20for%20Sustainability%20Management%20in%20the%20United%20Nations.Phase%20II.pdf (accessed on 10 August 2023).
13. United Nation. A Framework for Advancing Environmental and Social Sustainability in the United Nations System. 2012. Available online: https://sustainabledevelopment.un.org/content/documents/2738sustainabilityfinalweb-.pdf (accessed on 10 August 2023).
14. United Nation. Framework for Advancing Environmental and Social Sustainability in the UN System. 2016. Available online: https://unemg.org/images/emgdocs/sustainabilitymanagement/Synthesis_report_shortened_June2016.pdf (accessed on 10 August 2023).
15. COM. Urban Agenda for the EU Pact of Amsterdam. 2016. Available online: https://ec.europa.eu/regional_policy/sources/policy/themes/urban-development/agenda/pact-of-amsterdam.pdf (accessed on 18 September 2023).
16. COM. The European Green Deal. 2019. Available online: https://eur-lex.europa.eu/resource.html?uri=cellar:b828d165-1c22-11ea-8c1f-01aa75ed71a1.0002.02/DOC_1&format=PDF (accessed on 3 October 2023).
17. COM. Communication from the Commission to the European Parliament, the Council, the European Economic and Social Committee and the Committee of the Regions a Strong Social Europe for Just Transitions. 2020. Available online: https://eur-lex.europa.eu/resource.html?uri=cellar:e8c76c67-37a0-11ea-ba6e-01aa75ed71a1.0003.02/DOC_1&format=PDF (accessed on 3 October 2023).
18. COM. Overview of Investment Guidance on the just Transition Fund 2021–2027 per Memeber State. 2020. Available online: https://commission.europa.eu/system/files/2020-02/annex_d_crs_2020_en.pdf (accessed on 3 October 2023).
19. COM. New European Bauhaus. Beautiful, Sustainable, Together. 2021. Available online: https://new-european-bauhaus.europa.eu/system/files/2021-09/COM(2021)_573_EN_ACT.pdf (accessed on 3 October 2023).
20. ASVIS. Agenda Urbana Per lo Sviluppo Sostenibile. 2017. Available online: https://asvis.it/public/asvis/files/AgendaUrbana.pdf (accessed on 19 October 2023).
21. PNRR. Piano Nazionale di Ripresa e Resilienza. 2021. Available online: https://www.governo.it/sites/governo.it/files/PNRR.pdf (accessed on 19 October 2023).
22. Lefebvre, H. *Right to the City*; Ombre Corte: Verona, Italy, 1968.
23. Ricci, L. Città pubblica e Nuovo Welfare. Una rete di reti per la Rigenerazione Urbana. *Ananke 92 (Torino, Italy)* **2021**, 93–98.
24. Caldarice, O. Urban Welfare in Italy: From Urban Standards to Urban Facilities. In *Reconsidering Welfare Policies in Times of Crisis*; Caldarice, O., Ed.; Springer: Berlin/Heidelberg, Germany, 2018.
25. D'Onofrio, R.; Trusiani, E. *Città Salute e Benessere*; FrancoAngeli: Milano, Italy, 2017.
26. Sallis, J.F.; Cerin, E.; Conway, T.L.; Adams, M.A.; Frank, L.D.; Pratt, M.; Salvo, D.; Schipperijn, J.; Smith, G.; Cain, K.L.; et al. Physical activity in relation to urban environments in 14 cities worldwide: A cross-sectional study. *Lancet* **2016**, *387*, 2207–2217. [CrossRef] [PubMed]
27. Barton, H.; Grant, M. A health map for the local human habitat. *J. R. Soc. Promot. Health* **2006**, *126*, 252–253. [CrossRef] [PubMed]
28. Ricci, L.; Mariano, C.; Marino, M. *Well-Being Cities and Territorial Government Tools: Relationships and Interdependencies*; LNCE Springer: Berlin/Heidelberg, Germany, 2023; in press.
29. Istat. Allegato Statistico Per la Commissione Parlamentare di Inchiesta Sulle Condizioni di Sicurezza e Sullo Stato di Degrado Delle Città e Delle Loro Periferie. 2017. Available online: https://www.istat.it/it/archivio/202052 (accessed on 20 May 2023).
30. Legambiente e Caritas italiana. *Territori Civili. Indicatori, Mappe e Buone Pratiche verso L'ecologia Integrale*; Edizioni Palumbi: Teramo, Italy, 2020.
31. Istat. Indicatori SDGs. 2022. Available online: https://public.tableau.com/app/profile/istat.istituto.nazionale.di.statistica/viz/SDGs_public_ottobre_2022/SDGs?publish=yes (accessed on 20 May 2023).
32. Istat. Rapporto BES 2021: Il Benessere Equo e Sostenibile in Italia. 2021. Available online: https://www.istat.it/it/archivio/269316 (accessed on 20 May 2023).
33. EIU. The Global Liveability Index 2022. In *Recovery and Hardship*; The Economist Group: London, UK, 2022.
34. Sole 24 Ore. 33ª Indagine Sulla Qualità Della Vita. 2022. Available online: https://lab24.ilsole24ore.com/qualita-della-vita/ (accessed on 30 May 2023).

35. STEP 2025. *Urban Development Plan Vienna*; Vienna City Administration: Vienna, GA, USA, 2014; Available online: https://www.wien.gv.at/stadtentwicklung/studien/pdf/b008379b.pdf (accessed on 5 June 2023).
36. Masterplan Gründerzeit. Handlungsempfehlungen zur Qualitätsorientierten. *Weiterentwicklung der Gründerzeitlichen Bestandsstadt; Magistrat der Stadt Wien, MA 21—Stadtteilplanung und Flächennutzung*. 2018. Available online: https://www.superblock.at/_files/ugd/5bb4e2_10129184b72f4a939212b16ed0598ef8.pdf (accessed on 5 June 2023).
37. PUG Bologna. *Approfondimenti Conoscitivi: Governance*; Comune di Bologna: Bologna, Italy, 2021.
38. PUG Bologna. *Approfondimenti Conoscitivi: Mobilità*; Comune di Bologna: Bologna, Italy, 2021.
39. PUG Bologna. *Approfondimenti Conoscitivi: Paesaggio*; Comune di Bologna: Bologna, Italy, 2021.
40. PUG Bologna. *Approfondimenti Conoscitivi: Patrimonio Abitativo*; Comune di Bologna: Bologna, Italy, 2021.
41. PUG Bologna. *Approfondimenti Conoscitivi: Popolazione*; Comune di Bologna: Bologna, Italy, 2021.
42. PUG Bologna. *Approfondimenti Conoscitivi: Servizi Alle Persone*; Comune di Bologna: Bologna, Italy, 2021.
43. PUG Bologna. *Approfondimenti Conoscitivi: Sistema Economico*; Comune di Bologna: Bologna, Italy, 2021.
44. PUG Bologna. *Strategie urbane: Abitabilità e Inclusione*; Comune di Bologna: Bologna, Italy, 2021.
45. PUG Bologna. *Strategie urbane: Attrattività e Lavoro*; Comune di Bologna: Bologna, Italy, 2021.
46. PUG Bologna. *Strategie urbane: Resilienza e Ambiente*; Comune di Bologna: Bologna, Italy, 2021.
47. PUG Bologna. *Strategie urbane: Generale*; Comune di Bologna: Bologna, Italy, 2021.
48. PUG Bologna. *Leggere il Piano*; Comune di Bologna: Bologna, Italy, 2021.
49. Zhou, L.; Hu, F.; Wang, B.; Wei, C.; Sun, D.; Wang, S. Relationship between urban landscape structure and land surface temperature: Spatial hierarchy and interaction effects. *Sustain. Cities Soc.* **2022**, *80*, 103795. [CrossRef]
50. Zhou, L.; Yuan, B.; Hu, F.; Wei, C.; Dang, X.; Sun, D. Understanding the effects of 2D/3D urban morphology on land surface temperature based on local climate zones. *Build. Environ.* **2022**, *208*, 108578. [CrossRef]
51. Street Life. Available online: https://streetlife.wien/ (accessed on 8 June 2023).

land

MDPI

Article

Placing Urban Renewal in the Context of the Resilience Adaptive Cycle

Lars Marcus [1,2,*] and Johan Colding [2,3]

[1] Department of Architecture and Civil Engineering, Chalmers University of Technology, 412 96 Gothenburg, Sweden
[2] Department of Building Engineering, Energy Systems and Sustainability Science, University of Gävle, Kungsbäcksvägen 47, 801 76 Gävle, Sweden; johan.colding@hig.se or johanc@beijer.kva.se
[3] The Beijer Institute of Ecological Economics, Royal Swedish Academy of Sciences, 104 05 Stockholm, Sweden
* Correspondence: lars.marcus@chalmers.se

Abstract: Resilience thinking provides valuable insights into the dynamics of complex adaptive systems. To achieve resilience in urban systems, it can be fruitful to delve into the intricacies of resilience processes. This paper theorizes about how the specific characteristics of resilient systems can be integrated into the spatial design of cities. Emphasizing the importance of the built form and spatial systems in maintaining order within urban processes, we focus on how adaptive renewal cycles can be applied to various systems and dimensions where urban change, adaptation, and renewal occur. The paper identifies key resilient system characteristics applicable to urban spatial form and contextualizes urban renewal within the adaptive renewal cycle—a framework originally developed to capture temporal and spatial ecosystem dynamics. We integrate insights within 'space syntax theory', theorizing about how cities renew themselves over space and time. We discuss instances of 'compressed resilience' and the challenges posed by the 'tyranny of small decisions' in urban planning and development. In conclusion, we identify future research directions in the theory of spatial morphology and resilient urban systems, emphasizing the need for a deeper understanding of the interplay between urban processes, urban form, resilience, and adaptive renewal.

Keywords: resilience; urban renewal; the adaptive renewal cycle; compressed resilience; space syntax theory; spatial morphology

1. Introduction

With their irregularity, cities are systems of complexity par excellence where various elements interact in unpredictable ways. The prominent urban theorist and activist Jane Jacobs made a significant contribution to the understanding of cities and their complex dynamics. Her work emphasized the organic, diverse, and dynamic nature of urban systems. Her most influential work, *The Death and Life of Great American Cities*, published in 1961, challenged the prevailing urban planning theories of the time [1]. Jacobs underscored the importance of diversity, community engagement, bottom-up growth, social interaction, and the concept of cities as self-organizing, dynamic ecosystems rather than rigidly planned structures.

The recognition of the striking similarities between urban systems and ecosystems [2], initially highlighted by Jacobs, was about a decade later rediscovered by C.S. Holling as a relevant framework for understanding urban system dynamics [3]. It is striking how Jacob's keystone ideas resemble the key principles of resilience building [4–6], where resilience signifies a typical property found in all systems of complexity, i.e., their ability to sustain steady states over long time periods [7].

While the concept of complexity is often associated with irrationality and chaos, it signifies that systems fluctuate between stable and unstable states. Therefore, complexity encompasses all the possibilities that unfold between these states as evolution progresses.

What is surprising about cities, as well as many other complex systems, given their multitude of intricate sub-systems, is perhaps not their unpredictability and irregularity, but rather that, over long time periods, they remain stable and predictable just like the climate undergoes long periods of stability, considering that the Earth has, ever since the ice age, been in an interglacial period known as the Holocene, representing a relatively stable and warm period over the past approximately 11,700 years.

Resilience thinking provides planning theory with a new way of understanding complex, dynamic, and non-linear social, economic, and ecological relationships. As a result, it represents an important sustainability framework in which sustainability is seen as a process, rather than an end-product. Resilience, in this sense, denotes the adaptive capacity needed to deal with continual change and renewal, often triggered by different types of social and ecological disruptions [4].

Urban resilience is a multidimensional concept that has garnered significant attention in theoretical research within various fields during the past two decades, such as in urban studies, geography, environmental science, and urban planning and design. Researchers have explored different aspects of resilience, including its various definitions, measurement, contributing factors, and implications for sustainable development. They often draw on interdisciplinary perspectives, integrating insights from ecology, engineering, sociology, architecture, and economics [2,8–14]. Notable contributions of pivotal resilience research include Berkes and Folke's work on social–ecological resilience [15] and the development of resilience assessment methodologies by Meerow, Newell, and Stults [16]. The importance of adaptive governance and planning in enhancing urban resilience is also emphasized in works such as Pickett et al.'s exploration of anticipatory governance [8]. Marcus and Colding [17] have moreover explored the temporal and spatial dynamics of urban resilience through the lens of spatial analysis tools and theories. Other noteworthy contributions include Adger's exploration of social resilience [18] and Cote and Nightingale's work on resilience and political ecology [19].

In this paper, we make the theoretical argument that resilience thinking could help us to enrich our understanding about how a complex adaptive system, such as a city, undergoes renewal and regeneration. It is essential to underscore that this paper primarily serves as an abstraction, delving into conceptual frameworks and theoretical underpinnings. By exploring the application of resilience thinking in the context of urban dynamics, we aim to contribute to the theoretical foundations that underlie our understanding of the intricate mechanisms involved in the renewal and adaptive evolution of cities. Considering the theme of the Special Issue to which this paper contributes, it might be useful to distinguish between 'regeneration' and 'renewal', with the former implying a comprehensive and transformative process. It may involve not only physical changes but also improvements in social and economic aspects. Renewal, on the other hand, and as we use the term, has a broader application, focusing on the extension and the enhancement of an urban system's existing state without implying a complete overhaul of it. Rather, incremental change is a distinct mark of city development [20].

While ample research has been devoted to urban resilience, there has been a noticeable gap in efforts to bridge urban studies with the resilience adaptive renewal cycle proposed by Holling [21] for the renewal and dynamics of ecological systems. This paper takes a theoretical approach, primarily situating the urban form and urban renewal within the context of the adaptive renewal cycle. Holling and Goldberg [3] highlighted striking parallels between ecological and urban systems. They underscored how these systems operate as interdependent entities, relying on a sequence of historical events, exhibiting spatial connections, and possessing a nonlinear structure. Both systems seem to possess significant internal resilience within a specific range of stability [17]. Furthermore, they did not perceive a city as a uniform structure; instead, they saw it as a spatial mosaic of social, economic, and ecological variables that are connected by a variety of physical and social dispersal processes" [3] (p. 227), or what later Gunderson and Holling [21] termed 'panarchy', representing a nested set of adaptive cycles operating at discrete spatial and temporal scales. The adaptive renewal cycle functions as an elucidative framework for comprehending the dynamics of ecosystem renewal with potential for providing insights also into urban renewal processes more generally. As proposed in this paper, this implies a translation of temporal process into spatial structure, which also will demand support from spatial morphology [22].

Article Outline

This paper is organized as follows: Sections 2–5 delineate the essential theoretical foundations of resilience thinking, crucial for understanding the adaptive cycle. Section 2 outlines the definition of ecological resilience utilized in this paper. Section 3 presents theories of ecosystem renewal, shaped by resilience analytical thinking, originating from Holling's seminal paper, 'Resilience and Stability of Ecological Systems' [23]. Expanding upon these concepts, in Section 4, we interlink them with the adaptive renewal cycle—a pivotal framework in resilience science initially designed to capture temporal and spatial ecosystem dynamics [23,24]. Section 5 delves into situations where resilience may be diminished in the management of complex adaptive systems. In Sections 6 and 7, we scrutinize the intersection of resilience theory with spatial morphology as more specifically formulated in 'space syntax theory' and extensions of it [17,25]. The paper concludes by highlighting potential future research directions within this analytical research domain.

2. Ecosystem Renewal Insights: A Theoretical Framework for Understanding Urban Renewal?

The order or steady state typically found in complex adaptive systems, notwithstanding their intrinsic unpredictability, is in the resilience literature described as a 'regime' within an 'attraction domain'. This steady state is not a state of absolute equilibrium, but a state allowing for fluctuation within the boundaries of the attraction domain [26]. Fluctuation in cities occurs within the boundaries of a dominant attractor state that is generated by subsystem changes occurring at multiple spatial and temporal scales, referred to as 'panarchies' [21]. Thus, the multi-scale dynamics of urban systems are continuously renewed within these fluctuations through a mixture of intentional planning policies, informal self-organized processes, or a combination of both, as they interact with each other. In this sense, they represent self-organizing, complex adaptive systems that hold the opportunity to be managed in resilient ways towards sustainability through processes of transformation, just like any complex adaptive system [27]. The renewal of ecological systems, just as in the case of social systems (but in contrast to the systems of spatial urban form), is dynamic, which we can observe at an elementary level during the reoccurring seasonal changes. The spatial and temporal dynamics of ecosystems can be found in the succession cycle, which addresses the renewal and recuperation of ecosystems after a disturbance, such as fires or storms. Such processes, intrinsic to complex systems, can be contained within a given regime in the form of a dynamic steady state. However, if the domain boundaries of a regime are exceeded, the system may flip into a new regime, i.e., lose resilience, limited by a new attraction domain that changes the properties of the system—often radically [28]. Such a new regime may, moreover, prove difficult or even impossible to reverse [28]. A typical example from ecological studies is how a lake can change from a clear water regime to a turbid regime [29], where both may prove equally resilient, but where the properties and functions fundamentally differ, for instance, the abundance and diversity of fish in the lake. Similarly, degraded forests can turn into desserts, savannahs into shrub lands [30], and coral reefs become algae-dominated [31]. Such shifts may, from the point of view of humans, prove devastating.

Although an entire city rarely experiences a regime shift, except in war situations, the subsystems of a city undergo continuous regime shifts. For example, urban renewal projects can bring about significant changes in the dynamics of a city's subsystems [32], and these changes can have both positive and negative consequences [33,34]. Urban renewal projects may, for instance, lead to increased property values and more attractive living conditions, but they may also lead to gentrification processes through the displacement of existing residents and/or the loss of historical and cultural heritage. Hence, while it may make economic sense to increase property values in a city, it may have negative social effects. Thus, the reduction in social resilience may occur as an economic efficiency tradeoff.

In response to climate change and the need to reduce reliance on fossil fuel transportation, urban renewal in the form of planning frameworks, such as 'transit-oriented development', 'smart growth', and 'new urbanism', is often promoted by urban planners and policymakers. However, the adaptive capacity of cities cannot rely on static urban development paradigms, since social, economic, and ecological challenges evolve over time, presenting new and often complex challenges that demand new innovative solutions [35]. Rather, and in theory, we see the need for more adaptive management of urban development based on more contextually based governance principles.

3. Resilience Categories: Engineering vs. Ecosystem

The conception of resilience by C.S. Holling makes an important distinction between ecological resilience and engineering resilience [26]. The latter is measured as the time it takes for a perturbed system to return to its original state of equilibrium, often illustrated by a ball in a bowl being released after lifting up the side of the bowl. In contrast, Holling illustrates the idea of ecological resilience as: "a landscape of hills and valleys with a ball journeying among them, in part because of internal processes and in part because exogenous events may flip the ball from one stability domain to another" [36] (p. 72). In this case, resilience is measured as the magnitude of disturbance that an individual attraction domain, or valley, can absorb before the ball flips over into another valley or domain of attraction. This presents a different universe constituted by, first, broad valleys where the system's dynamics can play out without the loss of its intrinsic functions, and second, a landscape with multiple equilibria, where ecosystems (because of both internal and external forces) can shift from one steady state to another with the serious effect that their basic properties and functions change, as illustrated in the example of the clearwater lake.

Hence, a steady state in ecosystems does not imply an absolute equilibrium but can exist within a frame of fluctuation, where these fluctuations essentially support the system by staying within its regime. Hence, the idea of resilience significantly differs from the conventional conceptions of stability, or rather, ecological resilience fundamentally differs from engineering resilience [23]. As a matter of fact, one of Holling's key messages—not least true for natural resource management, but potentially equally important for urban management and development—is that keeping an ecosystem or resource management system close to an absolute equilibrium may over time undermine the resilience of the system, i.e., it may wear down the ridges defining the attraction domain so that the ball is more likely to flip into another domain, often with devastating results [37].

Adaptation is an important property of ecological resilience that makes it possible for an ecosystem to undergo change over time while retaining structure and function, and in extension, not only to change but to adapt, which can be described as a process of learning and involving ecological memory [36]. In the ecological conception of resilience, the system does not just bounce back after disturbance, but neither does it fluctuate aimlessly; what it does is to change and adapt. Whereas the ball in the bowl returns to its original position, we do not regain the same forest after its recovery from a fire; it is likely to have adapted. There is, for example, evidence of plant species subjected to repeated fires becoming distinctly more combustible, which helps the survival of their plant community, but is not true for species in communities not subject to a fire [36]. Keeping this analogy in mind, we may also see how cities after fires, for instance, the Great Fire of London in 1666 [38], are rebuilt in the same spatial form, but in brick rather than wood to better endure the next fire.

4. Connecting Processes: The Adaptive Renewal Cycle

C.S. Holling's work laid the foundation for the understanding of adaptive cycles within ecological systems, which can be related to the broader concept of adaptive renewal.

It is not only relevant for ecosystems but is intended as a framework for the understanding of complex adaptive systems generally, from cells to ecosystems and societies. Such all-embracing aims call for caution, but this means that the model was created from the perspective of general systems theory [39], the study of systems per se, with the aim to disclose the role of resilience in systems in principle, not explain the mechanisms of individual systems. Therefore, the adaptive renewal cycle has proved illuminating for a wide range of systems in many fields. At the same time, the reason for caution when comparing natural and human systems remains, as underlined by Holling himself [21].

In Holling's model of ecological succession, ecosystem dynamics are driven by the interaction of four basic functions: exploitation, conservation, release, and reorganization [36]. The first two come close to earlier conceptions of this cycle, often referred to as Clementian succession, where a progressive change in the composition of a community of organisms, either by colonization by species of a disturbed area or an established community, over time leads to a conditional stability identified as a climax community [40]. There are two stages in this progression, where 'exploitation' concerns a process of rapid colonization of disturbed ecosystems by r-strategists, which are species good at capturing dispersed resources, which is why it is often called the r-stage. The other stage, 'conservation', concerns a slower process of resource accumulation by late succession species associated with climax communities [41], which typically builds and stores increasingly complex structures. During the slow sequence from exploitation to conservation, connectedness and stability increase in the system, and nutrients and biomass are slowly accumulated. This stage is referred to as the K-stage. The first two stages are, moreover, referred to as the fore-loop of the cycle.

Holling then introduced two novel stages that are distinct additions to and are even in conflict with the Clementian model. The first is a stage that he called release, or the omega stage (Ω), which erratically occurs when the conservation phase has created an elaborated tightly bound organization that has become what in systems terms is called over-connected. The system then becomes fragile and vulnerable to disturbances, which makes it easily triggered into rapid change when its stored biomass is released, and the tight organization lost. This means an abrupt internal destruction of the system caused by an external disturbance, such as fires, storms, or disease. But this rapid process of destruction to the organization also creates opportunity for the fourth stage in the cycle, reorganization. In this stage, also called the alpha stage (α), the materials released in the previous release stage are again mobilized and made accessible for a new stage of exploitation.

In summary, the stability and productivity of the system are found in the slow exploitation and conservation sequence, whereas its resilience is determined by the last two phases, also referred to as the back-loop. Hence, reoccurring disturbances (or incremental changes and disruptions) are an important part of ecosystem succession and evolution and are crucial for sustaining resilience and integrity in such systems. This also means that, if disturbances are blocked out, e.g., through fire prevention, the ecosystem will over time accumulate biotic material and become over-connected, which creates an opportunity for large-scale perturbations with extensive destruction that can flip the system into a new attraction domain that profoundly changes its properties [15]. To evade this, creating conditions for the back-loop step in the succession cycle is critical for shifting the system into a new stage of exploitation rather than to change into a new regime. In Figure 1, this possibility is marked X.

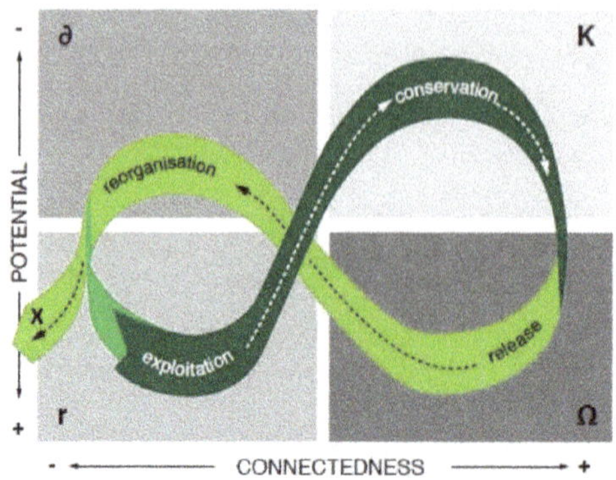

Figure 1. The adaptive renewal cycle. The figure illustrates how ecosystems evolve over time. The y-axis represents the amount of potentially stored nutrients, while the x-axis represents the degree of connection among the variables. The 'r-phase' denotes the establishment/exploitation of unvegetated land, such as land cleared by fire or by a heavy storm. During this stage, the land has significant potential for the redevelopment of biodiversity, reaching the 'K-phase'. The 'omega phase' symbolizes disturbance, involving the loss of energy, such as heat during a forest fire. In the 'alpha phase', the renewal of the system occurs after the disturbance. If the right conditions (nutrients, pioneering species, etc.) are not accessible in the alpha phase, the system risks tipping into a new stability domain, marked with an 'X' in the figure.

5. Optimization and Compressed Resilience

Efficiency is often viewed as a positive attribute in the governance of many systems, including economic, industrial, natural resource management, and organizational systems. However, it is important to balance efficiency with resilience to ensure that systems can respond and adapt to changing circumstances and avoid what can be referred to as 'compressed resilience', where options are successively reduced. This requires a shift in mindset from maximizing output to building the capacity for change and adaptation [42]. To promote resilience, resource managers and policymakers need to be aware of the potential trade-offs between efficiency and adaptability and to design policies and strategies that take both factors into account. This, among other things, includes adopting resilience principles, such as balancing strategies for managing connectivity and redundancy, building social networks and trust, and promoting learning and innovation [42].

As is generally known in natural resource management, planning that prioritizes optimization and efficiency often occurs at the expense of long-term adaptation or transformation [37]. To achieve a greater efficiency, many systems are streamlined to eliminate redundancies, meaning that there is little or no backup in place if the system as a whole or an element or component of the system fails. This can make the system more vulnerable to shocks and disturbances of different types and origins, like natural disasters, economic downturns, or equipment failures, which can have far-reaching and long-lasting consequences [21]. An illustrative case in natural resource management is the production efficiency in agriculture that often occurs at the expense of long-term agricultural resilience by degrading soil conditions and making the production system, in the long run, more vulnerable to pest and disease outbreaks as well as climate variability [43].

In urban development, a similar argument can be made, for example, for the neighborhood unit concept, in which new housing areas tend to be optimized for a particular size of inhabitants in regard of public and commercial services, but also for apartment sizes as well as for the measurements of individual rooms in apartments. This makes it hard to rearrange and restructure such legacies. Another well-known and illustrative example of this is building and construction legacies associated with the Olympic games [44]. Another example is modern traffic planning and its emphasis on individual road designs for different traffic modes and speeds. Whereas there are obvious arguments for such optimization, it does also create the vulnerability of change, such as changing demographics or new traffic technologies.

Hence, a narrow focus on efficiency and optimization can reduce resilience through both the creation of tightly coupled systems [45] and the lack of redundancy [46]. Thus, when a system is optimized to facilitate a specific function, it can become extremely difficult and costly to change. Tight coupling in the K-phase of the adaptive renewal cycle refers to a situation where the components of a system are highly connected, interdependent, and operate in a synchronized manner to provide a certain function. A lack of redundancy generally makes a system more vulnerable to unexpected events, such as natural disasters or economic downturns, which can cause cascading failures [23,47]. While tight coupling can increase efficiency in normal conditions, it also means that a disturbance or failure in one part of the system can quickly spread and affect the entire system, leading to greater chances for management failures (Figure 2). We refer to this situation as 'compressed resilience' in the adaptive renewal cycle, denoting a phase with the progressive loss of a system's ability to adapt to changing circumstances and recover from shocks or disruptions. Uncontrolled city growth represents a prime example of compressed resilience. A prime case is when local policymakers try to attract new inhabitants to their municipality to increase the tax base [48]. Over time, and due to changing circumstances, such a policy may lead to long-term social, ecological, and economic problems.

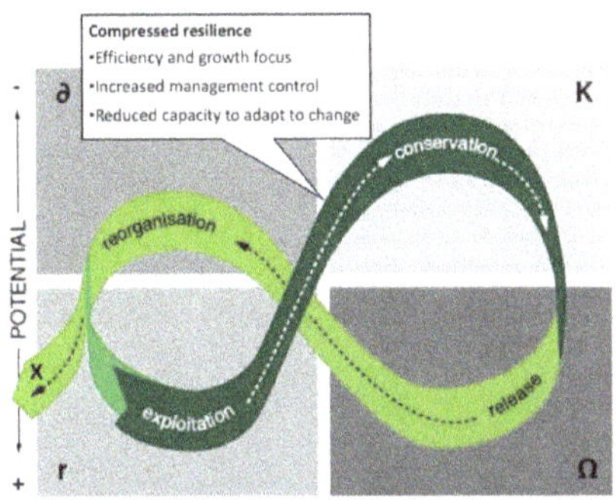

Figure 2. Compressed resilience illustrated through the framework of the adaptive renewal cycle. As the progression of the forward loop continues over time, the capacity to adapt to change and disturbance is declining—a situation that, in this paper, we refer to as 'compressed resilience'.

Tight coupling, however, may at the same time yield tangible benefits, such as urban densification for tackling climate change, which is endorsed by several international organizations, including the Intergovernmental Panel on Climate Change (IPCC) [49]. Densification is an important planning strategy to avoid urban sprawl, representing a leading land-use problem in many city regions. Hence, 'smart growth planning', often fused under the heading 'smart growth' or 'transit-oriented development', is a common urban design strategy for mitigating urban sprawl [50]. Yet, densification can, in some cases, also lead to situations of compressed resilience. With the tight coupling of the built urban form, often through taller buildings and packing more people into smaller spaces, comes an increasing risk of more severe consequences of disturbances [51,52]. Open spaces, such as plazas and parks, provide places of refuge during severe heat waves, new pandemics, for stress reduction, and other ecosystem services. Therefore, it is important for urban planners and policymakers to consider the potential risks associated with densification and incorporate measures that promote long-term resilience and adaptive capacity [53].

Compressed resilience also means that cities have fewer opportunities to renew themselves in a cost-effective way. One example of this is the restoration of skyscrapers, which often are the hallmark of dense urban areas and house many businesses and residential units. While some demolition and replacement of such buildings may be necessary for climate-proofing, studies indicate that there are significant opportunities to reduce carbon emissions by retrofitting them rather than constructing new ones [52]. However, the high cost of restoration may lead to a lack of investment in damaged areas or a reluctance to rebuild at all. This can have long-term economic and social impacts on the affected area and on the city itself. An emerging example is areas that are faced with sea-level rise due to climate change, leading to climate-induced gentrification [54,55].

A typical response for managers to deal with cases of compressed resilience is to resort to the imposition of new rules to implement more of the same policy solution, increasing the level of efficiency (Figure 2). This type of response is something that the economist Alfred E. Kahn warned about and coined as 'the tyranny of small decisions.' Kahn specialized in regulation and deregulation policies and presented the tyranny analogy to represent a situation in which a number of decisions, individually small in size and time, result in non-optimized and socially undesirable outcomes [56].

6. A Spatial Morphology for the Adaptive Renewal Cycle

Cities are rarely only dense or dispersed; they constitute landscapes of continuous variations in density, so that each location in principle is unique, allowing for parallel trajectories to develop. This not only concerns the local properties of individual locations, but more importantly, is applied to each locations' accessibility to other locations at different radii. This emphasis on relative location (location in relation to other locations, often defined by infrastructure) rather than absolute location (location on the surface on the Earth given by nature) is essential for understanding cities as well as for creating policies for sustainable urban development.

To understand the role of the adaptive renewal cycle in an urban context, it is crucial to translate such dynamics into the spatial urban form to fully understand how such processes play out in urban spaces and, by extension, to be able to inform urban planning and design how to create the spatial settings that can support and sustain them. For this, we need a spatial morphology through which urban space can be described and analyzed. In this paper, we provide a brief outline of such morphology [17], for which we identify three basic variables. First, 'accessibility' from each location to all other locations in an urban landscape, which typically is facilitated by a street network. This should not be seen as a neutral support system but defines the relative location of each location in a city, creating a greater or lower accessibility to the rest of the city, which can also be measured at different radii. Second, the 'density' of the built fabric, which allows for variations in the volume of activity at different locations by means of the built floor space. This can also be measured as accessibility to the floor space, meaning how much floor space one can access within a certain radius of a location, preferably measured following the street network. Third, the 'differentiation' of urban space into individual parcels and their adherent property rights, which creates opportunity for different forms of proprietorship (e.g., public, private, common, and their mixture) to develop individual strategies that, in turn, and in principle, can support social and economic diversity in urban space. Lessons from sustainable resource management suggest that policy efforts should concentrate on creating a diverse array of property rights frameworks tailored to the cultural, economic, and geographic contexts [57,58].

A spatial morphology based on resilience thinking offers an alternative to conventional top-down urban planning by adherence to a given set of resilience principles, such as redundancy and response diversity [46]. Framed within the context of the adaptive renewal cycle and beginning with the front-loop step, evolving from the r-stage to the K-stage, we argue that this phase is a transition from a stage that demands high spatial accessibility in the street network, high spatial differentiation in the parcel system, and moderate spatial density in the built fabric. This phase also provides 'spatial capital' [17] because it offers spatial support to the need for proximity between a large group of diverse actors, which is characteristic of the r-stage. These actors can, for example, be firms and business enterprises, people that meet and bond in social and public spaces, or actors facilitating different types of cultural amenities. In socio-economic terms, this can be interpreted as a need for rapid exchange between small, agile, and innovative firms and other types of actors [36]. From the r-phase, it proceeds to a state demanding lower accessibility and lower differentiation but increased density, due to the need for efficient accumulation and concentration within a few large firms, as typical for the K-stage, meaning firms or various urban service providers that are less agile and innovative, but highly competitive due to their accumulated capital. In resilience terms, it describes a shift from a stage of high resilience to a stage of low resilience, because the K-stage becomes increasingly exposed to disturbances, in part due to the system's reduction in diversity, where the loss of every actor has considerable effects, and in part due to the system's lower accessibility, which means less opportunities for invigorating input. But, at the same time, the potential for increased activity is improved due to increased density, or spatial volume, which affords a greater leverage to all action, including mistaken. Together, this amplifies the system's instability and vulnerability, which may prove devastating in times of disturbances of a more severe kind, such as an economic recession.

In the transition from the K-stage to the Ω-stage, connections are broken, and resources dispersed, not least because changes in the spatial form supporting it destabilize the resilience of the system. In spatial terms, it means a shift from the low spatial accessibility of the new regime to the need of increased accessibility, since actors, infrastructure, and other resources are unbound and need new connections. For similar reasons, it also demands an increase in spatial differentiation, since these actors and resources, previously connected in large organizations and clusters of spatial urban form, now are dispersed as individual entities in need of new spaces of their own to form spatial form clusters that promote the close interaction of these individuals. The demand for spatial density, on the other hand, plummets because many activities and resources were lost in the destruction of the previous organizations.

This Ω-stage forms the basis for a phase of reorganization, a shift from the Ω-stage to the α-stage, where the landscape of low spatial accessibility and differentiation, as well as low density, sees a surge in demand for all these variables. This is due to the process of self-organization of the dispersed actors, who, step by step, increase both their connections and differentiation. This is accelerated in the transition from the α-stage to the r-stage, where the organization of these activities both intensifies and proliferate, which increases the demand for spatial forms that facilitate diversity and exchange as well as volume. This rapid shift from the K-stage back to the r-stage also implies a transition from low spatial support for resilience to high support, as in general, all the spatial variables are inverted. Finally, the r-stage leads over to the slow transition into the K-stage where the cycle began, the front-loop evolving from exploitation to conservation, and the cycle starts over.

7. Synthesizing Resilience Theory and Urban Form: A Space Syntax Perspective

The adaptive renewal cycle is a representation of a process through time, whereas the spatial morphology mentioned above is a representation of spatial structures. However, up to this point in our discussion, we treated our spatial variables as part of a process through time so that spatial form changes as the process changes, which seems unlikely due to the extremely high costs of rebuilding the spatial urban form. It also undermines the argument that the spatial form can build resilience in urban systems since it changes only slowly over time and therefore, is able to control faster urban processes. For instance, whereas some processes change continuously, such as the intensity of the movement of people in a city, which typically varies over the course of the day, other aspects of this process remain quite constant, such as the relative distribution of movement in space. This constancy is due to the spatial form, in particular street networks, the configuration of which changes very slowly.

A fundamental property of the spatial urban form is not only that it is a slow variable that can influence faster variables, but that it, due to its very spatiality, can harbor several processes at the same time, that is, many trajectories occurring in parallel over time and in different places. This concerns not only different processes in different places but also different sequences of the same process, such as the different stages of the adaptive renewal cycle. For instance, we often see retail firms shift location as they grow, from the back street to the main street and into the mall, since each location is more suited to the different phases of the firm's growth. This ability to carry several processes in parallel is of course fundamental for cities, but as argued by British geographer Doreen Massey, we have an inclination to see space as a dimension of time—as in our treatment of the adaptive renewal cycle above—which she sees as part of our Western mode of thinking broadly, in which we continuously let time dominate over space [59]. Given the fundamental spatiality of cities, we should look for spatial support for the adaptive renewal cycle in the urban landscape as a whole and not in one and the same place—this would be to try to turn space into time.

To fully address this issue, it is important to realize that adaptive renewal cycles take place on several levels, as emphasized by Gunderson and Holling [21], since resilient systems consist of a hierarchy of interconnected systems, known as a 'panarchy', which is crucial for understanding how complex systems function and adapt to change over time. The panarchy model recognizes that systems exist at different scales, with each level of the hierarchy influencing the behavior and resilience of the systems on the other levels. At the same time, each level of the hierarchy operates on a different timescale, with faster cycles of change occurring at the lower levels and slower, more stable cycles of change occurring at the higher levels. The panarchy model can help us to better comprehend the dynamic relationships between different elements of an urban system and the interdependence between these elements at different levels of the hierarchy. This understanding is essential to predict how a system will change over time and how it will respond to various disturbances, shocks, or other types of changes that may occur in a city.

While high levels of connectivity can facilitate recovery after disturbances in ecosystems, it can, as argued, also spread disturbances much faster. Therefore, 'modular' or compartmentalized structures and patterns in ecosystems, such as in food webs, may offer resilience by isolating and retaining the impacts of a perturbation within a single module, and thus, minimize impacts on the other modules in the system [60]. Modularity calls for increased attention on how to configure and structure spatial urban form. For example, the integration of green spaces or other open public areas could help to mitigate the environmental stressors and alleviate urban challenges, making them modular elements in an otherwise highly connected urban fabric.

This challenges us to take another step in our translation of resilient systems to spatial form, again to fully understand how they play out in urban spaces, and by extension, how to support them through informed urban planning and design. In space syntax research, Hillier has presented a theoretical argument about the relation between the spatial form of cities and the evolution of society [61]. Urban street networks, which are the primary components of the spatial form for the distribution of accessibility in cities, often present a duality, he argues. In part, they are a foreground network covering the city, a kind of super-grid of high accessibility that facilitates quick connections and supports socio-economic exchange and innovation. In part, they are a background network in the shape of patches of sub-ordinated streets in this super-grid that are less accessible and therefore create more segregated spaces, which in contrast facilitate socio-cultural continuity and reproduction (Figure 3) [62]. Acknowledging that this may be manifested in different ways in different cities, the spatial form of cities may in principle be described as an entity that fundamentally supports both processes. In Hillier's words: "We call the first the generative use of space since it aims to generate copresence and make new things happen, and the second conservative since it aims to use space to reinforce existing features of society" [63] (p. 150).

Importantly, the processes supported by these networks, generation and conservation, are close to the processes identified by Holling and colleagues as fundamental for the sustainability of adaptive complex systems:

"The adaptive cycle [. . .] shows two very different stages. The front-loop stage, from r to K, is the slow, incremental phase of growth and accumulation. The back-loop stage, from Ω to α, is the rapid phase of reorganization leading to renewal. [. . .] It is as if two separate objectives are functioning, but in sequence. The first maximizes production and accumulation; the second maximizes invention and reassortment. We have no theorem to prove it, but our intuition suggests that any complex system, if it is adaptive, must generate these two phases in sequence, at some scale" [21] (p. 47).

In our argument, in cities, such phases of a process will by necessity require support from the physical structure of the spatial urban form. Hence, it is interesting that Hillier, in a similar manner, explains his idea of the foreground and background networks:

"by some as yet unknown process [...], cities of all kinds, however they begin seem to evolve into a foreground network of linked centres at all scales, from a couple of shops and a cafethrough to whole sub-cities, set into a background network of largely residential space" [63] (p. 139).

Holling and colleagues first seemed apt to think about the process they address in terms of time. Fearing the engineering principle of optimization, they underlined how: "The two objectives cannot be maximized simultaneously, they occur only sequentially" [21], but subsequently, they conceived a road forward, which we interpreted as being spatial:

"attempting to optimize around a single objective is fundamentally impossible for adaptive cycles, although optimizing the context that allows such a dynamic might be possible" [21] (p. 47).

We quote Hillier, again, in support of our interpretation of the context that Holling and colleagues refer to as the spatial form:

"The foreground structure, the network of linked centres, has emerged to maximise grid-induced movement, driven by micro-economic activity. Micro-economic activity takes a universal spatial form and this type of foreground pattern is a near-universal in self-organised cities. The residential background network is configured to restrain and structure movement in the image of a particular culture, and so tends to be culturally idiosyncratic, often expressed through a different geometry which makes the city as a whole look spatially different" [63] (p. 150).

Figure 3. The foreground network (dark grey), according to Hillier, facilitating socio-economic exchange and renewal through high accessibility, and the background network (light grey), facilitating socio-cultural continuity and reproduction through low accessibility. The pattern is produced by measuring the betweenness centrality for all street segments in the complete street system. Source: Hillier and Vaughan [61].

Again, different processes, or different phases of processes, can take place simultaneously in cities, albeit in different places. At a closer scrutiny, the large and high-resolution analysis of a street network in Figure 2 make us realize that we are not really referring to only two networks, but to a continuous hierarchy of spaces, from the most local to the most global, and that each street segment in a street system has unique properties, with these segments producing an exceptionally rich landscape of locations together. It may actually bring to mind an agricultural landscape, where the experienced farmer, knowing the land, recognizes exactly what crops work best and where.

Hence, we argue that the spatial urban form has a direct influence on urban systems and therefore, in principle, can be used to direct, through its design, such systems towards increased resilience. In the spatial morphology framework mentioned above, we also add the variables of spatial density and differentiation to the variable of accessibility primarily discussed by Hillier, which together offer a rich palette for the analysis of the spatial form of cities, as a means to increase our understanding of its role in building resilience and, in turn, inform practice in urban development.

8. Conclusions and Future Directions

As theorized in this paper, resilience thinking offers a novel approach to enhance our understanding of complex adaptive systems, with cities serving as prime examples. To foster resilience in urban systems, it is imperative to enhance our comprehension of the key characteristics inherent in resilient systems, translating these insights into the spatial urban form and understanding the processes of urban renewal. By integrating the adaptive renewal cycle and aligning it with spatial morphology and 'space syntax theory', we proposed in this paper several theoretical postulates that elucidate how cities undergo renewal, across both space and time. Additionally, we translated these postulates into spatial form, trying to present a more comprehensive understanding of their manifestation in urban spaces. While further research is essential to explore urban renewal through the analytical lens of resilience, future investigations in this domain should be oriented towards:

- Exploring the potential trade-offs between efficiency and adaptability for the development of design strategies that take both factors into account.
- The development of strategies for managing connectivity and redundancy, including improving understanding regarding the role of the modularity of urban spaces to mitigate environmental stressors and alleviate other types of urban challenges.
- Improving the understanding of both the generative use of space, which aims to generate co-presence and make new things happen, and the conservative use of space, which aims to use space to reinforce the existing features of society.
- Processes in cities that will need support from the spatial urban form and that confer resilience.

Author Contributions: Conceptualization, L.M. and J.C.; methodology, L.M. and J.C.; formal analysis, L.M. and J.C.; writing—original draft preparation, L.M. and J.C.; writing—review and editing, L.M. and J.C. All authors have read and agreed to the published version of the manuscript.

Funding: Lars Marcus's work in this paper was funded by the Department of Architecture and Civil Engineering, Chalmers University of Technology, Gothenburg, Sweden. Johan Colding's work was funded by a special grant from the J. Gustaf Richert Foundation (SWECO) within the project 'Studies of resilience and compressed adaptability in urban densification' (Studier av resiliens och komprimerad anpassningsförmåga vid förtätning i den byggda miljön). Colding's work was also funded by the Department of Building Engineering, Energy Systems and Sustainability Science, University of Gävle, and by the Beijer Institute of Ecological Economics.

Data Availability Statement: Data are contained within the article.

Conflicts of Interest: The authors declare no conflict of interest.

References

1. Jacobs, J. *The Death and Life of Great American Cities*; The Failure of Town Planning; Vintage: New York, NY, USA, 1961.
2. Colding, J. 'Ecological land-use complementation' for building resilience in urban ecosystems. *Landsc. Urban. Plan.* **2007**, *81*, 46–55. [CrossRef]
3. Holling, C.S.; Goldberg, M.A. Ecology and Planning. *J. Am. Inst. Plann.* **1971**, *37*, 221–230. [CrossRef]
4. Berkes, F.; Colding, J.; Folke, C. (Eds.) *Navigating Social-Ecological Systems*; Cambridge University Press: Cambridge, UK, 2001.
5. Biggs, R.; Schlüter, M.; Schoon, M.L. *Principles for Building Resilience: Sustaining Ecosystem Services in Social-Ecological Systems*; Cambridge University Press: Cambridge, UK, 2015.
6. Colding, J.; Colding, M.; Barthel, S. Applying seven resilience principles on the Vision of the Digital City. *Cities* **2020**, *103*, 102761. [CrossRef]
7. Van Der Leeuw, S. Land degradation as a socionatural process. In *The Way the Wind Blows: Climate, History and Human Action*; Mcintosh, R., Tainter, J., Mcintosh, S., Eds.; Columbia University Press: New York, NY, USA, 2000.
8. Pickett, S.T.A.; Cadenasso, M.L.; Grove, J.M. Resilient cities: Meaning, models, and metaphor for integrating the ecological, socio-economic, and planning realms. *Landsc. Urban. Plan.* **2004**, *69*, 369–384. [CrossRef]
9. Vale, L.J. *The Resilient City: How Modern Cities Recover from Disaster*; Oxford University Press: Cary, NC, USA, 2005; 392p.
10. Alberti, M.; Correa, C.; Marzluff, J.M.; Hendry, A.P.; Palkovacs, E.P.; Gotanda, K.M.; Hunt, V.M.; Apgar, T.M.; Zhou, Y. Global urban signatures of phenotypic change in animal and plant populations. *Proc. Natl. Acad. Sci. USA* **2017**, *114*, 8951–8956. [CrossRef]
11. Colding, J.; Barthel, S. The potential of 'Urban Green Commons' in the resilience building of cities. *Ecol. Econ.* **2013**, *86*, 156–166. [CrossRef]
12. Samuelsson, K.; Giusti, M.; Peterson, G.D.; Legeby, A.; Brandt, S.A.; Barthel, S. Impact of environment on people's everyday experiences in Stockholm. *Landsc. Urban. Plan.* **2018**, *171*, 7–17. [CrossRef]
13. Masnavi, M.R.; Gharai, F.; Hajibandeh, M. Exploring urban resilience thinking for its application in urban planning: A review of literature. *Int. J. Environ. Sci. Technol.* **2019**, *16*, 567–582. [CrossRef]
14. Sharifi, A. Resilient urban forms: A macro-scale analysis. *Cities* **2019**, *85*, 1–14. [CrossRef]
15. Berkes, F.; Folke, C. Linking social and ecological systems for resilience and sustainability. In *Linking Social and Ecological Systems*; Cambridge University Press: Cambridge, UK, 1998.
16. Meerow, S.; Newell, J.P.; Stults, M. Defining urban resilience: A review. *Landsc. Urban. Plan.* **2016**, *147*, 38–49. [CrossRef]
17. Marcus, L.; Colding, J. Toward an integrated theory of spatial morphology and resilient urban systems. *Ecol. Soc.* **2014**, *19*, art55. [CrossRef]
18. Adger, W. Social and ecological resilience: Are they related? *Prog. Hum. Geogr.* **2000**, *24*, 347–364. [CrossRef]
19. Cote, M.; Nightingale, A.J. Resilience thinking meets social theory: Situating social change in socio-ecological systems (SES) research. *Prog. Hum. Geogr.* **2012**, *36*, 475–489. [CrossRef]
20. Colding, J.; Gren, Å.; Barthel, S. The Incremental Demise of Urban Green Spaces. *Land* **2020**, *9*, 162. [CrossRef]
21. Gunderson, L.H.; Holling, C.S. *Panarchy: Understanding Transformations in Human and Natural Systems*; Island Press: Washington, DC, USA, 2002.
22. Hillier, B.; Hanson, J.; Peponis, J. What do we mean by building function? In *Designing for Building Utilisation*; Powell, J.A., Cooper, I., Lera, S., Eds.; E & F.N. Spon Ltd.: London, UK, 1984; pp. 61–72. Available online: https://discovery.ucl.ac.uk/id/eprint/15007/ (accessed on 8 December 2023).
23. Holling, C.S. Resilience and Stability of Ecological Systems. *Annu. Rev. Ecol. Syst.* **1973**, *4*, 1–23. [CrossRef]
24. Holling, C.S. Understanding the Complexity of Economic, Ecological, and Social Systems. *Ecosystems* **2001**, *4*, 390–405. [CrossRef]
25. Hillier, B.; Leaman, A.; Stansall, P.; Bedford, M. Space Syntax. *Environ. Plan. B Plan. Des.* **1976**, *3*, 147–185. [CrossRef]
26. Schultze, P. *Engineering Within Ecological Constraints*; National Academies Press: Washington, DC, USA, 1996; 222p.
27. Olazabal, M. Resilience, Sustainability and Transformability of Cities as Complex Adaptive Systems. In *Urban Regions Now & Tomorrow: Between Vulnerability, Resilience and Transformation*; Deppisch, S., Ed.; Springer Fachmedien: Wiesbaden, Germany, 2017; pp. 73–97. [CrossRef]
28. Folke, C.; Carpenter, S.; Walker, B.; Scheffer, M.; Elmqvist, T.; Gunderson, L.; Holling, C.S. Regime Shifts, Resilience, and Biodiversity in Ecosystem Management. *Annu. Rev. Ecol. Evol. Syst.* **2004**, *35*, 557–581. [CrossRef]
29. Scheffer, M.; Carpenter, S.R. Catastrophic regime shifts in ecosystems: Linking theory to observation. *Trends Ecol. Evol.* **2003**, *18*, 648–656. [CrossRef]
30. Folke, C.; Kofinas, G.P.; Chapin, F.S. (Eds.) *Principles of Ecosystem Stewardship*; Springer: New York, NY, USA, 2009; Available online: http://link.springer.com/10.1007/978-0-387-73033-2 (accessed on 8 December 2023).
31. Nyström, M.; Folke, C.; Moberg, F. Coral reef disturbance and resilience in a human-dominated environment. *Trends Ecol. Evol.* **2000**, *15*, 413–417. [CrossRef]
32. Batty, M. *Cities and Complexity. Underst Cities Cell Autom AgentBased Models Fractals*; MIT Press: Cambridge, MA, USA, 2005.
33. Glass, R. Aspects of Change. In *The Gentrification Debates*; Routledge: New York, NY, USA, 2010.
34. Pearsall, H. Moving out or moving in? Resilience to environmental gentrification in New York City. *Local. Environ.* **2012**, *17*, 1013–1026. [CrossRef]

35. Rice, J.L.; Cohen, D.A.; Long, J.; Jurjevich, J.R. Contradictions of the Climate-Friendly City: New Perspectives on Eco-Gentrification and Housing Justice. *Int. J. Urban. Reg. Res.* **2020**, *44*, 145–165. [CrossRef]
36. Holling, C.S. The resilience of terrestrial ecosystems; local surprise and global change. In *Sustainable Development of the Biosphere*; Clark, W.C., Munn, R.E., Conway, G.R., Eds.; Cambridge University Press: Cambridge, UK, 1986; pp. 292–317.
37. Holling, C.S.; Meffe, G.K. Command and Control and the Pathology of Natural Resource Management. *Conserv. Biol.* **1996**, *10*, 328–337. [CrossRef]
38. Hanson, J. Order and structure in urban design: The plans for the rebuilding of London after the Great Fire of 1666. *Ekistics* **1989**, *56*, 22–42.
39. Von Bertalanffy, L. The History and Status of General Systems Theory. *Acad. Manag. J.* **1972**, *15*, 407–426. [CrossRef]
40. Abercrombie, M.; Hickman, C.J.; Johnson, M.L. *A Dictionary of Biology*; Routledge: New York, NY, USA, 2017; 255p.
41. Clements, F.E. Nature and Structure of the Climax. *J. Ecol.* **1936**, *24*, 252–284. [CrossRef]
42. Folke, C.; Carpenter, S.R.; Walker, B.; Scheffer, M.; Chapin, T.; Rockström, J. Resilience Thinking: Integrating Resilience, Adaptability and Transformability. *Ecol. Soc.* **2010**, *15*, 9. [CrossRef]
43. Bennett, E.; Carpenter, S.; Gordon, L.; Ramankutty, N.; Balvanera, P.; Campbell, B.M.; Cramer, Q.; Dolwy, J.; Folke, C.; Karlberg, L.; et al. Skip to Navigation Login Subscribe | Sponsor | Submit | Donate | Sponsors and Partners | About Volume 5 | Issue 5 | Page 65-75 | Oct 2014 Toward a More Resilient Agriculture Idea Lab Features Perspectives Visionaries Your Solutions Community Global Toward a More Resilient Agriculture | Solutions. *Solutions* **2014**, *5*, 65–75.
44. Gratton, C.; Preuss, H. Maximizing Olympic Impacts by Building Up Legacies. *Int. J. Hist. Sport.* **2008**, *25*, 1922–1938. [CrossRef]
45. Grote, G. Rules Management as Source for Loose Coupling in High-Risk Systems. In *Resilience Engineering Perspectives*; CRC Press: Boca Raton, FL, USA, 2008; Volume 1.
46. Elmqvist, T.; Folke, C.; Nyström, M.; Peterson, G.; Bengtsson, J.; Walker, B.; Norberg, J. Response diversity, ecosystem change, and resilience. *Front. Ecol. Environ.* **2003**, *1*, 488–494. [CrossRef]
47. Comes, T.; de Walle, B.V. Measuring Disaster Resilience: The Impact of Hurricane Sandy on Critical Infrastructure Systems. In Proceedings of the 11th International ISCRAM Conference—University Park, College Township, PA, USA, 18–21 May 2014; 2014.
48. Leitner, H. Cities in pursuit of economic growth: The local state as entrepreneur. *Polit. Geogr. Q.* **1990**, *9*, 146–170. [CrossRef]
49. Kii, M. Reductions in CO_2 Emissions from Passenger Cars under Demography and Technology Scenarios in Japan by 2050. *Sustainability* **2020**, *12*, 6919. [CrossRef]
50. Gren, Å.; Colding, J.; Berghauser-Pont, M.; Marcus, L. How smart is smart growth? Examining the environmental validation behind city compaction. *Ambio* **2019**, *48*, 580–589. [CrossRef]
51. Iglesias, V.; Braswell, A.E.; Rossi, M.W.; Joseph, M.B.; McShane, C.; Cattau, M.; Koontz, M.J.; McGlinchy, J.; Nagy, R.C.; Balch, J.; et al. Risky Development: Increasing Exposure to Natural Hazards in the United States. *Earths Future* **2021**, *9*, e2020EF001795. [CrossRef]
52. Al-Kodmany, K. Green Retrofitting Skyscrapers: A Review. *Buildings* **2014**, *4*, 683–710. [CrossRef]
53. Villagra, P.; Rojas, C.; Ohno, R.; Xue, M.; Gómez, K. A GIS-base exploration of the relationships between open space systems and urban form for the adaptive capacity of cities after an earthquake: The cases of two Chilean cities. *Appl. Geogr.* **2014**, *48*, 64–78. [CrossRef]
54. Bouzarovski, S.; Frankowski, J.; Tirado Herrero, S. Low-Carbon Gentrification: When Climate Change Encounters Residential Displacement. *Int. J. Urban. Reg. Res.* **2018**, *42*, 845–863. [CrossRef]
55. Blok, A. Urban green gentrification in an unequal world of climate change. *Urban. Stud.* **2020**, *57*, 2803–2816. [CrossRef]
56. Kahn, A.E. The Tyranny of Small Decisions: Market Failures, Imperfections, and the Limits of Economics*. *Kyklos* **1966**, *19*, 23–47. [CrossRef]
57. Hanna, S.; Folke, C.; Mäler, K.G. Property Rights and the Natural Environment. In *Rights to Nature: Ecological, Economic, Cultural, and Political Principles of Institutions for the Environment*; Island Press: Washington, DC, USA, 1996.
58. Ostrom, E. *Understanding Institutional Diversity*; Princeton University Press: Princeton, NJ, USA, 2005; Available online: https://www.jstor.org/stable/j.ctt7s7wm (accessed on 8 December 2023).
59. Massey, D. Space-Time, "Science" and the Relationship between Physical Geography and Human Geography. *Trans. Inst. Br. Geogr.* **1999**, *24*, 261–276. [CrossRef]
60. May, R.M. Will a Large Complex System be Stable? *Nature* **1972**, *238*, 413–414. [CrossRef]
61. Hillier, B.; Vaughan, L. The city as one thing. *Prog. Plan.* **2007**, *67*, 205–230.
62. Hillier, B. Spatial sustainability in cities: Organic patterns and sustainable forms. In *Proceedings of the 7th International Space Syntax Symposium*; Koch, D., Marcus, L., Steen, J., Eds.; Royal Institute of Technology (KTH): Stockholm, Sweden, 2009; p. 1. Available online: http://www.sss7.org/Proceedings_list.html (accessed on 8 December 2023).
63. Hillier, B. The Genetic Code for Cities: Is It Simpler than We Think? In *Complexity Theories of Cities Have Come of Age: An Overview with Implications to Urban Planning and Design*; Portugali, J., Meyer, H., Stolk, E., Tan, E., Eds.; Springer: Berlin/Heidelberg, Germany, 2012; pp. 129–152. [CrossRef]

land

MDPI

Article

Counteract Soil Consumption through Ecosystem Services and Landscape Restoration for an Efficient Urban Regeneration

Celestina Fazia [1,*], Kh Md Nahiduzzaman [2], Baqer Al-Ramadan [3,4], Adel Aldosary [3] and Francesca Moraci [5]

1 Faculty of Engineering and Architecture, Kore University of Enna, 94100 Enna, Italy
2 School of Engineering, The University of British Columbia (UBC), Okanagan, Kelowna, BC V1V 1V7, Canada; kh.nahiduzzaman@ubc.ca
3 Architecture & City Design Department, KFUPM, Dhahran 31261, Saudi Arabia; bramadan@kfupm.edu.sa (B.A.-R.); asdosary@kfupm.edu.sa (A.A.)
4 Interdisciplinary Research Center for Smart Mobility and Logistics, King Fahd University of Petroleum & Minerals, Dhahran 31261, Saudi Arabia
5 DArTe Department, Mediterranea University of Reggio Calabria, 89124 Reggio Calabria, Italy; fmoraci@unirc.it
* Correspondence: celestina.fazia@unikore.it; Tel.: +39-3283719351

Abstract: Soil consumption, marked by the expansion of artificial land cover for residential, productive, and infrastructural purposes, is a concerning trend in Italy, as revealed by the Copernicus land monitoring program. The issue is exacerbated by agricultural intensification and urbanization, particularly affecting regions like Lombardia and Piemonte. However, Sicilia, Abruzzo, and Lazio experience notable increases in processes of abandonment and re-naturalization. Data from Ispra highlights the need for in-depth study, especially in regions like Sicilia, where contrasting phenomena occur. This study utilizes Ispra data to monitor and formulate strategies for mitigating soil consumption and safeguarding ecosystem services. The research aligns with objectives related to combating climate change and facilitating the ecological transition of territories. The complexity of land consumption, influenced by interdependent factors, is evident in the achieved results. Effective strategies for containment and re-naturalization involve the implementation of town planning regulations and multi-level behavioral pathways. This study aims to identify contextual actions that can reduce land consumption, promote de-impermeabilization, and encourage re-naturalization, focusing on enhancing ecosystem services in land use activities. Thus, it focuses on understanding the contributions of ecosystem services, landscape restoration and green infrastructure on climate mitigation, and a reduction in land consumption in urban regeneration processes. As well, through open-source systems, it is important to monitor in real time the trend of the quantity of factors and variables and the state of the environment, and the reasons to intervene with systemic strategies and actions constitutes another lens of focus.

Keywords: soil consumption; ecosystem services; landscape restoration; urban regeneration

1. Introduction

Soil consumption, as is well known, is the increase in artificial land cover. This occurs by transforming an area that was originally natural or semi-natural into areas to be used for residential, production, and infrastructure in general. In Italy, Ispra's National System for Environmental Protection monitors soil consumption by annually updating the relevant National Map [1,2]. The Copernicus land monitoring program—such as the Corine Land Cover, the high-resolution layers of the Pan-European and Local component—returns the state of the land and landscape. The data that emerge are worrying. If we consider only the indicator referring to fluxes relative to land cover changes during 1960–2018, it emerges that processes of agricultural intensification mainly affect Umbria, Lombardia, Valle d'Aosta, and Piemonte [3–5]; on the other hand, urbanization phenomena mainly concern Lombardy, Veneto, and Sicily; greater increases in cultivation abandonment or renaturation processes are recorded in Sicilia, Abruzzo, and Lazio [3,4]. These data deserve in-depth study, especially if we consider that some regions are particularly affected by different and apparently contrasting interchange flows; this is the case for Sicily [4,5], which will be the subject of a specific study.

This paper attempts to understand what actions should be proposed to reduce soil consumption, and encourage de-impermeabilization and renaturalization, paying attention to ecosystem services to promote fruition activities in the peri-urban system. All this can be achieved by identifying appropriate urban planning procedures and tools [4], and defining the contribution of the ecological–environmental dimension can be achieved through ecosystem services, landscape enhancement for ecological transition and urban regeneration, and through urban transformation management tools. An important contribution to reducing land consumption, resilience, and combating climate change in the broader domain of urban regeneration processes can be achieved by ecosystem services and intelligent landscape restoration. A strategically planned network of natural and semi-natural areas with other environmental features, the so-called green infrastructure, is a new supply of green spaces and solutions that must be managed to provide a wide range of ecosystem services. This 'warping' includes green spaces (or blue if it includes aquatic ecosystems) and other physical features found in terrestrial (including coastal) and marine areas, in rural and urban contexts. The United Nations 2030 Agenda [5–23] presents a comprehensive framework to suggest the implementation of blue–green solutions and achieve the Sustainable Development Goals. It is the most effective tool for delivering ecological, economic, and social benefits through natural solutions, which, among other things, helps us to understand the value of the benefits nature provides at no cost by avoiding reliance on expensive or artificial infrastructure, when nature already has the cheapest and most durable solutions. The European Commission's 2013 Communication [5] recommended the strengthening of green infrastructure based on the simple and agreeable principle that the protection and enhancement of natural processes and nature, and the benefits that human society derives from it, are intimately related to integrated spatial planning and development; not being a constraint on spatial development, green infrastructure promotes natural solutions only if these represent the best option. Investments in green infrastructure and landscape restoration projects have a high return and cost–benefit ratios in favor of the environmentally sustainable community. Such investments are also intercepted within the ecosystem services that are part of the territorial capital and identity of the European countries.

Green infrastructure will serve as a vital complement to diminish the carbon footprint of transportation and energy provision, thereby enhancing opportunities for more seamlessly integrating land use, ecosystems, and biodiversity into policy and planning. Green infrastructure solutions can contribute significantly to the development of green transport corridors, harnessing the potential of healthy ecosystems to sustainably mitigate carbon emissions, for instance.

Climate change, coupled with land fragmentation resulting from growing anthropization and infrastructure development, exacerbates cities' inherent susceptibility to various risks and disasters. This heightened vulnerability increases cities' exposure to natural disasters and extreme weather events, including storm surges, landslides, floods, avalanches, and forest fires, which lead to loss of life and significant damage.

The impact of these events on the environment and human society can often be reduced by using green infrastructure solutions. They become the safety 'drivers' by intercepting protective forests [5], in mountainous areas, riparian forests, alluvial plains functional to the environmental network, barrier beaches, and coastal wetlands. These can be implemented in combination with physical infrastructure for disaster reduction, e.g., river protection work.

2. Materials and Methods

2.1. Agenda 2030 in Europe and Italy, Land Target and Sustainability

Land and soil, as mentioned, are among the key components in the EU's natural resource package, and attention is high. Yet, every year, more than 1000 km^2 of land is subject to soil consumption for housing, industry, roads, or recreational activities [6–9]. It irreversibly compromises the quality by reducing the content of organic matter, in addition to the risk of soil contamination, which directly affects the health of the residents [6–21]. The provision of appropriate measures to implement green infrastructure in the planning and management process will help reduce the loss of ecosystem services, and improve and restore soil functions.

Land management for agriculture and forestry has an important impact on the condition of natural capital; agricultural policy and rural development provide tools and measures to encourage the development of green infrastructure and enhance highly valued nature in peri-urban areas. Large-scale direct support to farmers prevents the abandonment and fragmentation of land, while agri-environmental measures (e.g., those aimed at maintaining and improving hedges, terracing, and dry stone walls for the preservation of the agricultural landscape), are funding that promote the coherence of Natura 2000, and the conservation and restoration of rural heritage features. According to this aim, the Commission has included further aspects of 'greening' in its proposals for the reform of the Common Agricultural Policy, including the condition that farmers receiving incentives must comply, i.e., to maintain existing permanent grassland on their farm and to ensure that 7% of the land under arable and permanent crops is an area of ecological interest [7,9–21].

On the need to curb soil consumption, there is a willingness and convergence of all countries worldwide. This goal is not always made explicit in specific measures, sometimes it is cross-cutting, and in some cases, it is associated with other missions, such as the conservation of ecosystem services and reforestation. Agenda 2030, promoted by UN member countries, identifies 17 goals and associated sustainable development strategies to be implemented by 2030. Goals indicating direct or indirect targets to curb soil consumption [8,9], preserve ecosystem services, and support actions to restore the natural landscape are almost half. These are also the prerequisites set by the Next Generation EU (NGEU) for implementing efficient urban regeneration, and the NRP in Italy aligns with them accordingly [8,9]. Goal 15 "Protect, restore and foster sustainable use of the Earth's ecosystem" is aimed at preserving life on earth, halting and reversing land degradation and halting biodiversity loss, protecting and restoring terrestrial ecosystems, combating desertification, and sustainably managing forests [8–18].

Forests, in particular, harboring 80 percent of terrestrial species of animals, plants, and insects, are at the center of debate and are the focus of numerous research initiatives, both as entities to be protected and as dimensions to be restored [8,9]. Efforts pursued globally and locally to sustain forest ecosystems and their social, economic, and environmental functions are critical, as forest loss results in land degradation, increased carbon emissions, decreased biodiversity, and reduced livelihoods in rural communities.

The reforestation system, which is in place in several countries, must protect and foster biodiversity production/reproduction processes, especially in man-made areas and interchange areas present at the urban scale, where vital relationships between species belonging to different ecosystems [8,21–54] take place. It will give us pause to know that forests, woodlands, and agricultural land account for more than half of the land, while the remaining part is occupied by urban settlements (about 20 to 25 percent, containing a minimal amount of urban green spaces), water bodies (4 percent), and less than 1 percent by swamps, sandy or alluvial lands, and uncultivated land [8]. Within these differentiations, however, biodiversity is decreasing, and what worries us is the rate of reduction, which is higher than at any other time in human and earth history [8]. We are concerned about the figure for ecosystem services, which is discussed below.

The SDGs 2023 Report, compiled by Istat (national institute of statistics) for the 2030 Agenda in Italy [8,9], assumes the 17 goals contained therein as indicators for sustainable development in Italy; the study shows heterogeneous results from the different Italian regions despite the fact that the efforts made and the funding provided under the PNRR, National Recovery and Resilience Plan [8,9], cover the entire national territory with few territorial discrepancies. In the autonomous provinces of Bolzano and Trento, more than 40 percent of the 17 indicators are met, and in Valle d'Aosta, 33.8 percent are met, followed by Lombardia, with 24.8 percent of measures achieving the highest level in terms of sustainability results [8,9]. In the central regions, nearly 18 percent of measures in Lazio and Marche achieve appreciable results. In the southern regions, indicator values are among the lowest. By 2022, protected areas cover 21.7 percent of the national territory and partially include, with an average of about 76 percent, all the key biodiversity areas chosen by Istat for monitoring [8–13,18,21]. The vegetation cover of mountainous areas is declining, while the losses mainly affect the islands and the northwest, especially in the range below 1000 m above sea level, where there is more soil consumption. There are aspects, as argued before, that are interacting with soil consumption, understood both as loss (degradation) of natural land and resources and as a process of artificialization, with associated risks to humans.

In the first case, the factors responsible for the loss or "impairment" of the land's own characteristics to guarantee wildlife species, vegetation, and biodiversity are many. One example is the spread of allochthonous species, which shows, for the first time, signs of slowing down. Another concerns the effects of fires and deforestation. These also affect land responsiveness to extraordinary weather and climate events. In 2021, indices of temperature and precipitation extremes for provincial capital cities appear to be increasing compared to 1981–2010 climate values. At the same time, the number of days without rain is decreasing. More frequent are forest fires and worrying are the effects on the land: there has been a 23.1 percent increase in the number of fires between 2020 and 2021, and a doubling of the forest area involved (Figure 1).

There are other factors influencing environmental quality, and in some cases, climate change, such as emissions from manufacturing and transportation activities, which have increased more intensely than those from households in the past two years (+6.4 vs. +5.7 percent) [8,9], partly due to the resumption of post-COVID-19 activities.

2.2. The Perspective

We aim to understand the relationship between the factors that contribute to making the territory resilient by improving environmental and energy performance and urban quality standards through actions and measures to contain land consumption, de-waterproofing, and sustainable use. The map (Figure 2) summarizes the methodological path used for the general exploration of the themes and of the emerging orientations through the review of the literature and the state-of-the-art; we focus on the objectives of Agenda 2030 in Europe and in Italy [8,10–13,18–23] and on sustainability targets for the landscape. From the comparison of global strategies emerge the strategies in place to monitor and preserve the soil and the trend structures towards irreversible consumption. The discussion intervenes in the relationship between climate neutrality objectives and zero land consumption in Italy and the relationship and role of ecosystem services. The analysis of the results leads to defining some assumptions: the phenomenon of sprawl is not easy to contain, the containment of land consumption can be implemented through integrated regeneration strategies, and ecosystem services contribute to landscape planning. The discussion follows with a focus on renaturalization interventions in Sicily and the conclusions with future implications or expected results.

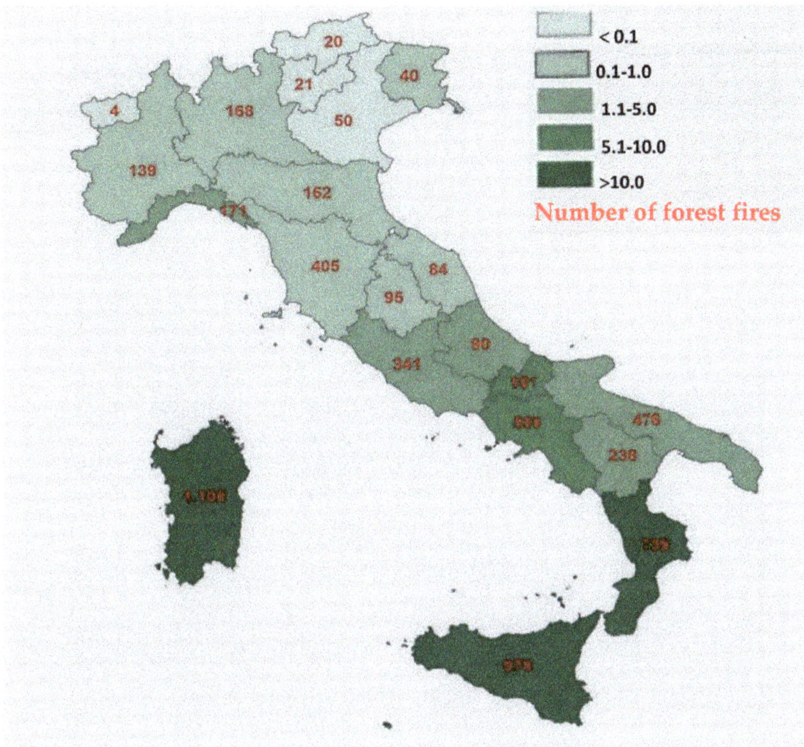

Figure 1. Forest area covered by fires per 1000 square kilometers and in red the number of fires, Source: Figure 13. Elaboration of SDGs 2023 Report on data from the Comando Carabinieri Tutela forestale (accessed on 10 September 2023).

Figure 2. Methodological steps of the study.

3. Results

3.1. Assumptions for Estimating Land Degradation

With Goal 15 "Protect, restore and promote sustainable use of the land ecosystem", Istat comes to measure land degradation through the implementation of a composite indicator. According to the UNCCD (United Nations Convention to Combat Desertification methodology) [9], land and land degradation is an extremely complex phenomenon, affected by many interdependent factors. While land quality can be reliably represented through biological, physical, or chemical parameters, it is also true that regarding its measurement, no single criterion has yet been defined that is capable of achieving a universal scientific consensus. The UN Statistical Commission has defined SDG indicator 15.3.1 [9] as the percentage of degraded areas in national territory; based on the methodology proposed by the UNCCD, it involves the application of the three sub-indicators: land cover and its changes over time, land productivity, and organic carbon content. Indicator 15.3.1 is partially covered, for Italy, by two statistical measures, both referring to particular aspects of land degradation: fragmentation of natural and agricultural land and land sealing/artificialization by artificial cover (soil consumption).

The measures were developed by Ispra, which also implements an overall indicator adapted to the Italian context. It uses as a source of information for the three sub-indicators: land use changes over the period 2000–2018 and land consumed over the period 2006–2021; productivity trajectory, estimated through the water use efficiency index (ratio of normalized difference vegetation index to evapotranspiration); and carbon changes, estimated through land use changes. At the national level, a percentage of degraded land is estimated at 17.2 percent, excluding water bodies; high values were recorded in Sardegna and Emilia-Romagna [9,10]. The measures integrated by Ispra are related to anthropogenic activities and the indirect effects of soil consumption (loss of habitat quality, presence of fire-ridden areas, land fragmentation, potential impact buffer of further consumption, areas with high and medium density of artificial cover [10,20,21], and increase in non-consumed areas and with an area of less than 1000 sq. m.).

The methodology considers only the most significant aspects of land degradation (changes in land use and land cover, land productivity, and the presence of organic matter); it does not cover other factors such as land compaction and contamination [11–13].

In order to overcome the limitations of the methodology indicated, Ispra is preparing the study of other indicators based on remote sensing, aimed at improving the measurement of land productivity and returning more reliable results on the level of degradation of the national territory [11–13]. All degradation factors considered were analyzed through a spatial overlay that provided the amount of area where degradation, over the 2016–2019 period (Figure 3), increased for one or more causes. The results obtained show an increase in the degraded land area of about 33,400 square kilometers. The Copernicus program, already active, returns the state of the Italian territory and landscape, and through the Land Cover Map [11–13], it represents the national reference for the elaboration of land and landscape analysis at high spatial resolution. Emerging data on land cover changes during 1960–2018 denote increasing land sealing and agricultural intensification. In Italy, Ispra monitors soil consumption by annually updating the relevant "National Map".

3.2. Climate Neutrality and the Concept of Zero Soil Consumption in Italy

The climate neutrality goals set for 2050 by the European Green Deal recognize that cities assume a key role in combating climate change, given that they consume more than 65 percent of the energy produced globally and that (the) cities themselves are responsible for 70 percent of CO_2 emissions [11–13,18–21]. This means that it is necessary to achieve the climate neutrality mission sought by the European Commission—with the document "Climate Neutral and Smart Cities"—by acting on anthropogenic factors. In this difficult context, urban transformation management tools have an important role: it is in urbanized contexts that the most important answers to the climate challenge can be found by directing plan actions to make the CO_2 concentrations produced by settlements neutral. Not only that, cities will have to reorganize and optimize functions within them by further limiting soil consumption and safeguarding ecosystems, protected areas, and areas that have not yet been artificialized. The situation is not simple. At the national level, the legislative framework is uneven, each region has legislated introducing specific measures related to context variables and different declinations of the concept of "soil consumption".

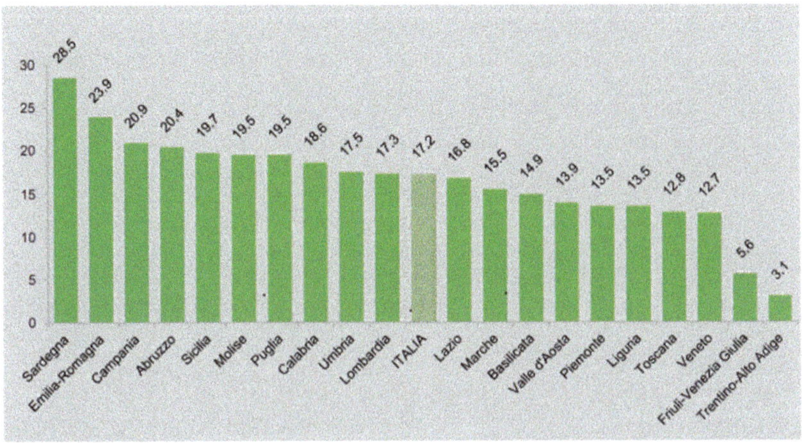

Figure 3. Percentage share of regional degraded land net of water bodies (UNCCD methodology for SDG 15.3.1 indicator) year 2019. Source: Ispra and SDGs Report 2023, ISTAT, Rome (accessed on 10 September 2023).

According to Ispra, in Italy, "soil consumption is a phenomenon associated with the loss of a fundamental environmental resource due to the occupation of originally agricultural, natural, or semi-natural surface area; soil consumption is, therefore, defined as a change from non-artificial land cover (non-consumed land) to artificial land cover (consumed land)" [12,31]. According to this understanding, land "consumption" is to be considered as such if the artificialization brought about by densification within an urban area or land conversion, new urban buildings and settlements, and road works and infrastructure affect non-artificial land cover [12,13]. Consequently, advocating zero soil consumption means encouraging the compacting of transformations within the urban area by providing for the restriction of new land use and soil consumption activities affecting areas that have not yet been artificialized.

In many new generation urban planning instruments, we find the concepts of "anthropisation" [12–14] and "urbanization" associated with the term "consumed" soil. The two terms "urbanized" and "soil consumption" are used by various disciplinary fields with different meanings, and are often anchored in a specific scientific and cultural matrix. The differences, as we shall see, are not insignificant.

The term "urbanized soil", proper to Corine Land Cover, refers to land use and introduces classification parameters related to all forms of anthropization, including roads, port areas, airports, and cemeteries. While "urbanization" is often associated with two conventional categories: the demographic one, linked to phenomena such as population increase in the areas defined as urban and proportion urban [12–17], and, on the other hand, the territorial one, based on indicators such as soil consumption, diffusion, and concentration; again, it is the Istat [12–15] study that proposes the terminological distinction.

In fact, within the new urban planning instruments related to some regional legislations, the interpretation of the concept of zero soil consumption is aimed at not allowing new areas of land transformations that have not already been provided for by the previous urban planning instrument.

The national law on "Containment of soil consumption and reuse of built-up land", which is still under discussion, understands "soil consumption" as the net annual increase in agricultural, natural, and semi-natural land area subject to sealing. To reinforce the precautionary principle, the term "containment" [13–17] of soil consumption has been replaced in some proposed amendments with the concept of "zeroing".

The June 2023 bill titled "Provisions on urban regeneration" intervenes exclusively in urbanized areas to prevent further expansion of cities, with the assumption that continuing to build outside urban centers costs too much in economic–social terms (public transportation, sewage networks, roads, and infrastructure in general), and for businesses, any economic return on investment is less and less profitable. Added to this is the consideration that it is necessary, as a priority, to tidy up the consolidated urban fabric by revitalizing cities [14] and making the current building stock more efficient and rational, given that half of it was built with techniques, materials, resources, and know-how from at least fifty years ago, with the prospect of dealing with the problems, needs, and risks associated with that period. The "National Program for Urban Regeneration", included in the Infrastructure Annex of the DEF, Document of Economy and Finance, aims to set criteria for the quality of design and implementation of interventions necessary to be met in the case of interventions intended to benefit from the economic contribution of the urban regeneration [14,16,17] fund (energy class A, seismic retrofitting, etc.). Of key importance are points 3 and 5, which provide:

- Realization of green areas and ecosystem services;
- Re-permeabilization of already sealed land, including through land renaturalization and reforestation for the purpose of hydrogeological risk mitigation in urban and peri-urban areas and the visual impact on the reference context.

In addition, for the social use of places, a "soil consumption equal to or less than the original lot, including infrastructure works" is recommended.

In Italy, in order to curb soil consumption, strategic and operational intervention has been envisioned through the new generation of general urban planning, implementation, and urban regeneration tools. In line with the regional regulations that support their legal effectiveness and within PNRR that indicates the accompanying economic measures for the feasibility of the interventions, the "territorialization" of the 'wide' and heterogeneous range of planning options should return by 2026 a complete picture of the actual implementation and achievement of the set objectives. The strategy is to contribute, thanks to the territorialization of interventions [15–17], to curb further depletion-preserving green areas and ecosystem services, restoring permeable lands within urban and peri-urban areas by renaturalizing and reforesting them for hydrogeological risk mitigation, etc. These are actions that will play a "central" role within the supply chain dedicated to urban regeneration, or "joint" because, acting synergistically and interconnected with other measures, they will provide indirect benefits and contributions [16,17] to other goals such as combating climate change, for example.

Ecological values, environmental quality, and cultural assets included in ecosystem services are fundamental to well-being and offer economic prospects in tandem. While the exploitation of natural resources is considered a threat to spatial development, it is also true that working with nature and in harmony with the landscape to provide essential goods and services through projects, renaturation of green infrastructure is economically beneficial and allows for the preservation of the physical and identity characteristics of local communities [17]. The benefits to communities are numerous, ranging from physical and socio-economic to psychological and emotional. It is the green infrastructure that establishes the best connections between the parties, establishing new and healthy balances and relating to urban areas [18,21–23]. Through urban food production and community gardens, the gap between food production and consumption is reduced, contributing to an increase in its perceived value. Any investment in green infrastructure, whether as a one-off intervention or in terms of large areas, has significant potential to strengthen regional and urban development by creating jobs and a circular economy [18,19]. There are many urban challenges and design solutions to be adopted within green infrastructure, the elements of which offer health benefits in cities, such as clean air and better water quality. Healthy ecosystems also reduce the spread of vector-borne diseases. The implementation of green infrastructure in urban areas creates a greater sense of community and helps combat social exclusion and segregation. Strategies that promote ecosystem-based approaches harness the adaptive forces of nature. They are among the most widespread, economically sustainable, and effective tools to combat the impacts of climate change and use green infrastructure to help people adapt to or mitigate the negative effects of climate change. An example of how a single natural capital measure can provide many benefits is the ecological restoration of floodplain forests. These can offer many benefits, such as preventing erosion, filtering water, and maintaining the water table. Forests also mitigate the effects of climate change by storing CO_2 [19,20,23] and providing bio-materials that can act as carbon substitutes, replacing carbon-intensive materials and fuels, as well as acting as a useful reservoir to store water and reduce the risk of flooding in human settlements. The restoration of floodplain forests is often cheaper in terms of one-off and maintenance costs than purely technical solutions such as the construction of dams and floodplain reservoirs. Since floodplain forest restoration measures are often connected to watercourses and the adjacent floodplain, together they succeed in ensuring connectivity and the conservation of protected fauna and birdlife.

Building resilience and improving our coastal flood defenses is also an important part of the land-use reduction strategy for urban resilience; the Alkborough Flats managed realignment project on the Humber Estuary, England [20,23], has delivered coastal flood protection benefits and reduced and deferred expenditure on artificial coastal defenses. The project is estimated to provide an annual flood protection benefit of EUR 465,000, as well as other benefits for wildlife and ecosystem services. The project cost EUR 11.8 million, but its effects are extraordinary: the implementation involved the restoration of tidal habitats affecting 440 hectares of agricultural land. What is more, the organization and management of flood and overflow flows within urbanized areas can be achieved through a complex of actions aimed at strengthening and optimizing the relationship between green infrastructure and watercourses. The Energy Performance of Buildings Directive [20–23] promoted the development and use of new materials and design solutions in the construction of buildings to reduce the significant level of greenhouse gas emissions from this sector. Agenda 2030 for Sustainable Development, in Goal 6—guarantee the availability and sustainable management of water and sanitation for all, and reuse of wastewater: a valuable practice in times of water scarcity—and in Goal 11—making cities and human settlements inclusive, safe, resilient, and sustainable—sets out these principles [20,23]. Green and blue infrastructure need to be linked. An example of efficient blue–green solutions to channel excess rainwater and reduce flooding are Green Roofs (GR) and Rainwater Harvesting (RWH) systems [20,23,50]. The installation of green roofs can promote urban biodiversity by providing a habitat for insects and birds and preserving biological diversity even in urban settings. Rainwater harvesting ensures that excess water is channeled and drained into special collection tanks; this process decreases the water load in the city's sewer system and is useful in areas prone to recurring flooding because increasing the permeable surface area of the urban layout facilitates drainage. In proper water management where water resources are scarce, rainwater is stored to be later reused for other purposes. The idea that sustainability is only an environmental and sectorial issue is being definitively overcome, and an integrated vision of the different dimensions of development and sustainability is being affirmed, which must pass through urban regeneration strategies in which each approach is functional to the systemic vision, hence, resilience, climate change mitigation, and the role of ecosystem services. In Seville, Spain, a young farmers' association managed an innovative LIFE project [20], defining a more sustainable soil management model. The project identified the vegetation cover types offering the best protection against erosion and focused on areas where improved soil retention capacity was an advantage in terms of soil resilience. On a larger scale, the project demonstrated that land cover change made the agricultural landscape more coherent and ready to respond to environmental challenges, in particular, climate change. Regarding action on water-related agri-environmental measures, the case study of Sint-Truiden, Belgium [20,23,50], is interesting, where measures were taken to protect the village from soil erosion and mud flooding through grassed watercourses, grassed buffer strips and retention ponds in the catchment area. The total cost of these measures was low compared to the cost of repairing and cleaning up the damage caused by mud floods in the study area and all the benefits induced, including improved quality, psychological well-being for the inhabitants, and protection of biodiversity.

3.3. *Ecosystem Services*

It is well known that Natura 2000 is an ecological network established by the European Habitats [21,33] and Birds [22,34] Directives. It was established to conserve and protect species and habitats that play an important role throughout the EU, while also providing many ecosystem services to human society. More than 27,000 Natura 2000 sites (terrestrial and marine) have been identified by the 27 Member States [22,23]. They cover an area of 1,219,403 km^2, corresponding to 17.5 % of the total land area of the EU, as well as 4 % of the marine waters under the jurisdiction of the Member States. The value of ecosystem services related to the Natura 2000 network has been estimated at EUR 200–300 billion per year [23–25]. Creating and consolidating the network means, above all, implementing the backbone of the EU's green infrastructure, as it represents a reservoir of biodiversity that can be drawn upon to repopulate and revitalize degraded environments by reducing ecosystem fragmentation and improving connectivity between Natura 2000 sites. This will be in line with the objectives of Article 10 of the Habitats Directive [23]. The Millennium Ecosystem Assessment (MA) [23,34] defined ecosystem services as those with "multiple benefits provided by ecosystems to humankind". Humans have changed ecosystems in the middle of a century with a speed unprecedented in previous periods; causes include the need for food, water, fiber/timber, and energy sources, habits and lifestyles, and production systems that in the most industrialized countries have changed abruptly by pandering to economic growth objectives to the detriment of the environment. Even the phenomenon of sprawl discussed below, brought about by new trends in settlement criteria, born out of the desire of families to live in the countryside surrounded by greenery [23–26] by building large artifacts on inexpensive lots but distant from the center and hardly or ineffectively connected to the urban center, has occurred in Europe and North America [23,32], and is a process of invasion of green spaces and low-density areas [24,30] that has engulfed potentially productive or high-nature areas with incidences of biodiversity.

This impact is causing an irreversible loss in biodiversity across the planet, and in particular, 60% of ecosystem services have been estimated to be compromised. The Millennium Ecosystem Assessment [23,24,34] identifies four categories of ecosystem services:

- Supply or provisioning: services that provide goods such as food, water, timber, fiber, fuel, etc;
- Regulatory: services that play a role in regulating climate, air and water quality, mitigate natural hazards such as erosion, or play a role in land formation or pollination;
- Supporting: genetic biodiversity conservation services and habitat creation;
- Cultural: useful services for cultural identity, aesthetic and recreational values, and intellectual enrichment.

The MA is tasked with identifying the state of ecosystems globally, assessing and monitoring the consequences of changes in ecosystems on human well-being by formulating scientific support for the identification of actions aimed at the conservation and sustainable use of ecosystems. The emerging data are worrying, so much so that the European Parliament came to pass the Nature Restoration Law (NRL) in June this year. As is well known, the NRL, calls for the implementation of a series of environmental restoration and restoration actions by 2030. These will be extended to 20 percent of European territory, both marine and terrestrial. At a later stage, by 2050, degraded ecosystems will be restored [23,24].

The earth's ecosystems—and the ecosystem services they provide, including diverse cultural [24–26], spiritual, and economic values—are fundamental to human life, contributing half of global GDP. However, there is a complex [25–27] crisis (so-called trivalent crisis) underway related to pollution, climate change and biodiversity loss. The trivalent crisis causes effects on humans and the environment; suffice it to say that in just four years, from 2015 to 2019, at least 100 million hectares of healthy and productive land were lost annually, and this impacted the lives of nearly 1.5 billion people [25].

4. Discussion

4.1. Urban Sprawl and Beyond—Data on the Phenomenon in Italy

Data from Eurostat's Land Use and Cover Area frame Survey [26] show that 7.8 percent of Italian land is affected by the phenomenon of anthropization, which is not in line with the European average of 4.6 percent. In fact, Italy, in 2013, ranked fifth in Europe [26], preceded by Malta, Belgium, the Netherlands, and Luxembourg. In the time span of almost a century, in terms of stock, Italian residential construction has experienced a consistent and continuous growth over time: from almost two million buildings already present before 1919 to almost thirteen million, excluding the demolitions that have occurred over the years [26]. But this reading, which refers only to the increase in soil consumption related to residential building, is not exhaustive. We must, in fact, consider that there are numerous other factors that determine forms of coverage in urban areas (business and commercial centers, production activities of all orders and scales of size, infrastructure, etc.) [27–29] increasing the level of anthropic use [27,30].

The 2006 European Environmental Agency defined sprawl as a physical pattern of low-density expansion of large urban areas at the expense of agricultural areas and from mixed land use and suburbs. Urban sprawl and urbanization are intimately related categories. Tangible repercussions of this phenomenon, in addition to the loss of agricultural land, are higher public expenditures for the provision of endowments and services and greater use of private vehicles; as a result, the sprawl of a city and its suburbs over rural areas is considered "unsustainable" [27–29,31,37]. The phenomenon of urban sprawl and the phenomenon of soil consumption are closely related, but they do not have the same meaning.

Sprawl is a form of low-density sprawl [28–31,37,43], while soil consumption is the consequence, but the two-way correspondence does not apply. Soil consumption, as mentioned above, in terms of loss of naturalness and biodiversity can result from other causes, such as deforestation with crop replacement and fires, for example.

In Italy, vegetation cover in mountainous areas is declining (−0.3 percent between 2012 and 2021, or about 4600 hectares per year), probably as a result of the devastating fires that have affected the peninsula in recent years. "Land losses" are concentrated in the Islands and the Northwest, especially in the belt below 1000 m above sea level, which is most affected by soil consumption for settlement purposes. Sicilia's percentage of degraded land is among the highest [29], reaching 19.7 percent (Table 1).

According to Ispra data, soil consumption is growing again, and in 2021, it reached 70 sq. km. of new artificial cover in a single year. The average is worrying, with 19 hectares per day, the highest value in the last ten years; it has been calculated that the speed of consumption reached 2 sq. m. per second. The artificialization of land through concrete affects 21,500 sq. km. of Italian land: of this, as much as 5400 sq. km., a territory as large as a region classified as medium-sized (such as Liguria or Piemonte) [29,30], is represented by construction, which accounts for 25 percent of the land consumed [29–32]. But that is not all. As mentioned, consumption, understood as the artificialization of land, may depend on anthropogenic processes, not necessarily related to "cement production" [29,31–33], meaning the construction of roads and general infrastructure and urbanization works.

Table 1. Soil consumption by province, Figures for the two cities of Palermo and Enna are highlighted in green, Regione Sicilia. Source: Ispra.

Regione Sicilia Provincia:	Soil Consumption 2006 [%]	Soil Consumption 2006 [Hectares]	Soil Consumption	Soil Consumption 2012 [Hectares]	Increase 2006–2012 [Net Annual Soil Consumption in Hectares]	Soil Consumption 2021 [Hectares]	Increase 2020–2021 [Net Annual Soil Consumption in Hectares]
Trapani	7.37	18.173	7.55	18.601	428.15	19.120	43.09
Palermo	**5.40**	**26.981**	**5.56**	**27.780**	**798.87**	**28.419**	**66.37**
Messina	5.77	18.727	5.87	19.044	317.08	19.572	30.30
Agrigento	5.43	16.540	5.65	17.197	657.60	17.603	27.41
Caltanissetta	4.51	9.607	4.64	9.889	282.20	10.209	36.13
Enna	**3.05**	**7.815**	**3.11**	**7.957**	**141.70**	**8.215**	**66.06**
Catania	7.35	26.104	7.68	27.279	1.175.13	28.118	59.06
Ragusa	9.75	15.744	10.15	16.399	654.97	17.116	96.52
Siracusa	8.53	18.002	8.83	18.650	647.18	19.217	62.23

4.1.1. Containing Soil Consumption and Regeneration through Planning and Landscape Design

Soil consumption of urban planning significance occurs by transforming an originally natural or semi-natural area into an area for residential, productive, and general infrastructure use.

The law on "soil consumption" approved in the House, applies to, in an "urbanized area", the part of the territory consisting of historic centers, built-up areas with continuity of lots for residential, industrial, and artisanal, commercial, office, service or tourist-receptive use, as well as areas equipped with equipment, services or technological facilities, urban parks, lots and unbuilt interclosed spaces [29,33–35] equipped with primary urbanization works.

In the urban planning domain of some regional laws, the definition of "urbanized territory" is unambiguous, but it is broad, not already polysemous, when its reference is made in the context of new urban planning instruments. Especially in some regional laws [29,30], it has an exquisitely technical configuration and an "open" temporal dimension, linked to the confirmation of existing urban planning provisions in the new urban planning instruments, the maintenance of building rights, and bonus incentives (house plan/earthquake bonus to set the building sector in motion).

The General Urban Plan—provided in Art. 32 of Law 24/17 of Emilia-Romagna—identifies the perimeter of the "urbanized territory", meaning not only the built-up areas with continuity and interclosed lots, but also the completion areas of the current plan conterminous to the urbanized territory and the undeveloped lots of urban implementation plans in progress. The same interpretation is given by RL (Regional Law) 19/02 of Calabria [30,43], the guidelines, and attached circulars.

Thus, according to the meanings described and some current regulations, in Italy land is:

- "consumed" if artificial land cover associated with the loss of a fundamental environmental resource intervenes, or has intervened, or if the naturalness of the land has been compromised by external or anthropogenic factors (fires, deforestation);
- "anthropized" if it has been transformed and its use is for human (anthropogenic) activities;
- Affected by "urbanization/urban sprawl" if there is a low-density extension of large urban areas at the expense of agricultural areas and mixed land use and suburbs;

- "urbanized" when it has pre-existing constructions of the urban planning instrument, but also all residual (unimplemented) areas of the current plan (B, C, D and F), and the undeveloped lots of detailed urban plans under implementation [30,43].

The trend of de-compaction or housing dispersion recorded in the last decade raises concerns. As pointed out by ISTAT, while between 1991 and 2001 there was a generalized trend of compaction (where densely built-up areas grow and suburban density decreases), a new progression of settlement dispersion emerged in the following decade. In 2011, consolidated built-up areas covered 6.7 percent of the territory [30,31]. Over the twenty years considered, the area grew by almost 2 percentage points, with higher progressions in the main urban realities, which represent almost one-fifth of the territory.

The result of this is the orientation to contain, preserve, limit, and reduce to zero the consumption of additional land by incentivizing "vertical" development with densification strategies and volumetric replacements [30,33,43] and promoting the reuse of the existing through special regulations and funding measures, such as regional l. no. 21 of 2008 "Norms for urban regeneration, in RL (Regional Law). no. 18 of 2019 of the Lombardia region "Simplification and incentive measures", but also the Budget Law of 2020 that allocated, from 2021 to 2034, EUR 8.5 billion [29,30,33] to be invested in regeneration projects starting from the suburbs with the aim of making investments to reduce marginalization and degradation.

In recent years, more and more green infrastructure projects have been implemented, including green interventions with a high impact and benefits for the environment, the community, and the landscape that are instrumental to mitigating the effects of climate change and land consumption [30]. There are tested results from best practices that show that a flexible, integrated, and concrete approach is the most cost-effective from an environmental and economic point of view. However, to further improve the functioning of strategies and maximize their benefits, they must assume interconnected and interdependent actions [30]. From the study carried out, it emerges that the benefits increase significantly when a minimum level of coherence is reached between the different scales, between the interventions and the contributions actually made to reduce land consumption, making the territory resilient by optimizing the responses in terms of regeneration by favoring measures aimed at implementing green infrastructures and enhancing ecosystem services. Only in this way, and measured against a broad territorial context, can the various strategies, combined in a green key, make a significant contribution to regional development, to combating the effects of climate change, to disaster risk management [30], and to improving agriculture/forestry and the environment.

4.1.2. The Contribution Made by Ecosystem Services

Recognition of the biophysical value of nature in economic terms took its first steps with the Millennium Ecosystem Assessment (MA) initiative (2005), followed by The Economics of Ecosystems and Biodiversity, TEEB [30–32], a study initiated in 2007 during the G8+5 meeting, European Mapping and Assessment of Ecosystems and their Services, MAES [30–32], and European projects such as Life+ MNG [30–32] that deepened the economic assessment system. The MA addressed the following topics: changes in ecosystems, ecosystem services, and the consequences for human well-being; the status of ecosystem services and human well-being; actions to conserve ecosystems; the effectiveness of decisions affecting ecosystems and the causes of ineffectiveness; and methodologies to strengthen the assessment of ecosystems and services.

The MA has established a scientific basis for taking ecosystems into account in decision making processes and has been credited with establishing the relationship between human well-being and the environment, highlighting the fact that the degradation of nature entails damage that has a significant cost. Acting on the environment is "complex and multidimensional" [30–32], and the MA proposes an analytical approach to analyzing ecosystems, their state, and their impacts, offering a common method of assessment in processes involving the UNCCD (United Nations Convention to Combat Desertification), the CBD (Convention on Biological Diversity), the Ramsar Convention on Wetlands, and the Convention on Migratory Species [30–32,50]. The motivation behind the economic estimation of ecosystem services is not to make resources tradable but to bring out the importance of the reciprocal influences of ecosystem services and community well-being; subsequently, at the national level, countries have produced their own schemes (https://openness.hugin.com/example/cices (accessed on 2 march 2024).

In order to simplify understanding, the CICES (Common International Classification of Ecosystem Services) [31,32,50] is, for example, the framework used in the EU MAES process, which was aimed at mapping ecosystem services on a European scale, in order to fulfil the commitments made in Action 5 of the EU Biodiversity Strategy to 2020 [31,32,50]. Examples include the missions of the TEEB process [31,32,50] to estimate the value of ecosystem services and integrate it into decision making, to invest in ecological infrastructure to make the value of nature visible, i.e., communicable in a clear way, to carry out measures and mapping to improve management, to identify the relationships between nature and human well-being, and to map protected areas.

Thus, in addition to systems for calculating and monitoring the area of soil consumed, the share of landscape restored through green infrastructure, of renatured area, and the contributions to climate mitigation made by ecosystem services must be considered; this information must be quantified and monitored by making it interfaceable with other systems referred to and in use, e.g., by Ispra to quantify soil consumption. The ecosystem services assessment process complements the mapping process and allows, by improving the knowledge framework of environmental resources, us to define the most appropriate management policies, and to bring out and know the value of natural capital. Many open-source computer models for valuing ecosystem services, such as ARIES, ARtificial Intelligence for Ecosystem Services, and InVEST, Integrated Valuation of Ecosystem Services and Tradeoffs tool [31,32,50], used in many programs and initiatives, are capable of geo-referencing and valuing ecosystem services in economic and biophysical terms.

Those analyzed in the preceding paragraphs are numbers that deserve a closer look, especially if we consider that some areas of the regions show that they have been particularly affected by urbanization processes—and soil consumption with urban relevance—already since the last few decades of the last century with urban sprawl towards peripheral areas [31–34,50]; the data that emerge are apparently contrasted with the amount of land degraded or compromised by factors independent of urban sprawl, and, thus, the transformation of woods and forests into productive cultivation activities, fires and others. Sicilia, for example, ranks among the regions with the highest percentage of degraded land, caused by productive conversions, deforestation, and fires, while still managing to maintain average sprawl.

If we consider Enna, the land consumed in the province of Enna as of 2006 was 3.05%, standing at 3.11% in 2012, according to Ispra data [31–34,50]. It will be necessary to understand what concrete actions to propose to reduce soil consumption, to encourage de-impermeabilization and renaturalization, to enhance and recover fragments of (diffuse) landscape within cities, to promote fruition activities in the peri-urban system, and to understand what can be the contribution of the ecological–environmental dimension rendered through landscape enhancement for ecological transition and through urban transformation management tools [32–34]. It will be necessary to realize that the "stability" of city fragments within the landscape depends on the ability of the landscape to contribute to urban circular metabolism, and that the model of urban resilience can be built through the "control" of the landscape and the stability of biodiversity and "urban ecosystems".

Ispra returned a snapshot of Italy, in the period 2000–2006 [32–34], in which peripheral and suburban areas grew four times faster [33,34] than urban centers. The phenomenon affected Europe, in which urban areas cover 4 percent of the surface area, but in Italy, the percentage figure was exceeded and the phenomenon of urban sprawl, or diffuse city, "implies that at least one quarter of the territory is directly involved" [34,35] in "urban use" [35]. The growth of suburban areas is not the only element characterizing the phenomenon of "urban sprawl" [36,43].

The environmental effects of "urban sprawl" include a reduction in the productivity of agricultural areas [37] and an increase in air pollution brought about by automobile use and the resulting road traffic congestion [38].

Even the vertiginous growth of intensive agriculture has had an impact on the territory and the rural landscape; alternative forms of agricultural activity linked to tourism, agritourism, for example, have had positive socio-economic–environmental effects both in terms of opportunities for valorization, greater knowledge of the cultural heritage, and in terms of impact—positive or negative—on the environment, on the balance of ecosystems and their services, on consumption models, and on the management of flows with respect to areas to be protected and resources to be preserved. The relaunch of tourism, and, therefore, of the economy of the territories, can take place considering the agricultural territory as an 'opportunity' for intelligent and sustainable development. Energy is a very significant indicator to measure the effectiveness of the measures envisaged: through the study of energy, we can identify the areas of strategic impact of the actions put in place for tourists' enjoyment of rural areas [39], in terms of increasing environmental awareness, thus contributing to the preservation of landscapes and responses to environmental challenges in the era of "ecological transition" [40].

4.2. Interventions of Renaturalization for Resilience, Focus Sicilia

The picture that emerges sees an ongoing crisis at the global level due to factors such as pollution, climate change, soil consumption, loss of biodiversity, and risk reduction in ecosystem services. These phenomena have different dimensions and are characterized by an acceleration in the speed of processes and changes that produce impacts directly on humans or indirectly involving other environmental factors in ways and cause–effect relationships that are difficult to understand.

According to the United Nations Intergovernmental Panel on Climate Change (IPCC) study, cities are the cause of more than 75 percent of the amount of CO_2 emissions estimated worldwide, and urban settlements are responsible for nearly half of the total greenhouse gas emissions produced [41–43]. It would seem to ascertain that this "picture", alarming in itself, will tend to get even worse. Some estimates speculate that the process of urbanization, in densely populated countries such as China and India, for example, will increase significantly. The number of residents within built-up areas will be on the rise: by 2050, 70 percent of the world's population will live in medium and large cities and metropolises, rising from the current 50 percent to 60 percent within a decade [42].

Absurdly, it would seem that a process of densification and growth proportional to the population's state of distress is underway: the population on a global basis will confirm the upward trend, especially [43] where there is widespread poverty.

To incentivize nature's reappropriation of areas taken from it, action must be taken at several levels, including reducing expansion areas and de-impermeabilizing lands through congruous and sustainable regulations and planning choices [44]. The process of "soil consumption" is difficult to stop. Some new-generation urban planning instruments have applied containment regulations to bring soil consumption to zero. What can be done, given that soil consumption is known to be the cause of increases in artificial land cover, and this can increase land vulnerabilities in light of climate change and interconnected factors that reduce responses in terms of increased resilience of cities? Nature needs to reclaim the places deputed to it. In many cities of the same region, for example, in Sicily, the causes of soil consumption are different; Enna and Palermo stand confirming the "consumption" trends of 2006, +0.16 for Palermo and +0.6 for Enna (Table 1).

The soil consumed [45] in the province of Enna (with a rural vocation) is probably linked to the cultivation transformations and the reconversion of productive types, while at the municipal level, the values are slightly higher (Table 2) because in lower Enna, the presence of the University Citadel [46] has determined a recent urban expansion [47] and the supply of student residences.

Table 2. Soil consumption in the province of Enna, Figure for the two city of Enna is highlighted in green. Regione Sicilia. Source: Ispra.

Regione Sicilia Provincia Enna Comune:	Soil Consumption 2006 [%]	Soil Consumption 2006 [Hectares]	Soil Consumption	Soil Consumption 2012 [Hectares]	Increase 2006–2012 [Net Annual Soil Consumption in Hectares]	Soil Consumption 2021 [%]	Soil Consumption 2021 [Hectares]	Increase 2020–2021 [Net Annual Soil Consumption in Hectares]
Agira	3.1	500	3.2	517.9	18.38	3.3	542	10.15
Aidone	1.7	349	1.7	351.47	2.3	1.7	363	0.90
Assoro	3.8	421	3.8	428.12	6.64	3.9	441	2.10
Barrafranca	5.0	270	5.1	274.86	5.21	5.4	288	1.37
Calascibetta	3.0	269	3.0	268.93	0.27	3.0	270	0.10
Catenanuova	12.6	141	12.7	142.29	1.5	13.1	146	0.14
Centuripe	3.1	530	3.1	529.74	0	3.1	543	1.24
Cerami	1.9	179	2.0	186.93	7.58	2.1	199	5.38
Enna	**3.6**	**1292**	**3.7**	**1325.98**	**33.68**	**3.8**	**1354**	**3.23**
Gagliano Castelferrato	3.3	182	3.3	183.46	1.35	3.3	191	6.14
Leonforte	3.7	312	3.9	323.81	11.42	3.9	328	3.77
Nicosia	3.2	688	3.2	702.08	13.85	3.4	754	18.51
Nissoria	3.3	203	3.3	204.63	1.73	3.4	213	2.16
Piazza Armerina	3.1	931	3.2	955.49	24.72	3.2	981	6.50
Pietraperzia	2.5	289	2.5	290.55	1.81	2.5	297	0.06
Regalbuto	2.4	404	2.4	411.59	7.66	2.5	420	1.33
Sperlinga	1.9	109	1.9	110.02	0.97	1.9	114	0.06
Troina	2.7	451	2.7	451.81	1.18	2.7	462	2.01
Valguarnera Caropepe	10.9	102	11.0	103.19	0.81	11.1	104	0.00

An urban reorganization of the territories will probably intervene. The current urban planning instruments in Sicilia [48], the PRG (General regulation Plan), drawn up in the implementation of RL (Regional Law) 71/78, will have to be replaced with the municipal General Urban Plan, PUG, which, in implementation of the principles of limiting soil consumption, regulates productive interventions in agricultural green areas. Not only that, RL, Regional Law, n. 19/20, provides a range of implementation tools by attributing exact legal configuration to urban regeneration interventions (regulated in Art. 33 of the recent RL). The new regulatory context solicits broad reflections, including the permanence of homogeneous zones in the two different domains—transformable areas and non-transformable areas—while pointing to paths of regulatory adaptation of the modalities in use until a few years ago.

In order to curb soil consumption in favor of urban regeneration, attention must also be turned to the recovery and redevelopment of degraded, disused, or abandoned areas; the redevelopment of urbanization in general and land endowments, connections, and contextual mobility; the use of urban equalization and compensation tools;, and the involvement of social partners in forms of participation. In order to achieve the set targets, it is necessary to ensure [17,48,54]:

- Consistent and reliable data;
- Improving the knowledge base and encouraging innovation;
- Providing financial support for regeneration projects aimed at combating land consumption through ecosystem services and landscape restoration;
- Integrating green infrastructure into policy implementation in key areas to trigger financing mechanisms across the EU [17,49,54].

Consistent and reliable data are essential for effective dissemination of actions. As part of the EU Biodiversity Strategy, together with the European Environment Agency, other Member State research bodies and agencies and other stakeholders have initiated monitoring activities and established interactive databases. There is a recognition that information is needed not only on land consumption, but also on the extent and condition of ecosystems, the services they provide, and the value of those services [49–54], so that ecosystem services are properly valued and then estimated with respect to the contribution they can make to resilience and the mitigation of the effects of climate change by promoting innovative solutions and approaches in spatial planning and decision making processes.

A minimum level of consistency should be encouraged in relation to the data used to inform these decisions, particularly for projects supported by EU funds. However, further efforts are needed to improve understanding of the links between biodiversity (species/habitat) and ecosystem condition (viability, resilience and productivity) and its capacity to provide ecosystem services.

The transnational continuity of geographical features such as mountain ranges (e.g., the Alps and Pyrenees), vegetation cover (e.g., forests) and river basins (e.g., the Rhine and Danube) [50–54], which are part of the EU's natural and cultural heritage and common identity, indicates that identity features, environmental problems, and solutions transcend national borders. For example, the European Green Belt initiative is an ecological network that aims to harmonize relations between human activities and the natural environment by increasing opportunities for socio-economic development in the area from the Barents Sea to the Black Sea [50–54]. In the TEN-T policy, already from its genesis, green infrastructure was conceived as an integral part of the proposed corridor projects. Strategies, to be effective, cannot have administrative limits, must be coordinated and joint, and must aim at a pan-European vision.

5. Conclusions

The policy documents and case studies cited, which are almost always empirical experiences, provide a useful reconnaissance to identify the most effective approach in terms of design solutions and strategies to be implemented in order to contain soil consumption through ecosystem services, landscape restoration, and its residual fragments within the existing city. In our research, we point out that this effort requires joint and integrated action and that this can take place within the urban regeneration processes envisaged by European planning, encouraged in the various fora by the various environmental protection organizations and scientific communities. All the initiatives mentioned have a high level of specialization and sectorization, and this is an obvious limitation. According to this approach, within the regeneration processes, it is necessary to define the individual actions (material and immaterial: de-impermeabilization/renaturalization, naturalistic engineering works for flood defense, implementation of green infrastructures in urban areas, ecological restoration of floodplain forests and woods in extra-urban areas) and the strategic impact areas (society, environment, climate, biodiversity, economy, etc.). Many factors are linked to certain environmental phenomena in a cause–effect relationship that has, at times, blurred boundaries even though they interact. Building resilience and improving our defenses against adverse events is an important point of the strategy to reduce land consumption, but it is clear that this strategy is a necessary but not sufficient condition. Specific action needs to be linked to something else.

It can be linked to green infrastructure (managing flows from floods and overflows can be achieved by strengthening and optimizing the relationship between green infrastructure and watercourses), to ecosystem services (economic estimation of the contribution of ecosystem services and benefits becomes a complementary support tool), and to punctual actions to restore landscape fragments in urban areas. The combination of the various measures leads to a reduction in landscape fragmentation and forest degradation, but it can also contribute to ensuring the conservation status of species and habitats and improving the provision of related ecosystem services, fostering a sense of community, and helping combat social exclusion and segregation. Clearly, all this needs to be monitored (using the tools described in the previous paragraphs), both to understand the trend state of the phenomena analyzed and to understand what the "response" of the interventions actually is in the implementation phase and measure it against the expectations set.

One of the aspects through which counteracting soil consumption generates positive effects is the protection of the land and the environment. This also has effects on the agricultural sector, as stopping the artificialization of the soil means addressing and promoting sustainable forms of use. In the management of their activities, "green" farmers use environmentally friendly practices, including composting of waste, use of renewable energy, recycling of materials, and the use and sale of zero-mile organic products. In this way, the sustainability "chain" takes shape, helping to preserve the environment and natural resources. At the same time, forms of rural tourism (rural farm) are encouraged [17,54], with a strong identity connotation, respecting the territory, the landscape and its peculiarities. Through the use of traditional techniques and practices, those involved in traditional agricultural production also offer opportunities to promote sustainable education by raising awareness of the environment [17,54] and encouraging an "intelligent and slow" use of the land. Through workshops and educational activities, users can put sustainable farming practices into practice, learn the importance of biodiversity, and reduce energy consumption [17,54]. These types of experiences contribute to forming more responsible citizens by making them aware of the importance of a sustainable lifestyle and offer numerous benefits to both guests and local communities.

The various modes of sustainable tourism also offer a significant experience, contributing to the enhancement of local cultures and the development of rural economies [17,54]. The 2030 Agenda, with its 17 goals, emphasizes the importance of implementing (Goal 11) 'Sustainable Cities and Communities'. There are numerous ecosystem services—regulating and producing, but also recreational and cultural—that can ensure the livability and stability of ecosystems by combining protection [17,54] and sustainable development.

But is the situation global? While this has been a global phenomenon, Europe has been not an exception. Since 2010, urban expansion in the continent has led to a significant conversion of agricultural land, driven by the surging demand for housing, infrastructure, and commercial development. This conversion has had far-reaching implications for the agricultural sector and the environment. The loss of fertile farmland has impacted food production and sustainability, posing challenges for Europe's food security. Factors like population growth, urbanization, and economic development contribute to this trend, with transportation network expansion further fueling the demand for land conversion. While urban expansion brings [17,54] economic growth and improved infrastructure, it also threatens food production, biodiversity, and environmental sustainability. Efforts such as land use planning, conservation measures, and promoting sustainable agriculture have been implemented to address these challenges. However, balancing the needs of urbanization with the preservation of agricultural resources on ecological–environmental fronts is essential for long-term food security and sustainable land use practices.

In addition, it is fundamental for future generations to understand the contribution of the ecological–environmental dimension in the fight against climate change and the ecological transition of cities. This can also be achieved through the enhancement of the landscape, the implementation of land de-impermeabilization criteria [17,54] and the management of meteoric water flows, the reintroduction of certain techniques of environmental restoration [17,54] and 'naturalistic' architecture, and the use of particular natural or recovered materials and native tree and shrub essences. The protection of landscape areas combined with some forms of sustainable use [17] can help reorganize and implement the offer of services and equipped green areas [17]. The forest favors the absorption of CO_2, and a network of urban gardens [17,54] could help the "urban landscape economy".

Author Contributions: Conceptualization, C.F., F.M., K.M.N., B.A.-R. and A.A.; methodology, C.F. and F.M.; validation, C.F., F.M., K.M.N., K.M.N., B.A.-R. and A.A.; investigation, C.F., F.M., K.M.N., B.A.-R. and A.A.; writing—original draft preparation, F.M., C.F. and K.M.N.; writing—review and editing, C.F. and K.M.N.; visualization, C.F.; supervision, F.M. The authors collaborated in the drafting of the following paragraphs, in order: Abstract and Section 1 are of all authors (C.F., K.M.N., B.A.-R., A.A. and F.M.); Sections 2, 2.1 and 2.2 are of C.F. and F.M.; Sections 3, 3.1 and 3.2 are of C.F.; Section 3.3

is of K.M.N.; Sections 4, 4.1.1 and 4.2 are of C.F.; Section 4.1.2 are of C.F., K.M.N. and F.M.; Section 5. is all authors (C.F., K.M.N., B.A.-R., A.A. and F.M.). All authors have read and agreed to the published version of the manuscript.

Funding: This research received no external funding.

Data Availability Statement: Data are contained within the article.

Conflicts of Interest: The authors declare no conflicts of interest.

References

1. Istat, Forme, Livelli e Dinamiche dell'Urbanizzazione in Italia. ISTAT. 2017. Available online: https://www.istat.it/it/files//2017/05/Urbanizzazione.pdf (accessed on 6 June 2023).
2. Munafò, M. (Ed.) Consumo di Suolo, Dinamiche Territoriali e Servizi Ecosistemici. Edizione 2022. Report SNPA 32/22. 2022. Available online: https://www.snpambiente.it/snpa/consumo-di-suolo-dinamiche-territoriali-e-servizi-ecosistemici-edizione-2022/ (accessed on 8 June 2023).
3. ISPRA. Corine Land Cover, Anni 2000, 2012 e 2018, Corine Land Cover—Italiano. Available online: https://www.isprambiente.gov.it/it/attivita/suolo-e-territorio/suolo/copertura-del-suolo/corine-land-cover (accessed on 8 June 2023).
4. ISPRA-SNPA. Carta Nazionale del Consumo di Suolo 2022, Rapporto Nazionale "Consumo di Suolo, Dinamiche Territoriali e Servizi Ecosistemici. Edizione 2022"—Italiano. 2022. Available online: https://www.isprambiente.gov.it/it/archivio/eventi/2022/07/presentazione-del-rapporto-nazionale-201cconsumo-di-suolo-dinamiche-territoriali-e-servizi-ecosistemici-edizione-2022201d (accessed on 20 June 2023).
5. COM(2012) 710 Final. COM/2013/0249_Communication from the Commission to the European Parliament, The Council, The European Economic and Social Committee and the Committee of the Regions, Green Infrastructure (GI)—Enhancing Europe's Natural Capital, EUR-Lex—52013DC0249—EN—EUR-Lex. Available online: https://europa.eu (accessed on 13 February 2024).
6. Bruxelles, 24.5.2019 COM(2019) 236 Final. RELAZIONE DELLA COMMISSIONE AL PARLAMENTO EUROPEO, AL CONSIGLIO, AL COMITATO ECONOMICO E SOCIALE EUROPEO E AL COMITATO DELLE REGIONI Riesame dei Progressi Compiuti Nell'attuazione della Strategia dell'UE per le Infrastrutture Verdi. Available online: https://eur-lex.europa.eu/legal-content/IT/TXT/PDF/?uri=CELEX:52019DC0236&qid=1562053537296#:~:text=Le%20infrastrutture%20verdi%20sono%20definite,ampio%20spettro%20di%20servizi%20ecosistemici (accessed on 13 February 2024).
7. Baldini, M. Le Infrastrutture blu e Verdi Nelle Città Aumentano la Connettività degli Habitat, in Le Infrastrutture blu e Verdi Nelle Città Aumentano la Connettività degli Habitat e Favoriscono la Biodiversità | AboutPlants. 2023. Available online: https://www.aboutplants.eu/notizie/paesaggio/Le-infrastrutture-blu-e-verdi-nelle-citta-aumentano-la-connettivita-degli-habitat (accessed on 30 October 2023).
8. Formez, Macro-Attività Realizzazione Azione di Sistema. Politiche Pubbliche e Rigenerazione Urbana: Un Quadro Ricostruttivo, Formez. 2021. Available online: https://formez.it (accessed on 5 October 2023).
9. Istat, Rapporto SDGs 2023. Informazioni Statistiche per l'Agenda 2030 in Italia, ISTAT, Roma. 2023. Available online: https://www.istat.it/storage/rapporti-tematici/sdgs/2023/Rapporto-SDGs-2023.pdf (accessed on 5 October 2023).
10. ASVIS, Le Infrastrutture Verdi Nelle Città a Dieci Anni dalla Legge Sugli Spazi Verdi Urbani, Position Paper, ASVIS, Dicembre 2023, Il Position Paper dell'ASviS a Dieci Anni dalla Legge sul Verde Urbano. Available online: https://asvis.it/public/asvis2/files/Pubblicazioni/PositionPaper/2023/PositionpaperASviSInfrastruttureVerdiDEF.pdf (accessed on 30 October 2023).
11. Bergantino, A.S.; Di Liddo, G.; Porcelli, F. A Relative Measure of Urban Sprawl for Italian Municipalities Using Satellite Light Images; Forthcoming on Applied Economics; Working papers SIET 2018–ISSN 1973-3208. Available online: http://www.sietitalia.org/wpsiet/2018_Bergantino.pdf (accessed on 30 December 2023).
12. Carraro, C. (Ed.) *Cambiamenti Climatici, Infrastrutture e Mobilità*; Ministero delle Infrastrutture e della Mobilità Sostenibili (MIMS): Roma, Italy, 2022; Cambiamenti climatici, infrastrutture, mobilità ‹Trasformazioni—Il Blog di Carlo Carraro›.
13. Ispra, Consumo di Suolo, Dinamiche Territoriali e Servizi Ecosistemici, Edizioni. 2018. Available online: https://climate.copernicus.eu (accessed on 6 June 2023).
14. European Commission, Copernicus Climate Change Service—C3S, and European Centre for Medium-Range Weather Forecasts—ECMWF. Heat Wave Days and Heat Related Mortality for Nine European Cities Derived from Climate Projections. 2021. Available online: https://cds.climate.copernicus.eu/cdsapp#!/software/app-health-urban-heat-relatedmortality-projections?tab=app (accessed on 10 September 2023).
15. Borie, A.; Micheloni, P.; Pinon, P. Forme urbane e siti di meandri. In *Casabella, I Terreni Della Tipologia*; 1985; number 509–510.
16. Ispra, Annuario dei dati Ambientali, Edizione 2012. ISPRA, Roma Annuario dei dati Ambientali 2012—Italiano. Available online: https://www.isprambiente.gov.it/it/pubblicazioni/stato-dellambiente/annuario-dei-dati-ambientali-2012 (accessed on 10 September 2023).
17. Territorial Agenda of the European Union 2020. Towards an Inclusive, Smart and Sustainable Europe of Diverse Regions. Informal Ministerial Meeting of Ministers Responsible for Spatial Planning and Territorial Development. 19 May 2011, Hungary. Available online: https://www.tandfonline.com/doi/pdf/10.1080/14649357.2012.707391 (accessed on 13 February 2024).

18. Reports, Studies and Review Documents Supported by the European Commission. Available online: http://ec.europa.eu/environment/nature/ecosystems/studies.htm (accessed on 13 February 2024).
19. European Commission. Brussels, 6.5.2013 COM(2013) 249 Final, See Case Examples of GI Creating Jobs in Table 2 of Commission Services Working Document (SWD(2013) 155 Final). Available online: https://eur-lex.europa.eu/resource.html?uri=cellar:d41348f2-01d5-4abe-b817-4c73e6f1b2df.0014.03/DOC_1&format=PDF (accessed on 13 February 2024).
20. European Commission, Communication from the Commission to the European Parliament, the Council, the European Economic and Social Committee and the Committee of the Regions. A Community Approach to the Prevention of Natural and Man-Made Disasters, COM(2009) 82 Final. *OJ L* **2003**, *1*, 65. Available online: https://eur-lex.europa.eu/LexUriServ/LexUriServ.do?uri=COM:2009:0082:FIN:EN:PDF (accessed on 13 February 2024).
21. AAVV, Natura, Paesaggio e Sviluppo del territorioNatura 2000. Provincia di Bolzano, Rete Natura. 2000. Available online: https://provincia.bz.it (accessed on 13 February 2024).
22. European Commission. Green Infrastracture. Available online: http://ec.europa.eu/environment/nature/ecosystems/docs/adaptation_fragmentation_guidelines.pdf (accessed on 13 February 2024).
23. Spano, D.; Mereu, V.; Bacciu, V.; Barbato, G.; Buonocore, M.; Casartelli, V.; Ellena, M.; Lamesso, E.; Ledda, A.; Marras, S.; et al. *I Cambiamenti Climatici in sei Città Italiane*; Fondazione Centro Euro-Mediterraneo sui Cambiamenti Climatici CMCC: Lecce, Italy, 2021; Available online: https://www.doi.org/10.25424/cmcc/analisi_del_rischio_2021 (accessed on 10 September 2023).
24. European Commission. Connecting Smart and Sustainable Growth through Smart Specialisation. 2012. Available online: https://www.researchgate.net/publication/312532217_Connecting_Smart_and_Sustainable_Growth_through_Smart_Specialisation_-_A_practical_guide_for_ERDF_Managing_Authorities (accessed on 13 February 2024).
25. Pavia, R.; Clementi, A. *Territori e Spazi Delle Infrastrutture*; Ed. Interspazi: Ancona, Italy, 1998.
26. Travisi, C.M.; Camagni, R.; Nijkamp, P. Impacts of urban sprawl and commuting: A modelling study for Italy. *J. Transp. Geogr.* **2010**, *18*, 382–392. [CrossRef]
27. Iaione, F.C. La città come bene comune. *Aedon* **2023**, *1*, 31–40.
28. Carruthers, J.I.; Ulfarsson, G.F. Does smart growth matter to public finance? *Urban Stud.* **2008**, *45*, 1791–1823. [CrossRef]
29. MA, Ecosystems and Human Well-Being: A Framework for Assessment. Available online: https://millenniumassessment.org (accessed on 2 February 2024).
30. FAO and ITPS. *Global Soil Organic Carbon Map (GSOCmap) Technical Report*; FAO: Rome, Italy, 2018; 162p, Available online: https://www.fao.org/3/I8891EN/i8891en.pdf (accessed on 2 February 2024).
31. Carruthers, J.I.; Ulfarsson, G.F. Urban sprawl and the cost of public services. *Environ. Plan. B Plan. Des.* **2003**, *30*, 503–522. [CrossRef]
32. Haines-Young, R.; e Potschin, M. Classificazione dei Servizi Ecosistemici. 2013. Available online: https://openness.hugin.com/example/cices (accessed on 10 September 2023).
33. Carruthers, J.I. The impacts of state growth management programmes: A comparative analysis. *Urban Stud.* **2002**, *39*, 1956–1982. [CrossRef]
34. Ispra, Consumo di Suolo, Dinamiche Territoriali e Servizi Ecosistemici. Rapporto ISPRA n.288, 2018. Consumo di Suolo, Dinamiche Territoriali e Servizi Ecosistemici. Edizione 2018—Italiano. Available online: https://www.isprambiente.gov.it/it/pubblicazioni/rapporti/consumo-di-suolo-dinamiche-territoriali-e-servizi-ecosistemici.-edizione-2018 (accessed on 10 September 2023).
35. Camagni, R.; Gibelli, M.C.; Rigamonti, P. *I Costi Collettivi della Città Dispersa*; Alinea: Firenze, Italy, 2002.
36. Calafati, A.G. Economia della città dispersa. *Econ. Ital.* **2003**, *1*, 215–230.
37. Bruederle, A.; Hodler, R. Nighttime lights as a proxy for human development at the local level. *PLoS ONE* **2018**, *13*, e0202231. [CrossRef] [PubMed]
38. Brueckner, J.K. Urban sprawl: Diagnosis and remedies. *Int. Reg. Sci. Rev.* **2000**, *23*, 160–171. [CrossRef]
39. Turri, E. *Il Paesaggio Come Teatro*; Marsilio Editore: Venezia, Italy, 1998.
40. Altinok, E.; Cengiz, H. The effects of Urban Sprawl on Spatial Fragmentation and Social Segregation in Instabul. In Proceedings of the 44th ISOCARP Congress, Dalian, China, 19–23 September 2008.
41. Haggblade, S.; Hazell, P.; Reardon, T. The rural non-farm economy: Prospects for growth and poverty reduction. *World Dev.* **2010**, *38*, 1429–1441. [CrossRef]
42. Jaeger, J.A.G.; Bertiller, R.; Schwick, C.; Cavens, D.; Kienast, F. Urban permeation of landscape and sprawl per capita: New measures of urban sprawl. *Ecol. Indic.* **2010**, *10*, 427–441. [CrossRef]
43. Moraci, F.; Fazia, C. Ambiguità degli effetti delle discipline regionali sul consumo di suolo zero e sulla verticalità in architettura/Ambiguous efects of regional regulations on zero soil consumption and on verticality in architecture. *Technè* **2019**, *17*, 78–95.
44. AAVV, Millennium Ecosystem Assesment, ISPRA, Millennium Ecosystem Assessment—Italiano. Available online: https://isprambiente.gov.it (accessed on 10 September 2023).
45. Papa, R.; Galderisi, A.; Vigo Majello, M.C.; Saretta, E. European Cities Dealing with Climate Issues: Ideas and Tools for a Better Framing of Current Practices. *TeMA-J. Land Use Mobil. Environ.* **2015**, *Special Issue ECCA 2015*, 63–80. [CrossRef]
46. Paulsen, K. The Effects of Land Development on Municipal Finance. *J. Plan. Lit.* **2013**, *29*, 20–40. [CrossRef]

47. Gibelli, M.C.; Salzano, E. (Eds.) *No Sprawl, Perché è Necessario Controllare la Dispersione Urbana e il Consumo di Suolo*; Alinea: Firenze, Italy, 2006.
48. Shi, M.; Janowicz, K.; Cai, L.; Mai, G.; Zhu, R. A socially aware huff model for destination choice in nature-based tourism. *AGILE GIScience Ser.* **2021**, *2*, 14. [CrossRef]
49. Maechler, S.; Boisvert, V. Valuing Nature to Save It? The Centrality of Valuation in the New Spirit of Conservation. *Glob. Environ. Politics* **2024**, *24*, 10–30. [CrossRef]
50. Mathew, P.V. Sustainable tourism development: Discerning the impact of responsible tourism on community well-being. *J. Hosp. Tour. Insights* **2022**, *5*, 987–1001.
51. Zhou, L.; Hu, F.; Wang, B.; Wei, C.; Sun, D.; Wang, S. Relationship between urban landscape structure and land surface temperature: Spatial hierarchy and interaction effects. *Sustain. Cities Soc.* **2022**, *80*, 103795. [CrossRef]
52. Barton, H.; Grant, M. A health map for the local human habitat. *J. R. Soc. Promot. Health* **2006**, *126*, 252–253. [CrossRef] [PubMed]
53. Khan, J.; Hildingsson, R.; Garting, L. Sustainable Welfare in Swedish Cities: Challenges of Eco-Social Integration in Urban Sustainability Governance. *Sustainability* **2020**, *12*, 383. [CrossRef]
54. Unione Europea, Verso un'Europa più Verde e Sostenibile. Available online: https://european-union.europa.eu/priorities-and-actions/actions-topic/environment_it (accessed on 13 February 2024).

land

MDPI

Article

Potential of Former Mill Race Corridors for Urban Regeneration Strategies—A Case Study from Podolínec in Prešov Region (Slovakia)

Juraj Illes [1,*], Katarina Kristianova [2], Viera Joklova [1] and Aida Shayegani [1]

1 Institute of Urban Design and Urban Planning, Faculty of Architecture and Design, Slovak University of Technology in Bratislava, Namestie Slobody 19, 812 45 Bratislava, Slovakia; viera.joklova@stuba.sk (V.J.); aida.shayegani@stuba.sk (A.S.)

2 Center of Landscape Architecture, Faculty of Architecture and Design, Slovak University of Technology in Bratislava, Namestie Slobody 19, 812 45 Bratislava, Slovakia; kristianova@stuba.sk

* Correspondence: juraj.illes@stuba.sk

Abstract: In the past, mill races were part of the urban structure of many towns in Slovakia. As regulated and artificially created waterways, they served to drive mills, rollers, or hammers. With the use of new sources of energy, they lost their functions, and most of them were dried, filled, or buried underground. In our research, we examine the former mill race corridor in Podolínec (Prešov region, Slovakia) and its potential to contribute to urban regeneration strategies. The research steps included the following, namely 1. Identification of the route of the mill race corridor according to historical sources, 2. survey of its current state and its spatial preservation in the urban structure, 3. evaluation of the possibilities of its new uses, which could increase the quality and attractiveness of urban public spaces. The research results show that the fragments of the corridor of the extinct mill race are still identifiable in the urban fabric, and in the cadaster, they are in public ownership and suitable for new uses. The fragments of the corridor of the extinct mill race in Podolínec represent a potential for strengthening the blue and green infrastructure, pedestrian and cycling greenways, and a potential for the presentation of cultural heritage values, which could contribute to the improvement of the qualities of the urban environment.

Keywords: mill race; blue and green infrastructure; greenways; cultural heritage; urban regeneration strategies; urban streams

1. Introduction

The transformations of the relations between the urban environment and the phenomenon of water play a significant role in the urban regeneration strategies [1,2]. Well-known examples of urban transformations of city–water relations are the spectacular transformations of embankments and former ports in the environments of large cities [3–6]. From an urban planning point of view, the contact of water and the urban structure is an important space, "a materialized symbiosis of the bipolar quality of the urban and natural structure" [7] (p. 107). The interactions of natural and artificial elements create a strong phenomenon of genius loci, the identity of a place [8]. In the current urban regeneration strategies in the context of water, a specific emphasis is put on the environmental and ecological aspects of water in an urbanized environment, adaptation to climate change, and issues of blue and green infrastructure [9–14]. Many urban regeneration projects focus on the revitalization and re-naturalization of rivers, streams, and water bodies, on restoring the environmental and ecological functions of water in urbanized areas, and on rainwater retention [15–24]. The revitalization and opening of watercourses and the restoration and return of water elements into the urban structure represent a significant contribution to the quality of urban landscapes and public spaces, and have a significant place-making potential [25–30]. Many works are devoted to the problem of opening covered streams or their parts [12,13,30–33]. Many urban streams have disappeared from the surface as a result of urban growth. Among the reasons was the gaining of space for the construction of roads and buildings or the use of stream water for the transport of wastewater. Spirn [34,35] points out that even buried streams can have important meanings for the identity of the place and the local community. Currently, efforts to uncover canalized streams and restore their beds are increasingly applied in urban blue–green space planning while recognizing the high ecological and social benefits of the opening, especially in supporting the resistance to the effects of climate change and the revival of the cultural ties with water [31].

In the past, along specific water channels, the mill races were part of the urban structure of many towns. They were historically built as artificial water channels or were created by regulating the original watercourses and their branches. The natural conditions, geomorphology of the terrain, hydrological conditions, and morphology of watercourses created various local prerequisites for the use of watercourses and for the construction of the mill races. The mill races were used as a source of energy to drive various kinds of mills, rollers, or hammers, that is, devices that used the power of the water flow through a water wheel to drive mechanisms enabling grinding, sawing, crushing ore and stone, forging iron, shaking and combing flax and hemp, or beating and felting woolen fabric into cloth. They enabled the use of water energy for the production of many material goods—flour, textiles, lumber, paper, or metal products. The mill races began to lose their importance with the invention of the steam engine and the use of coal, the accessibility of which also began to be ensured by the development of railway transport. Later, with the technological development of water turbines, some mill races and old mills were adapted for electricity production. However, they could not compete with other sources of electricity production, because they could not provide sufficient capacities. When the mill races lost their economic importance and became perceived as an obstacle to modern urban development, many of them were filled or buried and channeled underground [36–38]. In Slovakia, the changes in ownership, expropriation, and nationalization of mills, and the collectivization of land after the end of World War II and during the period of nationalization and industrialization of the national economy after 1948, as well as approaches towards anti-flood measures and water management in the second half of the 20th century, specifically contributed to the disappearance of the water mills and mill races [39].

Nowadays, society is once again aware of the benefits of the presence of water in an urbanized environment and, at the same time, of the importance of cultural heritage values. Various examples, specifically from towns from neighboring countries, with similar conditions as towns in Slovakia, show that the preservation and revitalization of the former mill races, new uses, and the presentation of the historical traces of their corridors are able to contribute to the urban regeneration strategies and increase the overall quality of the urban environment, as well as to the esthetics and attractiveness of public spaces.

Many examples show that the mill races play important roles in contemporary urban regeneration strategies. In Gdańsk (Poland), the Radunia Canal is involved in the contemporary life of the city; it flows under the modern shopping center and is presented in its central hall [40,41]. In Toruń (Poland), the preserved unique mill race system built by the Teutonic Knights in the 13th century contributes to the values for which Toruń is a UNESCO World Heritage City [42]. In Germany, the successful implementations of urban regeneration strategies, focusing on the transformation of brownfields into residential areas or into parks, such as, for example in Offenburg, Freital, or Erfurt, include the revitalization of the mill races [43–45]. Several examples show that also dry and filled corridors can be reused. The case of Plzeň (Czech Republic) shows that it is possible to return the water to the part of the dried and filled corridor of a former mill race [46]. In Prešov (Slovakia), a part of the dry and filled corridor serves as a cycling route [47]. The extinct mill races represent a hidden heritage, and its interpretation through various cultural activities is an opportunity to create awareness and promote a not well-known history as, for example, in Lancaster (UK) [48].

The revitalization of the corridors of the mill races represents challenges and opportunities for urban landscapes. Mill race corridors offer the potential for place-making, strengthening the ecosystem services of blue and green infrastructure, the potential for the creation of pedestrian and cycling routes, greenways, and recreational areas [49], and also possibilities of energy use [38]. The mill races represent historical and cultural heritage, and the presentation or interpretation of their values, even those that have disappeared, can enhance the quality and attractiveness of urban public spaces for both residents and visitors [36,37].

In this research, we examine the corridor of the former historical mill race in Podolínec, a small town in the Prešov Region in Slovakia, and its potential to contribute to contemporary urban regeneration strategies. The main aim is to evaluate the possibilities of its revitalization, its new uses, or the possibilities to interpret its extinct heritage values.

2. Materials and Methods

2.1. The Study Area

Podolínec is a small town located in the northwestern part of the Prešov Region of Slovakia, in the Stará Ľubovňa District (Figure 1).

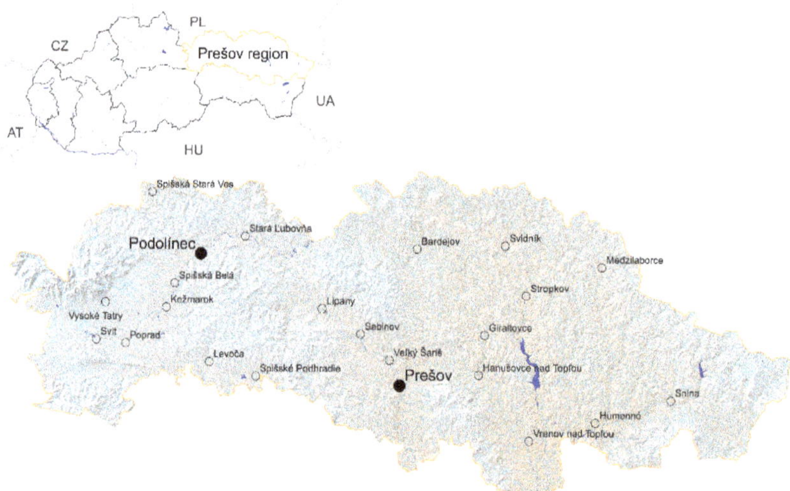

Figure 1. Prešov region in the Slovak Republic and the location of the town of Podolínec.

The Prešov Region is located in the northeastern part of Slovakia. It borders Poland in the north and Ukraine in the east. It is one of the eight Slovak administrative regions. It has an area of 8972.8 km^2 and 808,090 inhabitants (as of 31 December 2023) [50]. It is administratively divided into 13 districts. The region has a predominantly mountainous landscape of the Eastern and Western Carpathians with an outcrop of the Eastern Pannonian Basin in the southern part. Due to the different altitudes of highlands and lowlands, the region's northern temperate-continental climate is divided into three climatic areas—cold, moderately warm, and warm. The upper sections of the main rivers of Hornád, Torysa, Topľa, Ondava, Laborec, and Poprad are located in the territory of the Prešov Region. The Poprad and Dunajec rivers form a part of the border with Poland and belong to the drainage area of the Baltic Sea. There are 665 municipalities in the region, 23 of which have the status of a town (Bardejov, Giraltovce, Hanušovce nad Topľou, Humenné, Kežmarok, Levoča, Lipany, Medzilaborce, Podolínec, Poprad, Prešov, Sabinov, Snina, Spišská Belá, Spišské Podhradie, Spišská Stará Ves, Stará Ľubovňa, Stropkov, Svidník, Svit, Veľký Šariš, and Vranov nad Topľou, Vysoké Tatry). The administrative center of the region is Prešov. With 82,286 inhabitants (as of 31 December 2023), it is the third-largest town in Slovakia [50].

Podolínec developed in the Poprad Basin, in the valley of the river Poprad. The first written record of Podolínec dates from 1236. The Slavic settlement, destroyed in 1241 and 1285 during the Mongol invasion of Hungary, grew fast and received city privileges in 1292 for welcoming German settlers from Silesia. In 1412, it obtained the status of a free royal town [51]. Today, the municipality with a population of 3055 (as of 31 December 2023) keeps its town status [52]. Its historic center is formed around a triangular main square surrounded by late Renaissance burgher houses, with a centrally placed early Gothic Roman Catholic church and Renaissance bell tower. It is surrounded by a partially preserved set of fortification walls. The historic center was declared a monument reserve in 1991 due to its historical values [53].

The mill race, built outside the fortification walls via the regulation of the arm of the river Poprad, powered a mill and served as a fortification moat. This area was an important part of the town, concentrating the economic activities that required water use. The mill race was dried and filled during the 2nd half of the 20th century [53].

In our research, we examined the potential of the remnants of the mill race corridor for urban regeneration strategies. The research included archival research to identify the route of the historical mill race according to historical sources and on-site research to survey the current state of the mill race corridor in the current urban structure as a basis for the evaluation of its potential for urban regeneration strategies.

2.2. *Methodological Steps of the Research*

The research consisted of 3 methodological steps:

1. Identification of the route of the mill race corridor from historical sources;
2. Survey and analysis of the current state of the mill race corridor in the current urban structure;
3. Evaluation of the potential of the mill race corridor for urban regeneration strategies—formulation of ideas for its possible use and recommendations for practice.

2.2.1. Identification of the Mill Race Corridor from Historical Sources

The first step of the research was the identification of the route of the mill race corridor from historical sources, in different historical periods. This first stage aimed to find out the gradual transformations of the mill race corridor in Podolínec in the past. In this step, primarily historical maps were used, but also literary and visual sources (archival documents, historical engravings, historical photographs).

The important sources were the military maps of the Habsburg Empire and historical cadastral maps: the First Military Survey of Königreich Ungarn (1782–1785), the Second Military Survey of Hungary (1819–1869), the Third Military Survey of the Habsburg Empire (1869–1887), and maps of the original cadastral record from 1871 (reambulated in the 1930s). Other map sources from later periods included topographic maps 1:25,000 (1955), 1:5000 (1958), 1:10,000 (1964), and 1:10,000 (approx. 1990). Online available map sources for Arcanum maps were used. This stage of the research was based on research in the Central Archive of Geodesy and Cartography and Cadastre of the Slovak Republic (ÚAGK), in archival sources of the State Archive in Prešov, including its workplaces, in municipal archives, and the Archive of the Monuments Board of the Slovak Republic and archives of its regional offices.

The research was also based on sources from the local community, local monographs, historical photographs, postcards, as well as stories and memories shared by local interest groups, e.g., on social networks.

2.2.2. Survey and Analysis of the Current State of the Mill Race Corridor in the Current Urban Structure

In the second step of the research, the state of the current existence of the mill race corridor in the urban structure of Podolínec was examined. For the verification of its current existence in the urban structure, current map sources, cadastral maps, and orthophoto maps were used, available online through Geoportal, which provides a database for the geographic information system and is part of the information system created and maintained by the Geodesy, Cartography, and Cadastre Authority of the Slovak Republic. The information about the state of the current existence of the mill race corridor was collected from the Basic Data Base for the Geographic Information System (ZBGIS). Other sources included Mapy.cz and Google Maps with the Streetview service. Publications and articles in the daily press, local periodicals, or posts on social networks following the current events in the town, current land-use planning strategies, land-use planning documentation, urban master plans, and documents on the principles of the protection of monument areas were examined.

For the assessment and detailed analysis of the current state of the mill race corridor in the urban structure, on-site research was crucial. During the on-site visit, the entire route of the mill race corridor was examined and documented via photographs.

Based on the survey, the following aspects were analyzed:

- degree of spatial preservation of the corridor in the urban fabric;
- degree of preservation of the technical objects along the corridor (e.g., mills, weirs, modifications of banks, etc.);
- legibility and ownership of corridor parcels in the parcel fabric of the current cadastral record;
- current and planned functional uses along the corridor in the urban fabric.

The analyses of these aspects are important for the purposes of evaluating the various possibilities of using the mill race corridor and its potential for urban regeneration strategies.

2.2.3. Evaluation of the Potential of the Mill Race Corridor for Urban Regeneration Strategies

The corridor of the mill race in Podolínec, which was analyzed in the previous stage in terms of the degree of its preservation, the preservation of technical objects, the nature of ownership relations, and the conditions of the current and planned functional use in the urban structure, was in this stage evaluated in terms of its potential for various possibilities of its use and valorization. The evaluation of its potential enabled the suggestions of conceptual ideas for the possible use of different parts of the mill race corridor and the formulation of recommendations for the local government to protect and valorize this heritage.

The evaluation of the potential of the heritage of mill races for the current urban regeneration strategies was based on the assumptions that mill races—those preserved, but also the traces of those that have disappeared—can mean various benefits for the urban structure and urban public spaces. These assumptions relied on the concepts of urban green and blue infrastructure and its ecosystem services [54,55] and the concepts of heritage interpretation and valorization [56–59]. The benefits of urban green and blue infrastructure and its ecosystem services, including the wide range of provisioning, regulatory, and supporting services, covering the ecological and environmental aspects up to the benefits of cultural non-material services contributing to human well-being, are described and assessed by a large number of publications [60–64]. The benefits generated by heritage interpretation and valorization can be both direct and indirect, derived from the satisfaction of its direct users and the whole community. They are of an economic and social nature, influencing the increase in tourism, trade, and services or creating a sense of a common identity and place attachment [65,66]. The possibilities of the application of these conceptual approaches—the green infrastructure and the heritage interpretation and valorization approach for formulating suggestions and conceptual ideas for the revitalization and new uses of mill race corridors, as well as the amount and variety of the gained benefits—are mainly influenced by the degree of the preservation of the mill race corridors. It is possible to strengthen the green and blue infrastructure functions of the preserved mill race corridors or their preserved fragments, which can bring multiple benefits of various types and forms of green and blue infrastructure into the urban environment. While the continuous lines of mill race corridors can form greenways or biocorridors, the fragmented parts can be used for small-scale elements of urban green and blue infrastructure, for example, rain gardens or bioswales. When it is impossible to find tangible remnants of mill race corridors in the urban fabric, the techniques and methods of intangible heritage interpretation can help commemorate the extinct heritage values [48]. An overview of the potential benefits resulting from the use and valorization of the mill race corridors, contributing to the qualities of the urbanized environment and the quality of life in urban settlements, is given in Figure 2.

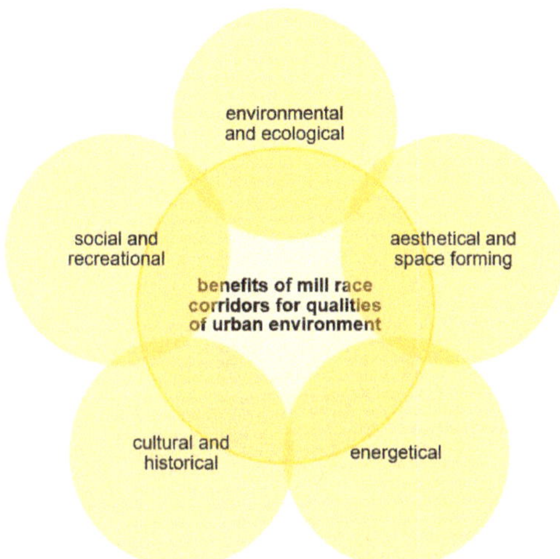

Figure 2. An overview of the possible benefits of mill race corridors for the overall quality of the urban environment.

3. Results

3.1. Identification of the Mill Race Corridor in Podolínec from Historical Sources

The map of the First Military Survey of the Hungarian Kingdom (1782–1785) captures the existence of the mill race in Podolínec, in the period of the end of the 18th century [67]. It shows that the mill race developed from the originally natural arm of the river Poprad. The map shows the location of the mill and the route of the small stream flowing around the fortifications (Figure 3). The map of the Second Military Survey of Hungary (1819–1869), which was carried out on the territory of the current Prešov Region from 1819 to 1827 [68], shows the mill race including the weir on the river Poprad and the mill (Figure 4). The mill race is depicted on the map of the Third Military Survey of the Habsburg Empire (1869–1887), which was carried out on the territory of the current Prešov Region from 1875 to 1876 [69]. The most useful source, depicting the details of the mill race corridor and details of the mill building, including the number of wheels is the cadastral map of the original cadastral record from 1871, reambulated in the 1930s [70]. It shows the corridor of the mill race in its last functional form in the first half of the 20th century (Figure 5).

Figure 3. The mill race and location of the mill (M) on the map of the First Military Survey. The mill race developed from the originally natural arm of the river Poprad [67].

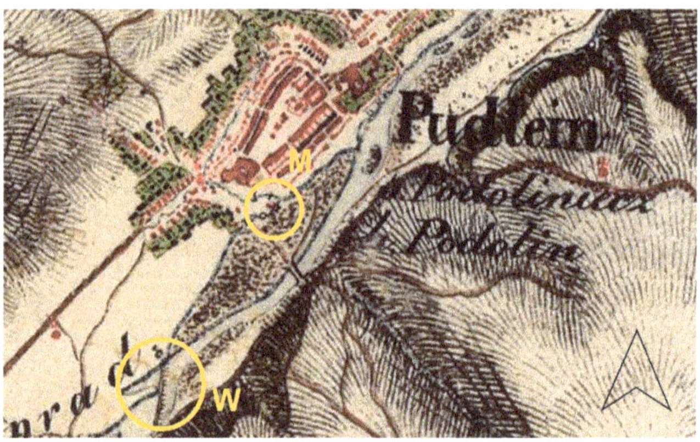

Figure 4. The mill race, location of the mill (M), and weir (W) on the river Poprad, depicted on the map of the Second Military Survey [68].

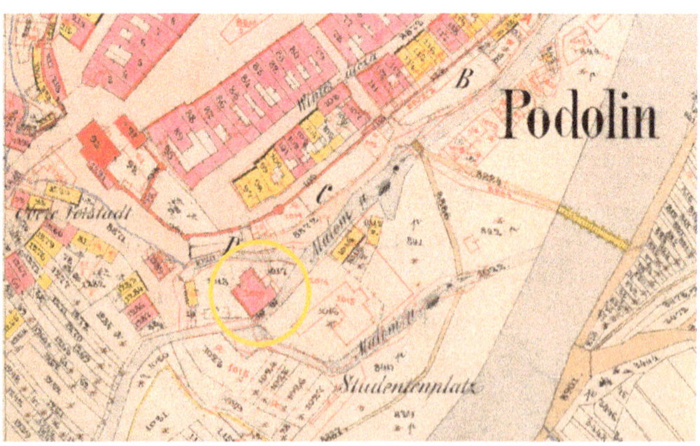

Figure 5. Detail of the mill race and its branches and the building of the mill, with two wheels, on the map of the original cadastral record from 1871, reambulated in 1930 [70] (map no. 20).

Photographs from the period of the first half of the 20th century have also been preserved (Figure 6).

Figure 6. View of the town from the river from the south, 1940s [53]. In the foreground, the branch of the mill race flowing into the river Poprad is highlighted.

On the topographic map from 1955, the corridor of the mill race is depicted with a water flow. One branch of the mill race, located near the mill and flowing into Poprad, has disappeared [71] (Figure 7a).

(**a**) (**b**)

Figure 7. (**a**) The mill race with flowing water on the topographic map from 1955 [71]. (**b**) On the topographic map from 1958, the corridor of the mill race is already dry, but the building of the mill is still preserved [72].

On the topographical map from 1958, however, the corridor of the mill race is already without water, and the water inlet to the mill race from the river Poprad has been removed. The mill building is still preserved [72] (Figure 7b). This situation persisted during the 1960s [73].

During the 1970s, the continuity of the corridor was disrupted by the development of family houses built in this part of the mill race corridor [74]. During the 1980s, the mill building was demolished and, in this part of the corridor, blocks of flats, a heating facility, and garages were built instead (Figure 8). The part around the outlet of the mill race to the river Poprad was filled and used for small production and warehouses [75].

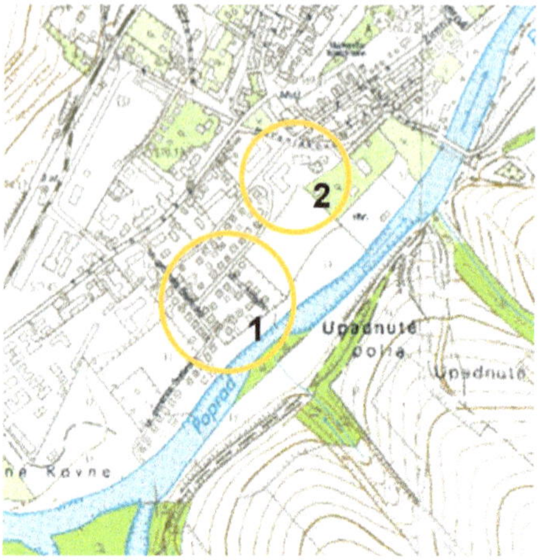

Figure 8. Development of family houses (1) and block of flats (2) in the 1970s and 1980s interrupted the mill race corridor depicted on the Basic Topographic Map from 1990. The mill has not been preserved [75].

3.2. *Survey and Analysis of the Current State of the Mill Race Corridor in the Current Urban Structure*

The survey and analyses of the current state show that the mill race corridor is partially preserved only in the form of fragments of parcels in the current cadastral record [76]. They are under public, municipal, or state ownership (Figure 9). The mill, weir, inlet, and outlet technical objects are not preserved. The public ownership and functional uses determined by the current land use plan [77]—green spaces and civic and commercial amenities—support the possibility of using the potential of the fragments of the corridor for urban regeneration strategies. The principles of the protection of the monument reserve [53] also set the requirement to preserve the corridor in the buffer zone (Figure 10) as an open space to keep the possibility of commemorating the mill race.

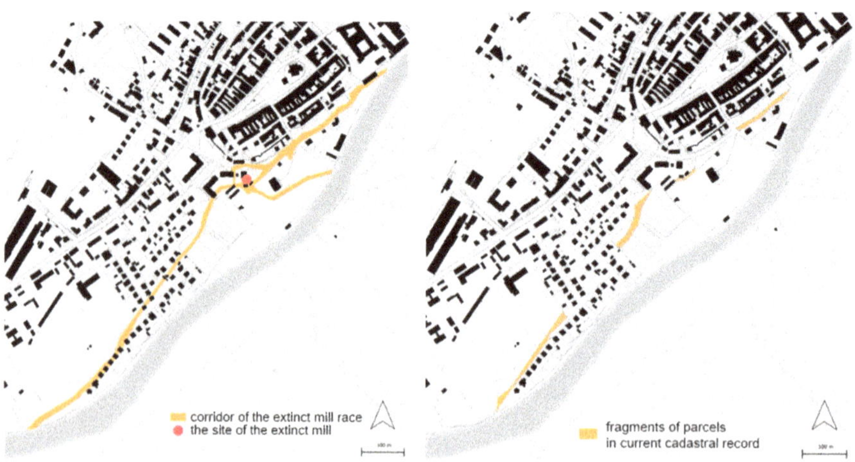

Figure 9. Corridor of the extinct mill race in Podolínec on the current cadastral map with the location of the place of the extinct mill and fragments of parcels of the mill race corridor under public ownership still identifiable in the current real estate cadaster.

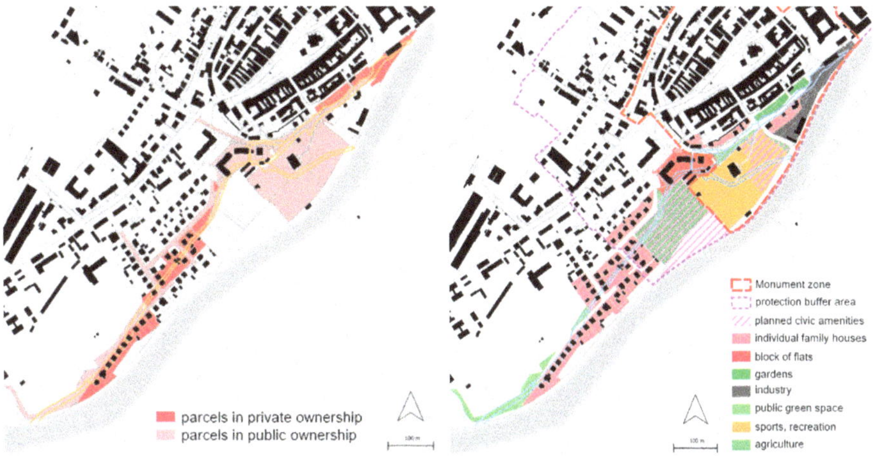

Figure 10. Analysis of the current ownership relations along the corridor of former mill race and analysis of the current functional uses in the urban structure along the corridor.

The corridor of the former mill race was examined during the on-site visit, and the accessible parts were documented via photographs (Figure 11). They capture the different types of the environment along the route and show the spatial possibilities for applying modifications that could commemorate the mill race and improve the quality of public spaces (Figures 11 and 12).

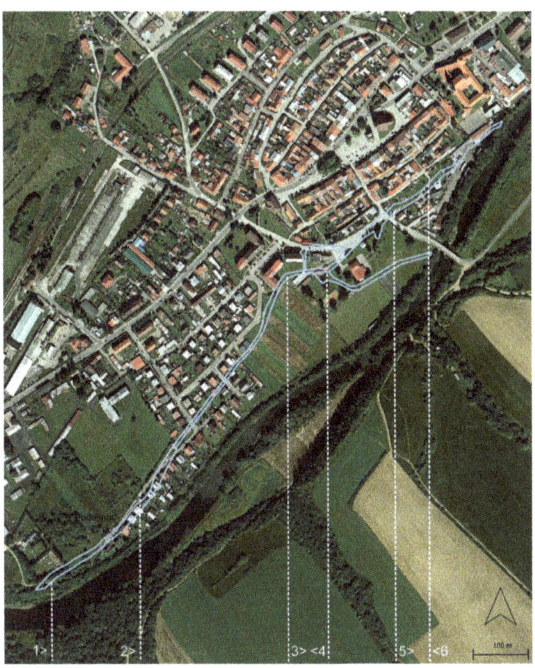

Figure 11. The corridor of the extinct mill race in Podolínec on the orthophoto map [78]. The orthophoto map allows observation of the occurrence of accompanying vegetation structures. The places documented by photographs (1)–(6) taken during the on-site survey are marked on the map.

Figure 12. Photos capturing different types of environments along the corridor of the extinct mill race. The route leads from the area of the former inlet of water to the mill race (**1**), along the road through the residential zone of family houses (**2**), through the public spaces of blocks of flats (**3**,**4**) to the area in direct contact with the historical center of the city and monument protection zone (**5**,**6**).

3.3. Evaluation of the Potential of the Mill Race Corridor for Urban Regeneration Strategies

Based on the survey and the analyses of the current state, the three parts of the corridor of the extinct mill race were identified, which, from the point of view of ownership relations and the spatial and functional conditions of the urban structure, can be used for strategies of urban regeneration:

1. the part by the river Poprad, at the former inlet to the mill race, continuing along the Street of Generál Štefánik, in the neighborhood of family houses (Figure 13);
2. the part forming public spaces around the blocks of flats and along Bernolákova Street (Figure 14);
3. the part in contact with the historical center of the town, south of Kláštorná Street (Figure 15).

These three parts were selected according to the analyses of the current state because they are the parts where the fragments of parcels of the former mill race corridor are preserved (Figure 9). These preserved fragments of parcels are under public ownership (municipality and state). They are surrounded by public spaces also under municipal or state ownership (Figure 10). The ownership rights are an important factor that allows the municipality to implement new ideas for using and presenting the heritage of the mill race corridor. The other parts along the route of the former mill race corridor are not suitable and not usable for redesign and revitalization interventions because they are under private ownership, built up by family houses and buildings of small industries. In these parts, the parcels of the former mill race corridor are not identifiable in the current cadastral record (Figure 10).

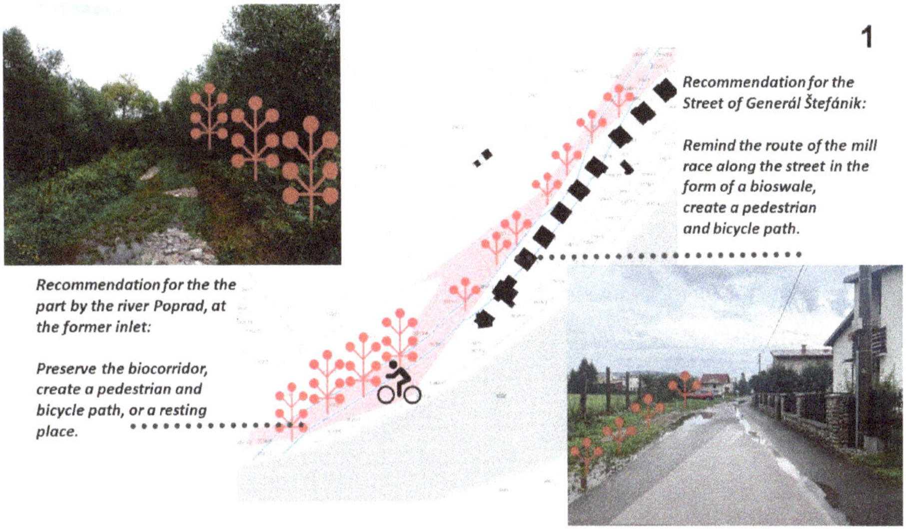

Figure 13. Evaluation of the urban regeneration potential and recommendations for the first part of the former mill race corridor by the river Poprad and along the Street of Generál Štefánik in the neighborhood of family houses.

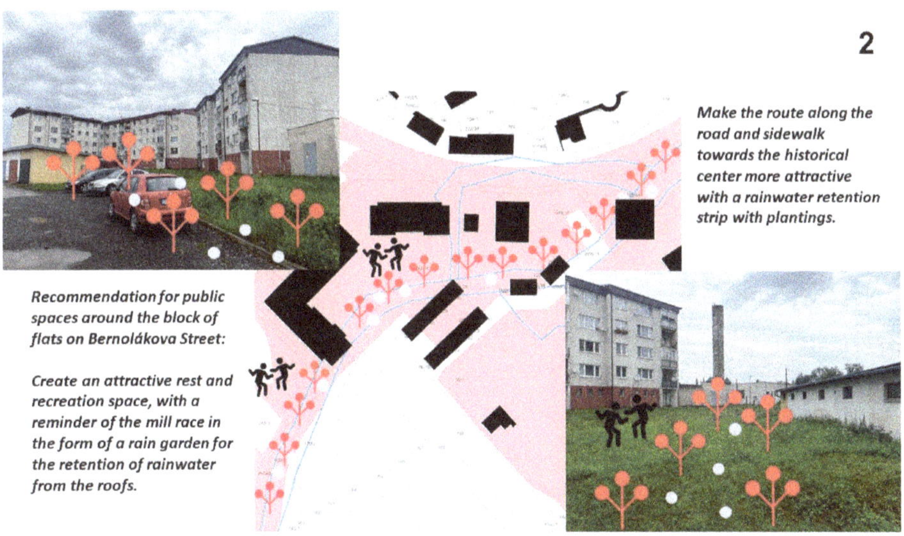

Figure 14. Evaluation of the potential and recommendations for the public spaces around the blocks of flats and along Bernolákova Street.

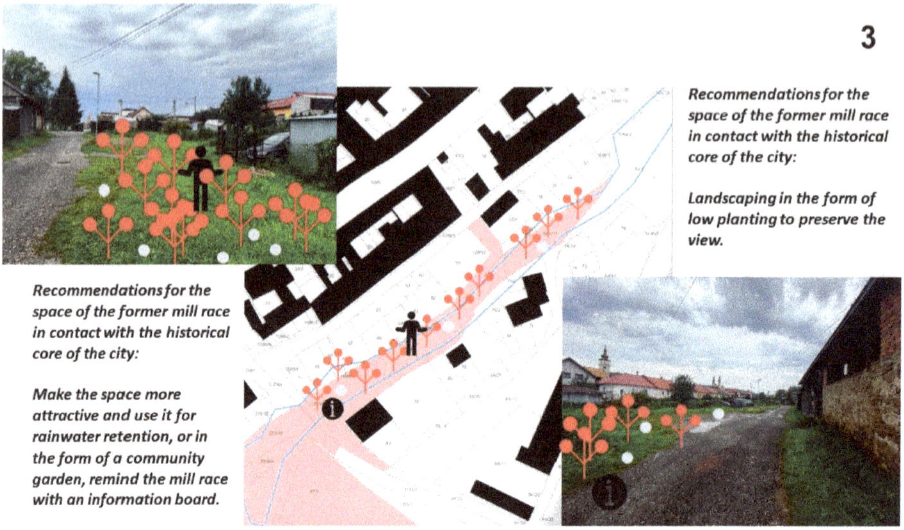

Figure 15. Evaluation of the potential and recommendations for the part in contact with the historical center of the town, south of Kláštorná Street.

For the three selected parts, recommendations were developed for the different possibilities of using the hidden potential of the heritage of the mill race, which could contribute to the urban regeneration strategies and the quality of public space improvement. The on-site investigation and photographs (Figure 12) document that the public spaces today are of low quality, do not offer a diversity of uses, and do not encourage people to meet and interact, socialize, rest, play, or participate in sports. The conceptual ideas for redesigning these public spaces (Figures 13–15) suggest possibilities for how to increase their quality and attractiveness, diversity of uses, place attachment, and local identity through valorizing the heritage of the former mill race corridor and implementing the current approaches of green and blue infrastructure planning. The conceptual ideas and recommendations reflect the specific conditions of the surrounding urban setting, land-use plan, and monument protection requirements and try to offer a variety of uses that could respond to the community's needs. The suggestion of various possibilities and variety of options can serve as a preliminary starting point for discussions and participatory processes with the local community, informing about the existence of the mill race heritage and stimulating the imagination of the possibilities of how to present or interpret the heritage in public spaces. Discussion and participation can verify the expectations and needs of the community, affirm their shared rights to the public spaces, and serve as a base for design guidelines and planning criteria or for preparing architectural design competitions for implementing these ideas.

3.3.1. Evaluation of the Potential and Recommendations for the Part by the River Poprad, at the Former Inlet, and Along the Street of Generál Štefánik

The recommendation for the site by the river Poprad, near the place of the former weir and inlet of water to the former mill race, is to preserve the biocorridor along the river, support the natural character of the area, create a pedestrian and cycle path, or a rest area, with a reminder of the mill race history. Along the Street of Generál Štefánik, it is possible to commemorate the route of the mill race in the form of a vegetated bioswale for soaking up rainwater and being planted with native plants supporting biodiversity (Figure 13). These design ideas could contribute to the urban regeneration strategies by valorizing the mill race heritage, strengthening the ecosystem services of green and blue infrastructure, and supporting the greenway walking and cycling connections and recreation activities.

3.3.2. Evaluation of the Potential and Recommendations for the Public Spaces around the Blocks of Flats and Along Bernolákova Street

The currently neglected and unused public spaces around the block of flats on Bernolákova Street need transformation to create an attractive relaxation and recreation area for residents, equipped with a children's playground, benches, or other elements according to the residents' needs. The recommendation for the revitalization is to remind people of the mill race, for example, in the form of a rain garden, for the rainwater retention from the roofs of apartment buildings and garages. For the design of the public spaces around the block of flats, usually, typified products of children's playgrounds, benches, and outdoor equipment are used. The mill race presentation in the space can strengthen the specific identity of the place and the sense of the inhabitants' place attachment.

The presentation of the mill race along Bernolákova Street, following the corridor of the former mill race towards the historical center, can be made more attractive with accompanying planted bioswale. Also in this part of the corridor of the former mill race, it is possible to check the possibilities for creating a cycle path (Figure 14).

3.3.3. Evaluation of the Potential and Recommendations for the Part in Contact with the Historical Center of the Town, South of Kláštorná Street

In contact with the historical core of the town, the corridor of the extinct mill race is preserved in the form of an open green space, with exceptional space-forming qualities and views. It is necessary to preserve this open space and not to build it up, in order to preserve the views towards the historical center. It is suitable to maintain the use of the space via current forms of community gardens, with low plantings. It is also possible to consider the return of water to the corridor, for example, in the form of rainwater retention. Since the area is located near the historical center visited by tourists, it would be suitable to remind them of the nonexistent mill race via an information board (Figure 15).

The Principles of the Protection of the Monument Zone also set out the demand to protect and not build up this open space to preserve the possibilities of reminding people of the heritage of the mill race. Unfortunately, this significant location cannot be connected with the river in continuation along the route of the former mill race to the place of the outlet. This part of the corridor was sold to a private owner and built up in the past period when the municipality and community were unaware of the values of the mill race heritage.

4. Discussion

The evaluation of the potential of the heritage of the mill races for the current urban regeneration strategies is based on the assumptions that the mill races—those preserved, but also the traces of those that have disappeared—can mean various important benefits for the urban structure and urban public spaces: environmental and ecological, social and recreational, cultural–historical, esthetical and space-forming, or energetic. The amount and variety of these benefits are mainly influenced by the degree of preservation of the mill races and their corridors.

The mill races preserved with the flow of water represent a valuable cultural and technical heritage, important elements of the green and blue infrastructure forming green corridors, greenways, cycling and walking tracks, connecting urban settlements with the landscape, and providing recreational opportunities, or in some cases, also sources of electric energy provision. The revitalization of the mill races, and opening and daylighting their covered parts offer great potential for the urban regeneration strategies. Many examples from the small towns in the neighboring regions show the values of the mill races and the contributions of their revitalization to the urban regeneration strategies. In Litovel (Czech Republic), called "Moravian Venice," the mill race flowing through the main square and under the town hall tower is the main attraction of the town [79]. The revitalization and restoration of the ecological functions and water quality in the mill races in Prostějov and Chrudim (Czech Republic) were highly appreciated by the residents [80]. In Slovakia, long-term efforts are underway to revitalize the mill race in Košice [81].

The case of Podolinec studied in this research represents an example of the extinct mill race, preserved only in the form of the dry and filled fragments of its corridor. Several studies and successful implementations show the possibilities of and approaches to how to reuse and valorize the remnants of the mill race corridors in the urban regeneration strategies [37,46,47,82,83]. In the cases where the mill races have been preserved only in the form of dry corridors and their fragments, it is possible to check the possibilities of reviving the water, as in the case of Plzeň [46], or options of creating greenways and green walking and cycling routes, as, for example, in the cases of Veľký Šariš or Prešov (Slovakia) [47,82]. The preserved parts of the spatial corridors of the former mill races, with accompanying vegetation, can form lines of biocorridors and a green skeleton of connections of the urban environment with the landscape. The relics of mills and technical objects carry cultural heritage values and can be presented in various forms along cycling or walking routes [37]. When only some parts or fragments of the spatial corridor of the former races are preserved, it is possible to verify their use for the small elements of green and blue infrastructure [83], for example, for rainwater retention. The fragments of the corridors of the former mill races can be presented in public spaces as elements that make the space attractive and recall its memory, cultural–historical, and technical values, as, for example, in the study of the courtyard of Krnov Castle [83].

These approaches inspired the examination of the mill race corridor in Podolínec and its potential for new uses, which could contribute to the urban regeneration strategies. The research results show that the fragments of the corridor of the extinct mill race are still identifiable in the urban fabric of Podolínec and the cadaster, and they are under public ownership. The research also shows that the evaluation of the possibilities of new uses of fragments needs to assess and reflect the specifics of the individual spatial and functional characteristics of the site and the surrounding urban landscape setting. While in the first part of the corridor, the natural character of the riverbank site supports recommendations to strengthen the biocorridor and greenway functions, in the second part around the block of flats, the emphasis is placed on using the mill race corridor to support the identity of active and attractive public spaces for residents, with the application of the current trends of rainwater retention. The recommendations for the third part reflect the specifics of the protected monument zone, with expected visits by tourists. The approach of the return of water in modified forms of water retention and the approach of heritage values' presentation and interpretation, which resonate in the proposed recommendations for all the parts of the former mill race, are based on the concepts of contributions of urban green and blue infrastructure and heritage valorization [54–59] to the urban regeneration strategies. The option of the regeneration of water flow is not recommended. The analyses show that the conditions of the current state of the mill race corridor in the urban fabric do not allow this option. When evaluating the benefits of mill races and issues of their regeneration, it is necessary to discuss the aspects of their impacts on the fluvial system of the water courses and aquatic organisms. Small dams and weirs used to divert and regulate the flow change the streambed and the inorganic and organic material transfer, and can act as a migration barrier for fish. The functional mill races need professional care for their flows and their maintenance is necessary, as well as regular revitalization interventions to improve their ecosystem functions.

The research tries to contribute to the raising of awareness about the values of the mill races and their corridors and the possibilities of their use in accordance with contemporary needs. Historically, the mill races in towns were mostly located on the edge of the urban structure, in industrial zones, and in the recent past, they were not the object of protection or valued for their values. The research results, acquainting the citizens of Podolínec with the races, many of whom do not know about the historical heritage of the mill race, could initiate discussions and participatory processes on how to use this heritage and improve the quality of the public spaces. The research can serve as an example for other municipalities to consider the potential of the mill race corridors, their spatial protection and preservation for possible public uses, and avoid the thoughtless sales of their parts as in the case of Podolínec or Prešov.

Many communities have become aware of the values that they have lost due to the disappearance of the mill races, and they try to remember this heritage in different ways [34,35]. In some cases, only the names and routes of streets remind people of the corridors of the mill races (Mill Street, Tannery Street, etc.). Even when the remnants of the mill race corridors have been built over, it is still possible to remember their heritage through presentation and interpretation in various forms, tangible or intangible, including virtual applications that educate about the lost cultural heritage [48,84].

5. Conclusions

The mill races are evidence of the civilizational and technical development and belong to a diverse and rich cultural heritage including their material and immaterial forms. The research draws attention to the potential of the former mill race corridors to improve the quality and attractiveness of urban public spaces, create pedestrian and cycling connections and greenways, and be significant elements of green and blue infrastructure, which is becoming a meaningful requirement in the context of the adaptation of the urban environment to climate change. The research contributes to the considerations of how to preserve the fragments or the cultural footprint of the mill races and how to present and use them, including their extinct and intangible values. The evaluation of the potential of the heritage of the mill races creates possibilities for the formulation of conceptual ideas and recommendations for local governments, how to protect and valorize this heritage, and how to use it in urban regeneration strategies.

The shortcomings of this study are that the research focused only on the case of one mill race. The research of several cases would allow a deeper insight into the issue, and the methodological approach, applied to the assessment of the current state and the possibilities of the use of the mill race corridor, was only from the perspective of architecture and urban and landscape planning, and aimed at evaluating the spatial and functional prerequisites for the use of the corridor in the urban structure. Further studies would benefit from the research on several cases, including cases with the assessment of the possibilities of restoring the water flow, and with applications of the interdisciplinary aspects of evaluation.

Author Contributions: Conceptualization, J.I. and K.K.; methodology, J.I. and K.K.; validation, J.I.; formal analysis, J.I.; investigation, J.I.; resources, J.I.; data curation, J.I.; writing—original draft preparation, J.I. and K.K.; writing—review and editing, J.I., K.K., V.J. and A.S.; visualization, J.I.; supervision, J.I., K.K. and V.J.; project administration, J.I.; funding acquisition, K.K. All authors have read and agreed to the published version of the manuscript.

Funding: This research received no external funding.

Data Availability Statement: The original contributions presented in the study are included in the article; further inquiries can be directed to the corresponding author.

Conflicts of Interest: The authors declare no conflicts of interest.

References

1. Kaushal, S.S.; McDowell, W.H.; Wollheim, W.M.; Johnson, T.A.N.; Mayer, P.M.; Belt, K.T.; Pennino, M.J. Urban Evolution: The Role of Water. *Water* **2015**, *7*, 4063–4087. [CrossRef]
2. Zingraff-Hamed, A.; Bonnefond, M.; Bonthoux, S.; Legay, N.; Greulich, S.; Robert, A.; Rotgé, V.; Serrano, J.; Cao, Y.; Bala, R.; et al. Human–River Encounter Sites: Looking for Harmony between Humans and Nature in Cities. *Sustainability* **2021**, *13*, 2864. [CrossRef]
3. Carta, M.; Ronsivalle, D. *The Fluid City Paradigm: Waterfront Regeneration as an Urban Renewal Strategy*, 1st ed.; Springer: Cham, Switzerland, 2016; 170p.
4. Joklová, V.; Furdík, J. Nábrežné stratégie v Bratislave-Petržalke. *ALFA* **2015**, *20*, 58–67.
5. Smith, H.C.; Ferrari, M.S.G. *Waterfront Regeneration: Experiences in City-Building*, 1st ed.; Routledge: London, UK, 2012; 256p.

6. Yıldız, R.; Şenlier, N.; Güzel, B.I. Sustainable urban design guidelines for waterfront developments. In Proceedings of the 2nd International Sustainable Buildings Symposium, Ankara, Turkey, 28–30 May 2015; pp. 487–494.
7. Bašová, S. Urbánne väzby historického jadra mesta a nábrežia. In Proceedings of the Bardkontakt: Problematika Mestských Pamiatkových Centier, Bardejov, Slovakia, 23–24 August 2011; pp. 107–113.
8. Vitková, Ľ. Sila génia loci. *Urbanita* **2014**, *26*, 10–13.
9. Pochodyła, E.; Glińska-Lewczuk, K.; Jaszczak, A. Blue-green infrastructure as a new trend and an effective tool for water management in urban areas. *Landsc. Online* **2021**, *92*, 1–20. [CrossRef]
10. Giannakis, E.; Bruggeman, A.; Poulou, D.; Zoumides, C.; Eliades, M. Linear Parks along Urban Rivers: Perceptions of Thermal Comfort and Climate Change Adaptation in Cyprus. *Sustainability* **2016**, *8*, 1023. [CrossRef]
11. Chou, R.-J. Achieving Successful River Restoration in Dense Urban Areas: Lessons from Taiwan. *Sustainability* **2016**, *8*, 1159. [CrossRef]
12. Jutrović, M.; Tomić Reljić, D.; Zovko, M.; Bubalo Kovačić, M.; Pereković, P.; Kamenečki, M. Potential for Applying Nature-Based Solutions to Urban Waterways: The Case Study of Medveščak and Črnomerec Streams in the City of Zagreb, Croatia. *Sustainability* **2023**, *15*, 9959. [CrossRef]
13. Gašparović, S.; Sopina, A.; Zeneral, A. Impacts of Zagreb's Urban Development on Dynamic Changes in Stream Landscapes from Mid-Twentieth Century. *Land* **2022**, *11*, 692. [CrossRef]
14. Jiang, Y.; Huang, J.; Shi, T.; Wang, H. Interaction of Urban Rivers and Green Space Morphology to Mitigate the Urban Heat Island Effect: Case-Based Comparative Analysis. *Int. J. Environ. Res. Public Health* **2021**, *18*, 11404. [CrossRef]
15. Baker, L.A.; Shanahan, P.; Holway, J. Principles for Managing the Urban Water Environment in the 21st Century. In *The Water Environment of Cities*, 1st ed.; Baker, L.A., Ed.; Springer: New York, NY, USA, 2009; pp. 275–289.
16. Booth, D.B.; Bledsdoe, B.P. Streams and Urbanization. In *The Water Environment of Cities*, 1st ed.; Baker, L.A., Ed.; Springer: New York, NY, USA, 2009; pp. 93–123.
17. Hill, K. Urban Design and Urban Water Ecosystems. In *The Water Environment of Cities*, 1st ed.; Baker, L.A., Ed.; Springer: New York, NY, USA, 2009; pp. 141–170.
18. Jurík, Ľ.; Pokrývková, J. Zadržiavanie vody v mestách—Teória a praktické riešenia. *Zivotn. Prostr.* **2018**, *52*, 42–48.
19. Just, T. Přírodě blízké úpravy vodních toků v intravilánech a jejich význam v ochraně před povodněmi. *Ochr. Přírody* **2010**, *6*, 15–17.
20. Just, T. *Ekologicky Orientovaná Správa Vodních Toků v Oblasti Péče o Jejich Morfologický Stav*, 1st ed.; Agentura Ochrany Přírody a Krajiny České Republiky: Prague, Czech Republic, 2016; 83p.
21. Golgher, A.; Callisto, M.; Hughes, R. Improved Ecosystem Services and Environmental Gentrification after Rehabilitating Brazilian Urban Streams. *Sustainability* **2023**, *15*, 3731. [CrossRef]
22. Theodosiou, G.; Panajiotidis, S. The Impact of Urban Land-Use Regimes on the Stream Vegetation and Quality of a Mediterranean City. *Hydrology* **2023**, *10*, 45. [CrossRef]
23. Lepeška, T. Vybrané krajinnoekologické problémy urbánnych povodí. *Geogr. Rev.* **2013**, *9*, 279–290.
24. Lepeška, T. Úloha ekohydrológie pri ochranę a posilnení ekosystémových služieb v urbánnych povodiach. In Proceedings of the Environmentálne Indexy, Agroenvironmentálne Opatrenaia a Ekosystémové Služby v Krajine, Bratislava, Slovakia, 27 November 2013; pp. 97–106.
25. Bašová, S.; Štefancová, L. Zážitková scéna urbánnej vegetácie nábreží. In Proceedings of the Krajinná Architektúra a Krajinné Plánovanie v Perspektíve, Bratislava, Slovakia, 1–2 October 2015.
26. Bašová, S.; Sopirová, A. Interactive spaces for contacts of cities with a watercourse. In Proceedings of the SGEM 2018, 5th International Multidisciplinary Scientific Conference on Social Sciences and Arts, Albena, Bulgaria, 26 August–1 September 2018.
27. Hanáček, T. Bratislava—Verejné priestory rieky. *ALFA* **2015**, *20*, 44–55.
28. Prominski, M.; Stokman, A.; Stimberg, D.; Voermanek, H.; Zeller, S.; Bajc, K.; Zheng, N. *River Space Design: Planning Strategies, Methods and Projects for Urban Rivers*, 3rd and enlarged ed.; Birkhäuser: Berlin, Germany; Boston, MA, USA, 2023; 356p.
29. Wantzen, K.M. River culture: How socio-ecological linkages to the rhythm of the waters develop, how they are lost, and how they can be regained. *Geogr. J.* **2022**, *190*, e12476. [CrossRef]
30. Hong, C.-Y.; Chung, E.-S.; Chang, H. The Right to Urban Streams: Quantitative Comparisons of Stakeholder Perceptions in Defining Adaptive Stream Restoration. *Sustainability* **2020**, *12*, 9500. [CrossRef]
31. Wantzen, K.M.; Piednoir, T.; Cao, Y.; Vazhayil, A.M.; Tan, C.; Kari, F.G.; Lagerström, M.; Gerner, N.V.; Sommerhäuser, M.M. Back to the surface—Daylighting urban streams in a Global North–South comparison. *Front. Ecol. Evol.* **2022**, *10*, 838794. [CrossRef]
32. Fitzsimmons, M.J. Rediscovering Nature: Daylighting an Urban Stream (Gwynns Run, Baltimore, Maryland). Master's Thesis, University of Maryland, College Park, ML, USA, 2004.
33. Koshaley, D.H. Developing Eligibility Criteria for Daylighting Streams as Applied to Dallas' Mill Creek. Ph.D. Thesis, The University of Texas, Arlington, TX, USA, 2008.
34. Spirn, A.W. Restoring Mill Creek: Landscape Literacy, Environmental Justice and City Planning and Design. *Landsc. Res.* **2005**, *30*, 395–413. [CrossRef]
35. Spirn, A.W. The Nature of Mill Creek: Landscape Literacy and Ecological Democracy. In *Pragmatic Sustainability*, 2nd ed.; Steven, A.M., Ed.; Routledge: London, UK, 2016; pp. 172–189.

36. Illéš, J. Mlynské náhony—Potenciál pre súčasné regeneračné stratégie mesta. In *Point for Science 3: Proceedings of Doctoral Research*; Legény, J., Rosinová, L., Benkovičová, L., Eds.; Spektrum STU: Bratislava, Slovakia, 2022; pp. 68–69.
37. Illéš, J.; Joklová, V.; Jaszczak, A. Traces of former mill races in Krnov: Possibilities of revitalization and interpretation. *ALFA* **2023**, *28*, 36–46. [CrossRef]
38. Illéš, J.; Kristiánová, K.; Joklová, V. Mill Races as Historical Sources of Hydropower and their Potential Today—Examples from Slovakia. In Proceedings of the WMCAUS 2022, Prague, Czech Republic, 5–9 September 2022.
39. Bartošíková, T. (Ed.) *K Dejinám Mlynov—Výskum a Mapovanie. Zborník zo Seminára K Dejinám Mlynov—Výskum a Mapovanie, Pamiatkový Úrad Slovenskej Republiky a Vodné Mlyny Oblazy, N.O., Online, 14 October 2021*; Pamiatkový úrad Slovenskej Republiky: Bratislava, Slovakia, 2023; 136p.
40. Forum Gdańsk/SUD Polska. 16 August 2019. *ArchDaily*. Available online: https://www.archdaily.com/909887/forum-gdansk-sud-polska (accessed on 23 June 2024).
41. Lorens, P.; Habier, M. Impact of large-scale urban interventions on contemporary city centers. In Proceedings of the 54th ISOCARP Congress 2018, Bodø, Norway, 1–5 October 2018.
42. Czarnecki, A.; Luc, M.; Lewandowska-Czarnecka, A. Wykorzystanie cech i zasobów środowiska w tworzeniu krajobrazu miast usytuowanych w dolinach rzecznych na przykładzie Torunia. In *Studia Krajobrazowe a Gingce Krajobrazy*; Chylińska, D., Łach, J., Eds.; Uniwersytet Wrocław: Wrocław, Poland, 2010; pp. 137–150.
43. Offenburg, Mühlbachareal 2016. Available online: https://faktorgruen.de/offenburg-muehlbachareal/ (accessed on 23 June 2024).
44. Die Geraaue—Stadtentwicklung im Rahmen der Buga. Erfurt.de—Das Offizielle Stadtportal der Landeshauptstadt Thüringens. Available online: https://www.erfurt.de/ef/de/leben/planen/projekte/buga2021/noerdliche-geraaue/index.html (accessed on 23 June 2024).
45. Abbruch Lederfabrik und Nachnutzung als Mühlenpark. Freital. Available online: https://www.freital.de/Unsere-Stadt/Informationen-zur-Stadt/Stadtentwicklung/Stadterneuerung/index.php?La=1&object=tx,3303.114.1&kat=&kuo=2&sub=0 (accessed on 23 June 2024).
46. Tolarová, I. Mlýnská Strouha je Parkem Roku. *Správa Veřejného Statku Města Plzně*. 25 October 2010. Available online: https://svsmp.cz/archiv/2010/mlynska-strouha-je-parkem-roku/ (accessed on 23 June 2024).
47. Hudačko, Ľ. V Prešove Máme Nový Cyklochodník Mlynský Náhon: Pribudnú aj Stojany na Bicycle. *Dnes 24 Prešov*. 21 May 2020. Available online: https://presov.dnes24.sk/v-presove-mame-novy-cyklochodnik-mlynsky-nahon-pribudnu-aj-stojany-na-bicykle-361992 (accessed on 23 June 2024).
48. Mill Race: Flow of Change. Cultural Programme, Lancaster. Available online: https://visitlancaster.org.uk/attractions/mill-race-flow-of-change/ (accessed on 23 June 2024).
49. Illéš, J.; Kristiánová, K. Rivers as backbones for urban and periurban recreation—Case studies from Košice and Prešov, Slovakia. In Proceedings of the Public Recreation and Landscape Protection—With Environment Hand in Hand? Křtiny, Czech Republic, 9–11 May 2023.
50. Prešovský Kraj—Characteristic of the Region. Available online: https://slovak.statistics.sk/wps/portal/ext/themes/regional/presovsky%20kraj/about/!ut/p/z1/jZFLb4JAFIV_iwu2zIEBZuhupBFpqBV5SGfToKFIw8Mglb9fqm6atNS7u8l3vtzcQyRJiWyyc1lkfdk2WTXur9J6S_w1n881AZdFGrzVKowTHtDngJHtBXBcsTSYD3DfNeGJZbyxA0ohKJH35PHHCNy (accessed on 4 May 2024).
51. Marhefka, P. *Putovanie Históriou Mesta Podolínec*, 1st ed.; Mesto Podolínec: Podolínec, Slovakia, 2002; 107p.
52. Moja Obec v Štatistikách—Podolínec. Available online: https://mojaobec.statistics.sk/html/sk.html?obec=SK041A526975 (accessed on 4 May 2024).
53. Pamiatková Rezervácia v Meste Podolínec—Zásady Ochrany Pamiatkového Územia. Krajský Pamiatkový Úrad Prešov. 2010. Available online: https://www.pamiatky.sk/page/zasady-ochrany-pr-podolinec/ (accessed on 4 May 2024).
54. de Oliveira, J.A.P.; Bellezoni, R.A.; Shih, W.Y.; Bayulken, B. Innovations in Urban Green and Blue Infrastructure: Tackling local and global challenges in cities. *J. Clean. Prod.* **2022**, *362*, 132355. [CrossRef]
55. Palliwoda, J.; Haase, A.; Suppee, C.; Rink, D.; Priess, J.A. Visions for development and management of urban green and blue infrastructure: A citizen's perspective. *Ecol. Soc.* **2022**, *27*, 2. [CrossRef]
56. Nowacki, M. Heritage Interpretation and Sustainable Development: A Systematic Literature Review. *Sustainability* **2021**, *13*, 4383. [CrossRef]
57. Uzzell, D.L. Creating place identity through heritage interpretation. *Int. J. Herit. Stud.* **1996**, *1*, 219–228. [CrossRef]
58. Meissner, M. The valorisation of intangible cultural heritage. In *Intangible Cultural Heritage and Sustainable Development: The Valorisation of Heritage Practices*; Heritage Studies; Springer: Cham, Switzerland, 2021; pp. 145–172. [CrossRef]
59. Riganti, P.; Nijkamp, P. Valuing Cultural Heritage Benefits to Urban and Regional Development. In Proceedings of the 44th Congress of the European Regional Science Association: Regions and Fiscal Federalism, Porto, Portugal, 25–29 August 2004. Available online: https://www.econstor.eu/handle/10419/117122 (accessed on 23 May 2024).
60. Jato-Espino, D.; Capra-Ribeiro, F.; Moscardó, V.; Bartolomé del Pino, L.E.; Mayor-Vitoria, F.; Gallardo, L.O.; Carracedo, P.; Dietrich, K. A systematic review on the ecosystem services provided by green infrastructure. *Urban For. Urban Green.* **2023**, *86*, 127998. [CrossRef]
61. Veerkamp, C.J.; Schipper, A.M.; Hedlund, K.; Lazarova, T.; Nordin, A.; Hanson, H.I. A review of studies assessing ecosystem services provided by urban green and blue infrastructure. *Ecosyst. Serv.* **2021**, *52*, 101367. [CrossRef]

62. Zhang, S.; Ramírez, F.M. Assessing and mapping ecosystem services to support urban green infrastructure: The case of Barcelona, Spain. *Cities* **2019**, *92*, 59–70. [CrossRef]
63. Haase, D.; Larondelle, N.; Andersson, E.; Artmann, M.; Borgström, S.; Breuste, J.; Gomez-Baggethun, E.; Gren, Å.; Hamstead, Z.; Hansen, R.; et al. A quantitative review of urban ecosystem service assessments: Concepts, models, and implementation. *Ambio* **2014**, *43*, 413–433. [CrossRef]
64. Luederitz, C.; Brink, E.; Gralla, F.; Hermelingmeier, V.; Meyer, M.; Niven, L.; Panzer, L.; Partelow, S.; Rau, A.L.; Sasaki, R.; et al. A review of urban ecosystem services: Six key challenges for future research. *Ecosyst. Serv.* **2015**, *14*, 98–112. [CrossRef]
65. La Bara, L.; Fiorani, G.; Litardi, I. Valorization of Historical and Cultural Heritage: A Strategy of Sustainable Growth. In Proceedings of the Conference Strategica—Challenging the Status Quo in Management and Economics, Bucharest, Romania, 11–12 October 2018.
66. Dümcke, C.; Gnedovsky, M. *The Social and Economic Value of Cultural Heritage: Literature Review*; European Expert Network on Culture (EENC): Barcelona, Spain, 2013; pp. 1–145. Available online: https://web.archive.org/web/20220120173612id_/https://www.interarts.net/descargas/interarts2557.pdf (accessed on 23 May 2024).
67. Königreich Ungarn (1782–1785)—First Military Survey. Available online: https://maps.arcanum.com/en/map/firstsurvey-hungary/?layers=147&bbox=2111707.6266405312,6019831.665236833,2140734.5881318273,6030399.084397258 (accessed on 4 May 2024).
68. Hungary (1819–1869)—Second Military Survey of the Habsburg Empire. Available online: https://maps.arcanum.com/en/map/secondsurvey-hungary/?layers=5&bbox=2114100.105625749,6021019.156812124,2128613.586371397,6026302.866392337 (accessed on 4 May 2024).
69. Habsburg Empire (1869–1887)—Third Military Survey (1:25,000) (1869–1887). Available online: https://maps.arcanum.com/en/map/thirdsurvey25000/?layers=129&bbox=2114252.9796822504,6020594.761835113,2128766.4604278984,6025878.471415326 (accessed on 4 May 2024).
70. *Cadastral Maps of the Original Cadastral Record from 1871, Reambulated in the 1930s (1:2880)*; Sp 159, map no. 19, 20, 22; Central Archive of Geodesy and Cartography: Bratislava, Slovakia, 1871.
71. Topographic Map 1:25,000. 1955. Available online: https://www.staremapy.sk/?zoom=16&lat=49.2533396488108&lng=20.532398158167698&map=SR1952 (accessed on 4 May 2024).
72. Topographic Map 1:5000. 1958. Available online: https://www.staremapy.sk/?zoom=16&lat=49.2533396488108&lng=20.532398158167698&map=TM5 (accessed on 4 May 2024).
73. Topographic Map 1:10,000. 1964. Available online: https://www.staremapy.sk/?zoom=16&lat=49.2533396488108&lng=20.532398158167698&map=SRTM10 (accessed on 4 May 2024).
74. State Map Derived 1:5000. 1980. Available online: https://www.staremapy.sk/?zoom=16&lat=49.2533396488108&lng=20.532398158167698&map=SMO52 (accessed on 4 May 2024).
75. Basic Topographic Map 1:10,000, Approx. 1990. Available online: https://www.staremapy.sk/?zoom=16&lat=49.2533396488108&lng=20.532398158167698&map=gkubazm10 (accessed on 4 May 2024).
76. Current Cadastral Map. Available online: https://zbgis.skgeodesy.sk/mapka/sk/kataster?pos=49.253751,20.533361,16 (accessed on 4 May 2024).
77. Územný Plán Mesta Podolínec. Available online: https://www.podolinec.eu/uzemny-plan-mesta/ (accessed on 4 May 2024).
78. Ortofotomozaika. 2022. Geoportál GKÚ. Available online: https://zbgis.skgeodesy.sk/mapka/sk/zakladna-mapa?pos=49.253751,20.533361,16 (accessed on 4 May 2024).
79. Augustinková, L.; Páclová, H. *Vodní dílo Nečíz v Litovli*; Agriprint Olomouc: Olomouc, Česká Republika, 2021; pp. 100–101.
80. Tomášková, H. Jak v Chrudimi Probíhala Revitalizace Vodních Toků v Urbanizovaném Prostředí. 2021. Available online: https://www.komunalniekologie.cz/info/jak-v-chrudimi-probihala-revitalizace-vodnich-toku-v-urbanizovanem-prostredi (accessed on 23 June 2024).
81. Tkac, S. Košice millrace, a unique historical waterwork has been bonding the technological advances and culture for centuries. In Proceedings of the 3rd Forum on Asian Industrial Heritage Conservation Bulletin, online, 27–28 October 2020; pp. 41–44.
82. Na Mieste Bývalého Mlynského Náhonu vo Veľkom Šariši sa Nachádza Nová Cykloaleja. Available online: https://www.kosiceonline.sk/na-mieste-byvaleho-mlynskeho-nahonu-vo-velkom-sarisi-sa-nachadza-nova-cykloaleja (accessed on 23 June 2024).
83. Kolarz, M. Studie Nádvoří Zámku Krnov. 2006. Available online: https://www.zamekkrnov.cz/o-spolecnosti/ (accessed on 23 June 2024).
84. Łabuz, R. The area of Kossak Square—A public space at the Młynówka Królewska Mill Race—In the Context of Its Historic, Cultural and Natural Assets. In *Topiarius 4 Landscape Studies*; University Rzeszów, Faculty of Biology and Agriculture: Rzeszów, Poland, 2017; pp. 117–129.

land

MDPI

Article

What Is the Perceived Environmental Restorative Potential of Informal Green Spaces? An Empirical Study Based on Visitor-Employed Photography

Jiayi Jiang [1,2,3], Hong Xu [2,4], Ruochen Ma [5,*], Shi Chen [1], Huixin Wang [5,*] and Ziang Zheng [6]

1 School of Architecture, Soochow University, No. 199 Ren-ai Road, Suzhou Industrial Park, Suzhou 215123, China; jyjiang@suda.edu.cn (J.J.); 20235241007@stu.suda.edu.cn (S.C.)
2 Soochow University—Suzhou Yuanke (SU-SY) Collaborative Innovation Center of Architecture and Urban Environment, School of Architecture, Soochow University, No. 199 Ren-ai Road, Suzhou Industrial Park, Suzhou 215123, China; yuanke@yuankest.cn
3 China-Portugal Joint Laboratory of Cultural Heritage Conservations Science Supported by the Belt and Road Initative, Soochow University, Suzhou 215123, China
4 Suzhou Yuanke Ecological Construction Group, No. 268 Dongping Street, Suzhou 215123, China
5 Graduate School of Horticulture, Chiba University, Chiba 271-8510, Japan
6 The Liverpool School of Architecture, University of Liverpool, 25 Abercromby Square, Liverpool L69 7ZN, UK; ziang.zheng@liverpool.ac.uk
* Correspondence: 24hd0401@student.gs.chiba-u.jp (R.M.); cefa5056@chiba-u.jp (H.W.); Tel.: +81-070-5518-8490 (R.M.); +81-070-3962-0399 (H.W.)

Abstract: Informal green spaces (IGSs) play an essential role in enhancing urban well-being by offering restorative environments, yet the impact of visitor behaviors on perceived restorativeness (PR) remains underexplored. This study investigates how different spatio-temporal behaviors influence PR in IGS, providing urban planners with actionable insights to optimize these spaces for better user experiences. Using a visitor-employed photography (VEP) survey and post-visit PR assessments, K-means clustering was applied to identify distinct visitor behavior patterns. Correlation analysis further explored the relationships between these patterns and PR; the results reveal three unique clusters of visitor behaviors—fast, extensive exploration; moderate, focused exploration; and slow, thorough exploration—each showing distinct impacts on PR. Visitors who engage in rapid, broad exploration perceive larger, navigable spaces as more restorative, while those focusing on specific or in-depth exploration emphasize psychological aspects like escape and fascination. These behavioral patterns demonstrate varying strengths in their association with restorative experiences; This study underscores the importance of integrating spatio-temporal behavior data with PR assessments, highlighting how the physical and psychological features of IGS influence visitor experiences. These findings offer critical insights for designing and managing IGS to accommodate diverse user needs and promote urban well-being.

Keywords: informal green space; visitor-employed photography; restorative environment; spatial–temporal patterns; human well-being; cluster analysis; landscape experience

1. Introduction

High-density urban areas are facing numerous challenges due to urbanization, such as environmental pollution [1,2], the urban heat island effect [3,4], and a growing disconnection between people and nature [5]. A substantial body of research highlights that exposure to natural environments provides significant physical and mental health benefits for urban residents, including stress reduction, attention restoration, positive emotional states, and disease prevention [6–8]. However, as urban density rises, per capita green space is decreasing, reducing access to nature in cities [9,10]. In this context, informal green spaces (IGSs) have garnered increased attention as potential solutions.

IGS refers to urban green spaces that are not formally planned and are characterized by spontaneous vegetation and irregular maintenance [11]. Introduced by Rupprecht and Byrne, examples of IGSs include vacant lots, brownfields, and the edges of streets or railroads [12]. Unlike formal green spaces, which are intentionally planned and maintained, IGSs tend to emerge in neglected urban areas, exhibiting a disorganized and heterogeneous appearance [11,13]. This natural disorganization, however, is exactly what grants IGSs unique ecological and social functions, supporting biodiversity and offering unstructured recreational opportunities [14–16]. Furthermore, IGSs enhance ecological resilience in urban development by providing essential green infrastructure services, such as regulating temperature, pollution control, and disaster mitigation [17,18]

Research on IGS varies in focus across different regions globally. In Europe, especially in Poland, several studies have shown that IGSs contribute significantly to the well-being of urban residents and ecosystem services [18,19]. In addition, Sikorska et al. state that IGS has the potential to reduce inequitable distribution of urban green space availability [15]. In Asian countries, particularly Japan and China, research on IGS has focused more on residents' perceptions. In Japan, studies often use questionnaires to investigate residents' views, usage patterns, and management preferences regarding IGS [20,21]. In contrast, Chinese researchers, such as Chen et al., have employed machine learning techniques to accurately measure and analyze residents' complaints about IGS on social media platforms [13]. Rupprecht et al. conducted two cross-cultural studies exploring residents' perceptions of IGS in Australia and Japan, finding that the functional role of IGS in the lives of respondents from both countries was quite similar [14,22]. However, how the respondents used and evaluated IGS was closely tied to their personal environmental preferences.

Urban residents hold mixed attitudes toward IGS. Some residents prefer the less manicured, more natural appearance of IGSs compared to formal green spaces and already use IGSs for activities such as children's play, dog walking, and barbecuing [23]. Meanwhile, the unofficial and unmanaged nature of IGS raises safety concerns for others, limiting the full realization of its potential [23,24]. The lack of formal protection for IGS further limits its potential as a valuable urban resource [25].

In densely populated urban environments where formal green spaces are limited, IGS can provide substantial social and environmental benefits, but its potential has not yet been fully realized [20]. To unlock the potential of IGS in high-density urban areas, it is crucial to better understand the specific contributions that IGS can make to human well-being [21]. Although most existing studies have focused on the spatial distribution [25,26], resident attitudes [13,14,27], and ecological functions like biodiversity and climate regulation [11,19,28], limited attention has been paid to its potential restorative effects on mental health. Herman et al., for instance, examined the emotional well-being of 20 participants using a portable electroencephalography device during their visits to IGS but found no significant difference in emotional states compared to visits to formal green space [17].

To further investigate the mental health benefits of IGS, this study introduces Perceived Restorativeness as a key indicator for measuring the psychological benefits provided by urban green spaces [29,30]. According to Kaplan et al.'s Attention Restoration Theory (ART), perceived restorativeness refers to the extent of psychological recovery that an individual experiences in a given environment, typically linked to reduced mental fatigue, restored attention, and improved mood states [31,32]. Numerous studies have demonstrated that individuals perceive higher levels of restorativeness in natural environments compared to urban settings [33–37]. Despite the potential for IGS to provide higher levels of perceived restorativeness due to its natural features, there is still a lack of empirical evidence to support this assumption [38]. Most existing studies on restorative effects have focused on traditional green spaces (e.g., parks and gardens), often employing methods such as on-site questionnaires, photographs, or video assessments [39,40]. However, these methods have limitations when applied to IGS, as they are typically not designated by management authorities or owners for public use, and their disorganized and uncertain nature makes them less visible to the public [25,41]. Many people have little direct experience with IGSs, and relying solely on these methods may fail to accurately capture their restorative potential or could underestimate their contribution to psychological well-being [21]. Considering this, this study aims to fill this research gap by validating the perceived restorative function of IGS through surveys of participants' experiences following their on-site visits, while considering the specific characteristics of IGS environments.

One distinctive feature of IGS is the absence of designated pathways, meaning that individuals' choices regarding route selection and the duration of their visit can significantly influence their green space experience, potentially leading to varying levels of psychological restoration [42,43]. Spatio-temporal behavioral patterns—referring to the trajectories and time-related dynamics of individuals' movements across space—capture how people navigate environments and interact with different areas [44]. GPS-based tracking systems provide an effective tool to capture such movement patterns, shedding light on how individuals explore and experience IGS over time [45,46]. This study introduces GPS-based spatio-temporal data to link movement patterns with perceived restorativeness, providing a dynamic approach to understanding the utilization of IGS.

Spatio-temporal analysis has been widely used in various fields to better understand movement patterns, space exploration behaviors, and visitor flows [46], to analyze public space use [47], and to monitor urban mobility [48,49]. In particular, spatio-temporal clustering techniques have been proven effective in identifying tourist characteristics and understanding how visitors use different spaces. For example, Liu et al. [50] utilized open GPS trajectory data in mountainous scenic areas to identify microscopic movement patterns, revealing key insights for managing tourist flows and resource allocation. Brian et al. [51] applied spatio-temporal analysis of GPS trajectory data to cluster and compare travel groups in a national park, helping optimize resource allocation and management strategies by identifying the time visitors spent in different areas. Research methods that employ GPS trajectory data to analyze visitor behavior within destinations have become relatively advanced [52]. By treating IGSs, which are less restricted spaces, as small-scale destinations and combining spatio-temporal behavioral data with perceived restorativeness assessments, it is possible to provide a more precise and dynamic analytical perspective. This approach allows for a more nuanced understanding of how different route selection in IGS environments influence visitors' restorative experiences, providing valuable insights into how these informal spaces can be better integrated into urban planning to enhance human well-being. Ultimately, this study will propose specific planning and management strategies for IGS based on the analysis results, aiming to better realize their potential as a valuable resource for enhancing the well-being of urban residents.

Although previous research has examined IGS across various urban contexts, there remains limited understanding of how visitor behaviors specifically influence their perception of restorativeness. This study seeks to bridge that gap by employing GPS-based spatio-temporal analysis in conjunction with psychological assessments of restorativeness, offering new insights into how different routes and behaviors within IGS impact mental well-being. The study aims to address the following key research questions:

(1) What are the spatio-temporal behavioral patterns of visitors within IGS?
(2) How do these behavioral patterns influence visitors' perceived restorative effects?
(3) How can the findings from these patterns inform the planning and management of IGS to optimize their use and enhance restorative experiences?

Through this research, we aim to provide a scientific basis for the integration of IGS into urban planning, offering new insights and practical recommendations for the allocation of limited natural resources in high-density urban areas.

2. Materials and Methods

2.1. Study Sites

Suzhou, located in the Yangtze River Delta region, is a significant central city known for its rapid economic growth and urbanization. By the end of 2023, the city had a permanent resident population of approximately 12.96 million, ranking first in Jiangsu Province in both population size and Gross Domestic Product (GDP). Suzhou Industrial Park (SIP), a leading industrial development zone within the city, is characterized by its significant industrial agglomeration and robust economic growth (Figure 1a). The informal green spaces (IGSs) within SIP have developed in response to rapid urban expansion, offering unique opportunities to examine the dynamics of urban green space formation and utilization due to their distinct locations and flexible usage patterns.

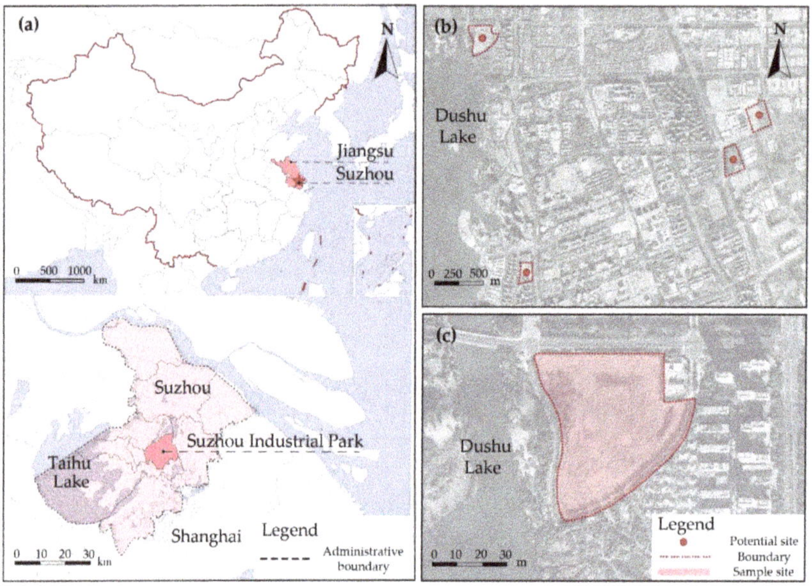

Figure 1. Study site: (**a**) Suzhou City, Jiangsu Province, China; (**b**) Four potential IGS sites; (**c**) The sample site.

In identifying potential study sites, we applied specific criteria based on the work of Sikorski et al. [15], the Code for the Design of Public Parks [53], and the findings of Jiang et al. [54]. The criteria included a substantial area size, a green space ratio of at least 65%, vegetation coverage exceeding 30% [11], and unrestricted public access. Based on these parameters, four potential IGS sites were identified (Figure 1b). Following a comprehensive evaluation involving on-site surveys, assessments of site longevity, and preliminary experimental data, a site adjacent to Dushu Lake was selected as the sample site. The site covers approximately 11.4 hectares, with 85.18% consisting of a mix of trees, shrubs, and grasses. It features multiple access points and open areas conducive to visitor exploration (Figure 1c).

Data collection was conducted on clear, windless days, including visitor behavior observation through hired photographers and a structured questionnaire survey. According to historical imagery, the site has existed for approximately 15 years and is classified as a Type I residential area [55].

2.2. Research Protocol

A total of 89 students majoring in landscape architecture and related fields participated in this study, including 75 undergraduates and 14 Master's students, with a gender ratio of approximately 1:1.5 (male to female). Participants were instructed to explore an informal green space (IGS) site and record their paths using a mobile app, followed by completing a questionnaire. The trajectory data were collected using the "Six Feet (version 4.202.23)" app, which recorded participants' GPS coordinates and timestamps in real time, as well as their routes and travel distances. The questionnaire gathered information on spatio-temporal behaviors, socio-demographic characteristics, usage of park and informal green spaces, and perceived environmental restorative qualities of the site. Given their academic background in environmental studies, participants were expected to provide informed and reliable feedback. Each participant was compensated with RMB 60 for their involvement.

2.2.1. Visitor-Employed Photography (VEP)

The visitor-employed photography (VEP) survey was conducted on two clear, windless mornings, 13 and 14 April 2024, from 9:00 to 12:00, with average temperatures of 20 °C and 21 °C, respectively, to avoid weather-related variability and ensure consistent conditions. Based on the methodology and guidelines suggested by Liang et al. [56] and Fefer et al. [57], we ultimately recruited 89 volunteers to participate in the VEP survey, ensuring a diverse sample size to capture a range of visitor behaviors.

Since the IGS lacked designated entry points, two fixed entrances along a secondary road were established to ensure participant safety and manage participant flow (Figure 2). To minimize entry bias, considering the positive correlation between physical fatigue and travel distance, and to ensure sufficient data collection while avoiding redundancy [58], multiple entrances at different locations were used as starting points for the VEP survey. Specifically, on 13 April, groups 1, 2, and 3 entered from Entrance 1, while on April 14, groups 4, 5, and 6 entered from Entrance 2, following the methodology of Jurisic et al. [59].

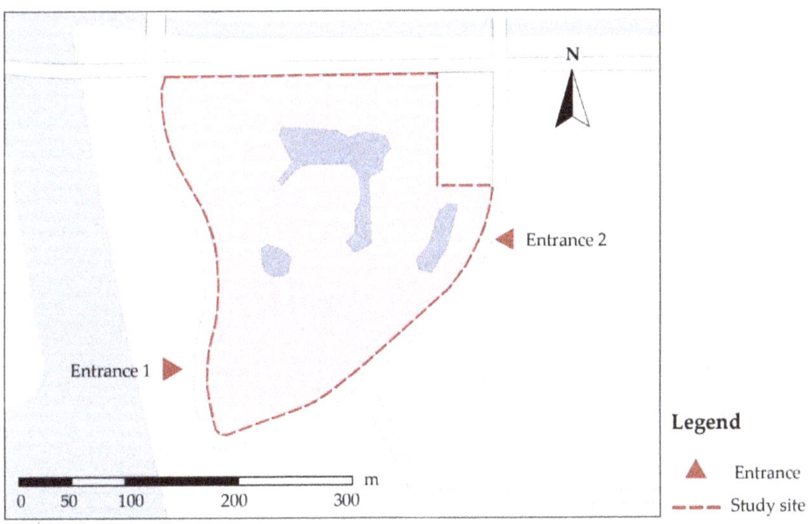

Figure 2. IGS Site and Entrances.

Participants used the "Six Feet" app, which they were required to download before the survey. After confirming the use of the AMAP, participants entered the site at staggered intervals of 3 min to avoid interactions. They were free to explore according to their preferences and could take photos ("footprints") with descriptions during their visit. The app automatically recorded their routes, GPS coordinates, and time spent, and participants uploaded these data upon completion for subsequent analysis.

2.2.2. Perceived Restorativeness Scale (PRS)

Following the site visit, participants immediately completed a questionnaire assessing the perceived environmental restorative qualities of the IGS. The questionnaire consisted of four sections: (1) demographic information (gender, age, educational background); (2) experience with parks and informal green spaces (frequency of visits, activity duration); (3) evaluation of perceived restorativeness scale (PRS) of IGS across 13 items [31,32], rated on a seven-point Likert scale (1 = not at all; 7 = totally) (Table 1 presents 13 items of PRS); and (4) willingness to revisit the site.

Table 1. Perceived Restorativeness Scale (PRS) [30,31].

Subscale Membership	Items
Being away	3.1 Being here is an escape experience. 3.2 Spending time here gives me a break from my day-to-day routine. 3.3 It is a place to get away from it all. 3.4 Being here helps me to relax my focus on getting things done. 3.5 Coming here helps me to get relief from unwanted demands on my attention.
Fascination	3.6 This place has fascinating qualities. 3.7 My attention is drawn to many interesting things.
Extent	3.8 I want to get to know this place better. 3.9 There is much to explore and discover here. 3.10 I want to spend more time looking at the surroundings.
Compatibility	3.11 This place is boring. 3.12 The setting is fascinating. 3.13 There is nothing worth looking at here.

Tables 2 and 3 provide the results of the reliability and validity analyses. To assess the internal consistency of the variables, Cronbach's alpha was calculated, yielding a value of 0.83. This exceeds the acceptable threshold of 0.7, indicating strong reliability and consistency within the scale. In terms of validity, factor analysis was performed using the Kaiser–Meyer–Olkin (KMO) Measure and Bartlett's Test of Sphericity via SPSS software v22.0 to examine the inter-item correlations. The KMO statistic was 0.88, within the desirable range of 0.8–0.9, signifying an adequate level of shared variance among the variables. Moreover, the Bartlett's Test result was significant at $p < 0.05$, confirming that the original variables are correlated, thereby affirming the structural validity of the questionnaire.

Table 2. Reliability analysis of Perceived Restorativeness Scale (PRS).

Cronbach's Alpha	N of Items
0.83	13

Table 3. Validity analysis of Perceived Restorativeness Scale (PRS).

Kaiser–Meyer–Olkin Measure (KMO) and Bartlett's Test		
KMO Measure of Sampling Adequacy		0.88
Bartlett's Test of Sphericity	Approx. Chi-Square	758.58
	df	78
	Sig.	0

2.3. Data Analysis

The data analysis consisted of two main parts: (1) analysis of spatio-temporal behaviors based on participants' movement trajectories within the informal green space (IGS), focusing on metrics such as visit duration, path length, area of coverage, and clustering characteristics; and (2) examination of differences in perceived environmental restorativeness across participant clusters derived from the trajectory data.

2.3.1. Spatio-Temporal Path Analysis

The spatio-temporal paths of participants were visualized using GPS data from the "Six Feet" app by ArcGIS 10.8. Following the methodology of Huang et al. [49], four core indicators were identified to represent participants' spatio-temporal behavior: trajectory duration, path length, coverage area, and coverage perimeter:

- Trajectory duration represents the total time a visitor spent in the study area, calculated as the difference between the start and end times of the GPS-recorded trajectory;
- Path length denotes the total distance traveled by each visitor, calculated as the sum of the distances between consecutive GPS coordinates;
- Coverage area refers to the projected area of a visitor's trajectory on the XY coordinate plane. The standard deviation ellipse was employed to encompass all trajectory points. Trajectory coverage maps were generated in ArcGIS based on the set of points for each trajectory;
- Coverage perimeter is the perimeter of the ellipse that covers all trajectory points, serving as an indicator of the spatial characteristics of the visitor movement patterns.

Clustering analysis was applied using four key indicators to classify participants' movement patterns within the IGS. Clustering is a widely used technique in spatial analysis and can be broadly categorized into partition-based, hierarchical, density-based, graph-based, and model-based methods. Among these, K-means clustering, a partition-based method, was chosen for this study due to its suitability for handling spatial data with overlapping clusters. K-means divides the dataset into K clusters by minimizing intra-cluster distances while maximizing inter-cluster distances. This method is particularly advantageous for detecting clusters that are not clearly separated, even when there is some overlap between them.

In contrast to density-based methods such as DBSCAN (Density-Based Spatial Clustering of Applications with Noise), which can merge overlapping clusters and are highly sensitive to parameter selection, K-means is more effective for identifying distinct movement patterns. Thus, K-means offers an optimal balance between computational efficiency and the ability to distinguish nuanced movement behaviors within the IGS environment.

The GIS clustering module was implemented using Python 3.9 with the Anaconda platform (version 4.0.15). The clustering algorithm was configured in Python to follow two criteria: (1) areas of high density, where cluster centers are surrounded by lower-density neighbors, and (2) large distances between dense data points to ensure clear separation between clusters. This approach allowed for effective categorization of participants' spatial behavior, facilitating a deeper understanding of their movement within the IGS.

To confirm the clustering results, we used a one-way analysis of variance (ANOVA) to determine the statistical significance of the clusters.

2.3.2. Cluster Difference Analysis

When the ANOVA results were significant at the $p < 0.05$ level, we conducted an analysis of the differences in questionnaire responses among the different clusters. We used descriptive statistics and comparative analyses to look at variations in demographic characteristics, experiences with green spaces, and willingness to revisit. Additionally, we conducted Pearson correlation analyses to explore the relationships between participants' perceived environmental restorativeness and their behavioral patterns within each identified cluster.

3. Results and Analysis

3.1. General Statistic

Table 4 summarizes the spatio-temporal trajectory indicators of 89 participants, including trajectory duration, path length, coverage area, and coverage perimeter. The data show that trajectory duration and path length are right-skewed, indicating a few visitors spent significantly more time and covered longer distances. Conversely, coverage area is left-skewed, with most visitors covering larger areas. The mean and median values for coverage perimeter are closely aligned, suggesting a more balanced distribution. The standard deviations indicate greater variability in trajectory duration and coverage area, reflecting diverse exploration behaviors among visitors.

Table 4. General statistical analysis of spatio-temporal trajectory indicators.

Title	Minimum	Maximum	Mean	Median	Std.
Trajectory duration (s)	0.00	10,732.00	3282.14	2294.00	2202.55
Path length (m)	505.02	4008.38	1402.22	1259.86	566.20
Coverage area (m^2)	10,545.99	86,040.46	56,813.50	66,135.84	22,696.50
Coverage perimeter (m)	455.54	1126.15	930.46	1008.00	169.84

3.2. Clusters Analysis

To understand the diversity in visitor behaviors, K-means clustering was applied based on four trajectory indicators: trajectory duration, path length, coverage area, and coverage perimeter. This clustering process resulted in three distinct clusters after several iterations. The validity of the clustering results was verified using analysis of variance (ANOVA), with all four indicators showing p-values less than 0.001, confirming the statistical significance and reliability of the clusters (Table 5).

Table 5. ANOVA results based on differences in respondents' spatio-temporal trajectories.

Indicator	Cluster 1	Cluster 2	Cluster 3	Test Score
Trajectory duration	2153.15	2589.78	6913.00	F = 125.25, $p < 0.001$ *** [1]
Path length	1224.76	1028.53	2293.55	F = 97.29, $p < 0.001$ ***
Coverage area	69,297.54	22,578.09	67,374.82	F = 165.73, $p < 0.001$ ***
Coverage perimeter	1023.41	675.83	1008.76	F = 156.46, $p < 0.001$ ***

[1] *** $p < 0.001$.

3.2.1. Cluster 1 Analysis

Cluster 1, comprising nearly half of the sample, exhibited the widest coverage area (69,297.54 m^2) and the largest perimeter (1023.41 m), while also having the shortest total browsing time (2153.15 s) and a moderate path length (1224.76 m). Figure 3 shows the trajectory maps for this cluster, revealing a pattern of fast, broad exploration, where participants moved quickly through the IGS with minimal stops or prolonged engagement in specific areas. The expansive paths suggest that visitors in this cluster prioritized covering as much ground as possible, focusing more on the overall environment rather than particular features. This behavior indicates a preference for rapid and extensive exploration, with limited in-depth interaction with the surroundings.

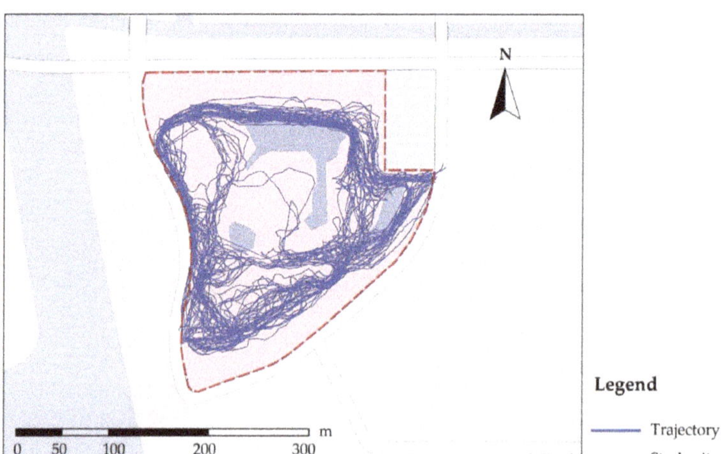

Figure 3. Trajectory maps for Cluster 1.

3.2.2. Cluster 2 Analysis

Cluster 2 exhibited a moderate browsing time (2589.78 s) but had the shortest path length (22,578.09 m^2) and smallest coverage area (22,578.09 m^2) and perimeter (675.83 m). Figure 4 illustrates the trajectory maps for this cluster, revealing that the paths are concentrated in specific sections of the IGS, particularly around key landscape features such as water bodies and shaded areas. These visitors appeared to focus on specific areas of interest rather than covering vast distances, indicating a targeted exploration approach where visitors concentrate on select features of the site rather than engaging in extensive exploration.

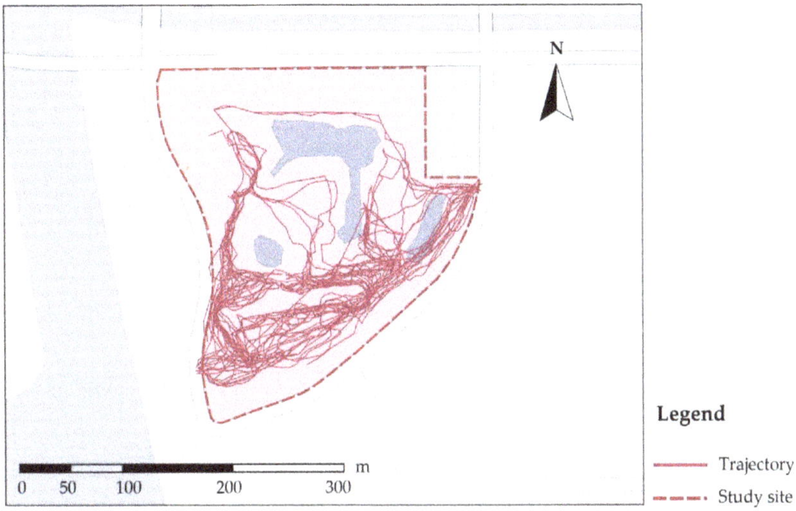

Figure 4. Trajectory maps for Cluster 2.

3.2.3. Cluster 3 Analysis

Cluster 3 exhibited the longest total browsing time (6913.00 s) and the longest path length (2293.55 m), with a substantial coverage area (67,374.82 m^2) and perimeter (1008.76 m). Figure 5 shows the trajectory maps for this cluster, revealing a comprehensive exploration pattern where visitors traversed a wide range of the IGS, taking time to engage with various areas. These results suggest a more deliberate, in-depth exploration style, with visitors taking their time to experience a broad range of features within the IGS. This cluster reflects a comprehensive engagement with the space, likely providing a more immersive environmental experience.

The clustering results highlight three distinct visitor behavior patterns: Cluster 1 reflects fast, extensive exploration across a wide area; Cluster 2 represents moderate, focused exploration of specific regions; and Cluster 3 indicates slow, thorough exploration of a substantial portion of the IGS. These patterns provide valuable insights into how different visitor types interact with green spaces, which can inform future management and design strategies to cater to varied preferences and behaviors. Differences in spatio-temporal behavior also align with variations in perceived restorativeness across clusters.

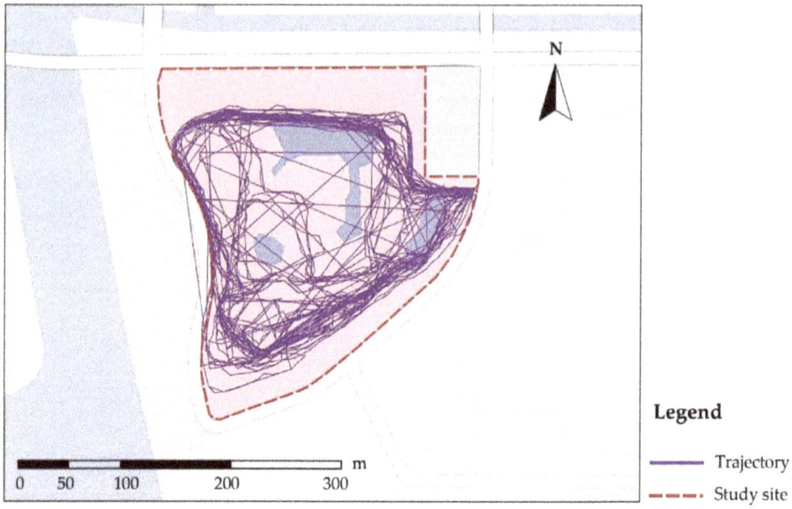

Figure 5. Trajectory maps for Cluster 3.

3.3. *Analysis of Cluster Differences*

3.3.1. Green Space Usage and Overall Impressions Across Clusters

We compared the green space usage habits and willingness to revisit across the clusters (Table 6). Significant differences were found in green space usage habits and willingness to revisit across clusters. Participants in Cluster 1 demonstrated a balanced approach regarding both the frequency of visits and willingness to revisit, showing moderate familiarity with informal green spaces. In contrast, Cluster 2 respondents visited formal parks more frequently but had relatively lower familiarity with informal green spaces. Meanwhile, Cluster 3 respondents had a higher frequency of visits to informal green spaces but were more uncertain about their willingness to revisit. These differences highlight the need for tailored informal green space management strategies to cater to the unique preferences of different user groups.

Table 6. Summary of green space usage and willingness to revisit by cluster.

Categories	Variables	Cluster 1	Cluster 2	Cluster 3	Total
Frequency of visiting parks	Almost never	5 (10.64%)	0 (0.00%)	5 (26.32%)	10 (11.24%)
	A few times a month	23 (48.94%)	18 (78.26%)	12 (63.16%)	53 (59.55%)
	Once a week	11 (23.40%)	2 (8.70%)	1 (5.26%)	14 (15.73%)
	More than 2–3 times a week	8 (17.02%)	3 (13.04%)	1 (5.26%)	12 (13.48%)
	Daily	0 (0.00%)	0 (0.00%)	0(0.00%)	0 (0.00%)
Average time spent per park visit	<15 min	8 (17.02%)	2 (8.70%)	2 (10.53%)	12 (13.48%)
	15–30 min	24 (51.06%)	9 (39.13%)	9 (47.37%)	42 (47.19%)
	30–60 min	11 (23.41%)	10 (43.47%)	7 (36.84%)	28 (31.46%)
	>60 min	4 (8.51%)	2 (8.70%)	1 (5.26%)	7 (7.87%)
Familiarity with informal green spaces in cities	Not very familiar	19 (40.43%)	10 (43.47%)	9 (47.37%)	38 (42.70%)
	Fairly familiar	27 (57.44%)	13 (56.53%)	10 (52.63%)	50 (56.18%)
	Very familiar	1 (2.13%)	0 (0.00%)	0 (0.00%)	1 (1.12%)
Frequency of visiting IGS	Almost never	19 (40.42%)	9 (39.13%)	4 (21.05%)	32 (35.95%)
	A few times a month	18 (38.30%)	6 (26.09%)	14 (73.69%)	38 (42.70%)
	Once a week	5 (10.64%)	4 (17.39%)	0 (0.00%)	9 (10.11%)
	More than 2–3 times a week	5 (10.64%)	4 (17.39%)	1 (5.26%)	10 (11.24%)
	Daily	0 (0.00%)	0 (0.00%)	0 (0.00%)	0 (0.00%)

Table 6. *Cont.*

Categories	Variables	Cluster 1	Cluster 2	Cluster 3	Total
Average time spent per visit to IGS	<15 min	23 (48.94%)	10 (43.47%)	7 (36.84%)	40 (44.94%)
	15–30 min	17 (36.17%)	12 (52.18%)	11 (57.90%)	40 (44.94%)
	30–60 min	6 (12.76%)	1 (4.35%)	1 (5.26%)	8 (9.00%)
	>60 min	1 (2.13%)	0 (0.00%)	0 (0.00%)	1 (1.12%)
Willingness to revisit	Strongly unwilling	0 (0.00%)	0 (0.00%)	0 (0.00%)	0 (0.00%)
	Unwilling	5 (10.64%)	3 (13.04%)	1 (5.26%)	9 (10.11%)
	Somewhat unwilling	5 (10.64%)	7 (30.44%)	8 (42.11%)	20 (22.47%)
	Neutral	13 (27.66%)	3 (13.04%)	1 (5.26%)	17 (19.10%)
	Somewhat willing	14 (29.79%)	6 (26.09%)	6 (31.58%)	26 (29.21%)
	Willing	6 (12.76%)	4 (17.39%)	3 (15.79%)	13 (14.61%)
	Strongly willing	4 (8.51%)	0 (0.00%)	0 (0.00%)	4 (4.50%)

3.3.2. Perceived Restorativeness Scale (PRS) Scores Across Clusters

Correlation analysis was conducted to examine the relationships between trajectory indicators (CI1–CI4: trajectory duration, path length, coverage area, and coverage perimeter) and perceived restorativeness variables (Table 1, 3.1–3.13) across clusters. Key correlations for each cluster are summarized below, highlighting differences in perceived restorativeness among respondents.

Table 7 highlights key differences in perceived restorativeness across the three clusters. Cluster 1 generally scored the highest in the "being away" subscale (3.1–3.5), particularly on items 3.4 (Mean = 5.47) and 3.5 (Mean = 5.43), indicating a stronger sense of escape from daily routines. This aligns with the "being away" component of ART, which refers to psychological distance from everyday demands. Visitors in this cluster, who explored larger coverage areas and had longer perimeters, seemed to achieve a greater sense of escape, likely facilitated by the opportunity to navigate wide, open spaces. Cluster 2 showed moderate scores overall, with lower ratings in fascination-related items like 3.6 (Mean = 4.17), but it matched Cluster 1 in item 3.9 (Mean = 5.04), indicating a similar perception of exploration opportunities. Cluster 3, despite the longest engagement time, scored lower in "being away" (3.3, Mean = 4.21), but had similar fascination scores to Cluster 1 (3.7, Mean = 4.79). It also scored higher on item 3.13 (Mean = 3.74), suggesting more mixed perceptions of the space's appeal.

Table 7. General statistical results of the three clusters.

Items	Cluster 1		Cluster 2		Cluster 3	
	Mean	Std.	Mean	Std.	Mean	Std.
3.1	4.98	1.23	4.78	1.38	4.68	0.86
3.2	5.34	1.40	4.87	1.39	4.89	1.25
3.3	4.91	1.35	4.22	1.50	4.21	1.20
3.4	5.47	1.22	4.83	1.52	5.00	0.73
3.5	5.43	1.25	5.17	1.13	5.11	0.85
3.6	4.68	1.37	4.17	1.37	4.58	0.94
3.7	4.91	1.38	4.48	1.53	4.79	1.00
3.8	4.72	1.36	4.30	1.46	4.37	1.22
3.9	5.04	1.35	5.04	1.23	4.79	1.15
3.10	4.89	1.48	4.57	1.10	4.68	1.30
3.11	3.60	1.36	3.48	1.31	4.11	1.29
3.12	4.38	1.45	4.26	1.15	4.42	1.27
3.13	3.28	1.42	3.48	1.10	3.74	1.33

Figures 6–8 present the correlation analysis of the variables for the indicators across three clusters. For Cluster 1, significant correlations were observed between trajectory indicators and perceived restorativeness variables. CI3 (coverage area) and CI4 (coverage

perimeter) showed a strong positive correlation (r = 0.97), indicating that larger coverage areas tend to have longer perimeters. Among the PR variables, a strong positive correlation was found between items 3.6 and 3.2 (r = 0.60), suggesting that areas perceived as more engaging are also seen as more restorative. In contrast, negative correlations were observed between perceived boredom (3.11) and fascination-related items (3.12, r = −0.51; 3.13, r = −0.83), reflecting differing perceptions of the space's restorativeness.

	CI1	CI2	CI3	CI4	3.1	3.2	3.3	3.4	3.5	3.6	3.7	3.8	3.9	3.10	3.11	3.12	3.13
CI1	1.00																
CI2	0.24	1.00															
CI3	0.01	0.30*	1.00														
CI4	-0.04	0.28	0.97**	1.00													
3.1	0.11	-0.08	0.00	0.01	1.00												
3.2	0.21	-0.05	-0.02	-0.06	0.72**	1.00											
3.3	0.14	-0.19	0.09	0.06	0.32*	0.34*	1.00										
3.4	0.27	-0.10	-0.14	-0.17	0.60**	0.80**	0.22	1.00									
3.5	0.10	0.01	-0.12	-0.12	0.64**	0.69**	0.26	0.65**	1.00								
3.6	-0.11	-0.12	0.06	0.09	0.50**	0.60**	0.43**	0.49**	0.55**	1.00							
3.7	-0.06	0.00	0.12	0.13	0.41**	0.47**	0.43**	0.38**	0.54**	0.68**	1.00						
3.8	0.04	-0.07	-0.14	-0.15	0.49**	0.69**	0.28	0.69**	0.52**	0.68**	0.63**	1.00					
3.9	0.15	0.01	-0.08	-0.11	0.54**	0.62**	0.46**	0.51**	0.52**	0.55**	0.62**	0.56**	1.00				
3.10	0.10	-0.01	-0.21	-0.26	0.48**	0.63**	0.39**	0.61**	0.58**	0.64**	0.59**	0.73**	0.66**	1.00			
3.11	0.07	0.08	0.12	0.10	-0.29	-0.43**	-0.16	-0.45**	-0.52**	-0.49**	-0.48**	-0.58**	-0.38**	-0.53**	1.00		
3.12	-0.02	-0.06	-0.20	-0.20	0.49**	0.64**	0.36*	0.62**	0.60**	0.66**	0.47**	0.77**	0.56**	0.74**	-0.51**	1.00	
3.13	0.04	0.06	-0.09	-0.09	-0.18	-0.44**	-0.09	-0.44**	-0.34*	-0.50**	-0.44**	-0.60**	-0.37*	-0.49**	0.83**	-0.47**	1.00

Correlation
1.0
0.5
0.0
-0.5
-1.0

*p<0.05;**p<0.01.

Figure 6. A correlation analysis of the variables for the indicators in Cluster 1.

	CI1	CI2	CI3	CI4	3.1	3.2	3.3	3.4	3.5	3.6	3.7	3.8	3.9	3.10	3.11	3.12	3.13
CI1	1.00																
CI2	0.71**	1.00															
CI3	0.25	0.66**	1.00														
CI4	0.39	0.72**	0.84**	1.00													
3.1	-0.15	-0.24	-0.16	-0.24	1.00												
3.2	0.36	0.13	-0.11	-0.09	0.53**	1.00											
3.3	-0.14	-0.06	-0.11	-0.29	0.04	0.03	1.00										
3.4	0.13	-0.01	-0.23	-0.25	0.83**	0.54**	-0.10	1.00									
3.5	0.01	-0.20	-0.34	-0.39	0.50*	0.15	0.05	0.65**	1.00								
3.6	0.24	0.15	-0.09	-0.15	0.57**	0.47*	0.17	0.66**	0.35	1.00							
3.7	0.26	0.16	0.05	0.07	0.50*	0.52*	0.01	0.52*	0.41	0.83**	1.00						
3.8	0.18	0.07	-0.04	-0.05	0.49*	0.41	0.11	0.42*	0.13	0.80**	0.83**	1.00					
3.9	0.17	0.05	0.01	-0.06	0.18	0.03	0.46*	0.17	0.43*	0.38	0.52*	0.55**	1.00				
3.10	-0.10	-0.22	-0.44*	-0.54**	0.31	0.11	0.51*	0.35	0.52*	0.54**	0.41	0.49*	0.59**	1.00			
3.11	0.00	-0.09	-0.01	0.06	-0.54**	-0.35	0.04	-0.55**	0.26	-0.72**	-0.63**	-0.67**	-0.20	-0.46*	1.00		
3.12	0.11	0.04	-0.15	-0.22	0.64**	0.56**	0.19	0.62**	0.33	0.80**	0.79**	0.86**	0.53*	0.54**	-0.66**	1.00	
3.13	-0.09	-0.10	0.17	0.13	0.51*	-0.44*	0.10	-0.57**	-0.28	-0.69**	-0.63**	-0.63**	-0.02	-0.37	0.87**	-0.65**	1.00

Correlation
1.0
0.5
0.0
-0.5
-1.0

*p<0.05;**p<0.01.

Figure 7. A correlation analysis of the variables for the indicators in Cluster 2.

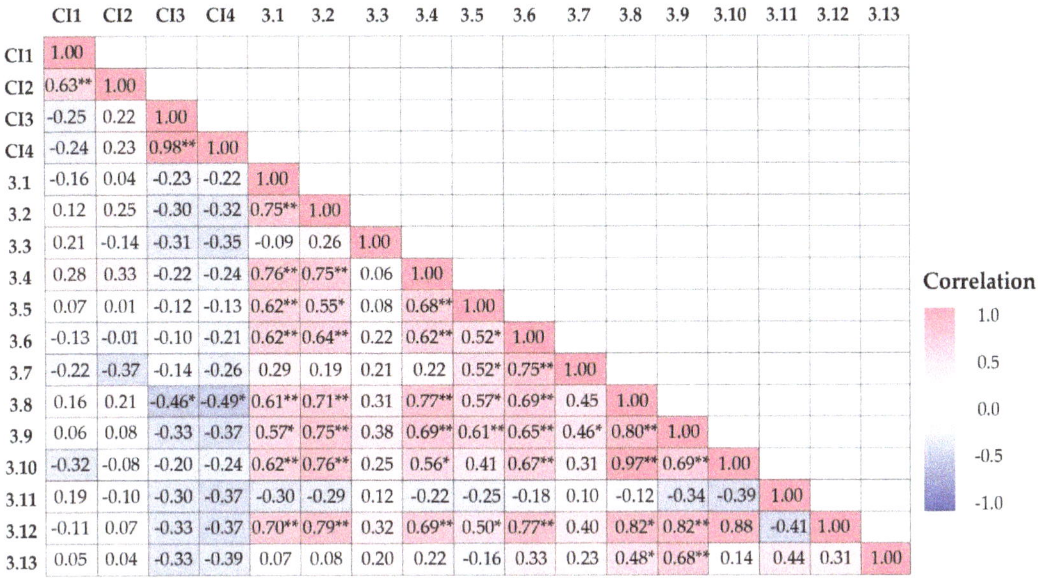

*p<0.05;**p<0.01.

Figure 8. A correlation analysis of the variables for the indicators in Cluster 3.

In Cluster 2, significant positive correlations were noted between CI2 (path length) and CI4 (coverage perimeter) (r = 0.72), indicating that longer paths correspond to larger coverage perimeters. Positive correlations between PR items 3.4 and 3.1 (r = 0.83) suggest that respondents in Cluster 2 associate restorativeness more strongly with feelings of escape. Negative correlations between perceived boredom (3.11) and restorativeness items like 3.12 (r = −0.66) further highlight contrasting views on the restorativeness of these spaces.

Cluster 3 demonstrated strong correlations between trajectory indicators CI3 (coverage area) and CI4 (coverage perimeter) (r = 0.98), consistent with findings in Cluster 1. Positive correlations were observed between PR items 3.6 and 3.8 (r = 0.71) (r = 0.71), suggesting a strong association between fascination and the desire for exploration. However, there were more varied correlations in Cluster 3, such as negative associations between some trajectory indicators and restorativeness items, indicating diverse perceptions of restorativeness within this cluster. This variability implies that while some visitors found the space fascinating and conducive to exploration, others may have perceived it as overstimulating or less compatible with their personal needs for restoration.

In summary, the correlation analysis highlights distinct psychological implications for each cluster based on their movement patterns and alignment with ART. Cluster 1 visitors benefit most from environments offering extent and fascination, while Cluster 2 visitors achieve restorativeness primarily through compatibility and psychological escape. Cluster 3 visitors experience mixed outcomes, with some benefiting from fascination, but others finding the environment less conducive to mental restoration. These insights suggest that tailoring the design and management of informal green spaces (IGSs) to meet these varied psychological needs could enhance their restorative potential for a broader range of users.

4. Discussion

This study investigates how visitors' spatio-temporal behaviors in informal green spaces (IGSs) influence their perceived environmental restorativeness. By integrating data from a Visitor-Employed Photography (VEP) survey and post-visit perceived restorativeness assessments, we identified distinct variations in visitor behaviors, explored how these patterns relate to perceived restorativeness, and proposed strategies to enhance the design and management of IGS for maximizing their restorative potential. The discussion is structured into three sections: differences in spatio-temporal behaviors among clusters, the relationship between these behaviors and perceived restorativeness, and strategic recommendations for IGS optimization.

4.1. Differences in Spatio-Temporal Behaviors Among Clusters

We identified three distinct clusters of visitor behaviors, each exhibiting unique spatio-temporal patterns. The application of Visitor-Employed Photography (VEP) proved to be particularly valuable in capturing spatio-temporal behaviors in IGSs, which are typically less structured and less familiar than traditional parks. VEP provides a means of documenting participants' subjective experiences, allowing for a deeper understanding of how specific landscape elements influence movement and engagement in informal green spaces. This method complements quantitative measures by offering a more nuanced perspective on visitor interactions with IGS.

Cluster 1 visitors demonstrated rapid, broad spatial exploration, covering large areas in a short time with the widest coverage area and perimeter, but the shortest total browsing time. This suggests a preference for open, navigable environments that encourage fast-paced exploration and provide a sense of freedom. These behaviors align with previous research indicating that expansive green spaces facilitate broader spatial movement and are perceived as more restorative due to their capacity to offer escape and fascination [12,31]. Furthermore, similar findings by Chang et al. [60] suggest that open and visually accessible spaces enhance feelings of safety and comfort, which may further encourage dynamic movement and exploration.

In contrast, Cluster 2 visitors displayed more focused, targeted exploration, with the shortest path length, smallest coverage area, and smallest perimeter, but moderate browsing time. This indicates a preference for specific landscape features within the IGS rather than general exploration, reflecting a more concentrated engagement with particular aspects of the environment. Visitors in this cluster are likely drawn to unique landscape elements or secluded spots that provide a sense of mental escape and stress relief. This pattern is consistent with Anderson and Minor [28], who emphasize the importance of distinctive environmental features in attracting and retaining visitor interest. Moreover, research by Fisher et al. [61] also suggests that specific landscape elements like water features enhance perceived restorativeness by promoting tranquility and escape from urban life.

Cluster 3 visitors engaged in slow, thorough exploration with the longest browsing time and path length, combined with a large coverage area and perimeter. This behavior suggests a more comprehensive engagement with the environment, where visitors take time to explore various facets of the space. Such visitors may seek diverse and rich environments that offer various stimuli, aligning with Sikorska et al. [15], who underscore the need for varied and immersive environments that promote prolonged engagement in green spaces. These findings contrast with studies that report quicker, less immersive interactions in similar settings [29], suggesting that certain landscape features may support deeper visitor engagement.

4.2. Relationship Between Behavior Patterns and Perceived Restorativeness

The correlation analysis between trajectory indicators and perceived restorativeness variables provides deeper insights into how spatio-temporal behaviors influence perceived restorativeness in IGSs. In Cluster 1, the strong positive correlation between coverage area (CI3) and coverage perimeter (CI4) suggests that larger, navigable spaces are associated with higher perceived restorativeness. Visitors in this cluster perceive environments that support extensive movement and exploration as more restorative. This finding is consistent with Kaplan's Attention Restoration Theory [31], which argues that environments offering fascination and the experience of being away enhance psychological restoration. Negative correlations between perceived boredom and fascination-related items further highlight contrasting perceptions, indicating that dynamic, stimulating environments are essential for restorativeness. Arnberger et al. [62] emphasize that environments perceived as fascinating are more likely to capture attention and promote mental recovery.

These results underscore the importance of spatial diversity within IGS in promoting restorativeness, especially for Cluster 1 visitors who prioritize movement and exploration. To cater to these users, urban planners should focus on maintaining larger, navigable areas with meandering paths and diverse vegetation to encourage dynamic engagement with the environment. This strategy is consistent with research highlighting the role of environmental diversity in enhancing user satisfaction [28].

In Cluster 2, significant correlations between path length and perceived psychological restorativeness—such as feelings of escape and relief from routine—suggest that for these visitors, the restorative experience is less dependent on the physical attributes of the space and more about the psychological benefits provided by specific interactions with features. Incorporating elements that promote mental escape, such as quiet zones or reflective spots, may be more effective for this group. These findings align with studies emphasizing the importance of secluded, contemplative spaces in enhancing psychological well-being [5,30]. Unlike research focusing on general green space use [13], the results here highlight the importance of strategically designed elements to maximize perceived restorativeness.

For visitors in Cluster 2, who prioritize psychological escape, urban planners should incorporate secluded zones and quiet areas to provide mental relief. Placing benches near water features or small, sheltered areas for quiet reflection could offer opportunities for solitude and contemplation, similar to strategies implemented in London's urban parks, where the inclusion of quiet areas enhances the restorative experience. Minimal interventions, such as directional signage, can help guide visitors without detracting from the informal, unstructured character of IGS.

Cluster 3 reveals a complex relationship between behavior patterns and perceived restorativeness, with strong associations between fascination and exploration but also some negative correlations between trajectory indicators and PR items. This indicates that visitors in this cluster have diverse expectations and preferences regarding restorativeness, with some favoring varied and stimulating environments and others preferring more straightforward, navigable spaces. These mixed results suggest that providing a balanced mix of features catering to different restorative needs can maximize the perceived benefits of IGS. This aligns with the findings by Anderson et al. [29], who found that a diverse range of green space attributes can cater to varied user preferences, enhancing overall satisfaction and perceived well-being.

To accommodate the varied preferences of visitors in Cluster 3, urban planners should design multi-functional spaces that allow for both active exploration and peaceful reflection. Walking trails that traverse varied landscapes and provide opportunities to pause for rest or reflection can effectively cater to this diversity, ensuring that the space meets a broader range of user needs.

4.3. Implications for Planning and Management of IGS

The findings of this study provide several strategic recommendations for urban planners and landscape architects, offering practical examples of how cities can integrate the insights from spatio-temporal behavior patterns to better design and manage IGS for diverse user needs.

First, cities should prioritize preserving spatial diversity within IGS, particularly when making updates or interventions [30,32]. Spatial diversity, including variations in terrain, vegetation, and open spaces, is essential for promoting exploration and enhancing the restorative qualities of IGS, especially for individuals who exhibit preferences for movement and discovery, as observed in Cluster 1. Rather than extensive modifications, an internal signage system could be implemented to highlight key features, such as wetland areas, main bird species, or different types of vegetation cover. For instance, the urban green space Amager Fælled in Copenhagen serves as an exemplary model by maintaining diverse vegetation and meandering paths that offer a range of natural landscapes, allowing visitors to engage with the environment freely. This approach effectively balances open space and natural diversity, thereby fostering an environment conducive to both dynamic exploration and psychological restoration.

Second, improving accessibility and navigability is crucial for those who value extensive exploration, with clear pathways and open spaces that encourage fluid movement throughout the IGS. Clear pathways and open spaces that facilitate fluid movement throughout IGS are important, but minimal management interventions—such as directional signage or limited vegetation control—can help maintain the informal nature of these spaces. Previous studies suggest that adding guiding signs along nearby roads can direct visitors to IGS [19,24]. Additionally, safety signs in wetland areas can enhance safety while maintaining the informal nature of the space. Cities such as London, where natural spaces like Hampstead Heath employ unobtrusive signage and minimal management, offer useful examples of how to balance navigability with maintaining the inherent wildness of the space. Similarly, safety signage in areas prone to hazards, such as wetlands, can enhance safety while preserving the informal nature of the environment. This supports findings from previous studies emphasizing the need for well-structured pathways to facilitate user engagement [24].

Third, incorporating restorative elements such as diverse vegetation, water features, and sheltered nooks can appeal to visitors seeking psychological escape and relaxation. The High Line in New York serves as an illustrative example of how strategic use of diverse vegetation and seating nooks can create restorative pockets within a densely urbanized area. For example, research by Fischer et al. [63] suggests that to enhance the landscape effects of IGS and provide a food source for wildlife, methods such as broadcasting flowers and grass seeds could be considered to help improve the ecological stability of IGS.

Finally, promoting awareness and inclusivity through community engagement and targeted programs is essential to increasing the familiarity and positive perception of IGS among all visitor groups. Public education initiatives aimed at highlighting the environmental and mental health benefits of IGSs, alongside community-led maintenance or stewardship programs, can strengthen the connection between residents and these spaces. This supports findings from Feltynowski et al. [41], which emphasize the role of community ties in the utilization and appreciation of green spaces.

5. Conclusions

This study highlights the importance of integrating spatio-temporal behavior data with perceived restorativeness assessments to enhance the understanding of the restorative potential of informal green spaces. The findings demonstrate that both physical attributes (such as coverage area and path length) and psychological experiences (such as feelings of escape and fascination) play crucial roles in shaping visitor experiences and their perceptions of restorativeness. Our findings reveal that visitors' engagement with IGS—characterized by varied spatial coverage, path length, and duration—significantly affects their perception of restorativeness through distinct behavioral patterns. These patterns operate through different dimensions of environmental experience, such as psychological escape, fascination, and exploration, offering critical insights for urban planners and landscape architects aiming to design IGSs that accommodate diverse user needs and enhance urban well-being.

From a planning and management perspective, the findings suggest that enhancing the spatial diversity, accessibility, and ecological features of IGSs can significantly improve their utilization and restorative potential. Preserving spatial diversity while adding minimal interventions, such as signage or clear pathways, could maintain the informal nature of these spaces while making them more navigable and appealing to a wider audience. Incorporating restorative elements like diverse vegetation, water features, and quiet spaces for relaxation could further enhance the psychological benefits of IGSs. Additionally, promoting public awareness and community involvement is essential for increasing engagement with these spaces. Targeted programs and community education can help improve the positive perception of IGSs, supporting their integration into urban environments as valuable resources for well-being.

Moreover, this study supports prior research on public attitudes toward IGSs, showing that while some individuals appreciate the unstructured nature of these spaces, others may have reservations about their safety or accessibility. This underlines the need for careful planning and management strategies to optimize IGS usage while maintaining their unique character. In particular, Visitor-Employed Photography (VEP) proved valuable in capturing authentic and reflective evaluations of IGSs, providing participants with the opportunity to engage with and assess these spaces in a more personal manner. This approach, while useful, also highlights potential biases, as it relies on subjective participant perspectives.

However, VEP, while effective in offering personal insights into participants' experiences, also introduces potential subjective biases. Participants may focus on elements that align with their preferences or prior beliefs, leading to selective representation of the IGS environment. These subjective perceptions may not fully reflect the broader user experience, which could influence the findings related to restorativeness. Future research could mitigate this bias by incorporating more objective measures, such as physiological data (e.g., heart rate or EEG) or additional mixed methods to balance personal impressions with more quantitative assessments.

Additionally, the sample primarily consisted of students within a narrow age range, which limits the generalizability of the findings. Future studies should expand to include a broader demographic, encompassing various ages, occupations, and cultural backgrounds, to capture a more comprehensive spectrum of perceptions and behaviors. In addition, the influence of seasonal and temporal variations on IGS usage remains underexplored; understanding these patterns could be pivotal in optimizing green space design for year-round engagement and restorative benefits. Future longitudinal studies could investigate how IGSs are utilized year-round, capturing the temporal changes and seasonal fluctuations that may affect their restorative potential. Understanding these patterns will be pivotal in optimizing green space design for year-round engagement and psychological benefits. Investigating these aspects further will refine strategies and provide more nuanced guidelines for maximizing the restorative potential of IGS in diverse urban contexts. Building on this, cross-cultural studies can help explore how different populations in diverse urban environments engage with IGS, adding a comparative dimension to the research. This could reveal cultural differences in the perception of restorativeness, as well as how environmental factors like vegetation, pathways, or facilities are valued across regions. Long-term monitoring of IGSs could also capture how perceptions of restorativeness evolve over time, particularly in the context of urban development and ecological changes.

In summary, this study provides a basis for future research to optimize IGS design and management strategies that integrate visitor behavior patterns and restorativeness perceptions, aiming to enhance user engagement and maximize restorative experiences in urban environments.

Author Contributions: Conceptualization, J.J. and H.W.; methodology, J.J. and S.C.; software, J.J. and S.C.; validation, J.J. and S.C.; formal analysis, J.J. and S.C.; investigation, J.J., R.M., S.C. and Z.Z.; resources, J.J. and H.X.; data curation, S.C.; writing—original draft preparation, J.J., R.M. and S.C.; writing—review and editing, J.J., R.M., H.X. and H.W.; visualization, S.C.; supervision, J.J.; project administration, J.J. and H.X.; funding acquisition, J.J. and H.X. All authors have read and agreed to the published version of the manuscript.

Funding: This research was funded by the National Natural Science Foundation of China (32301646), MOE (Ministry of Education in China) Project of Humanities and Social Sciences (23YJCZH089), and Soochow University—the Suzhou Yuanke(su-sy) Collaborative Innovation Center of Architecture and Urban Environment (SY2022003).

Data Availability Statement: The data presented in this study are available on request from the corresponding authors due to privacy.

Acknowledgments: The authors acknowledge all the participants and administrators in this study.

Conflicts of Interest: Author Hong Xu was employed by the company Suzhou Yuanke Ecological Construction Group. The remaining authors declare that the research was conducted in the absence of any commercial or financial relationships that could be construed as a potential conflict of interest.

References

1. Yuan, C.; Shan, R.; Zhang, Y.; Li, X.-X.; Yin, T.; Hang, J.; Norford, L. Multilayer urban canopy modelling and mapping for traffic pollutant dispersion at high density urban areas. *Sci. Total Environ.* **2019**, *647*, 255–267. [CrossRef] [PubMed]
2. Chen, S.; He, P.; Yu, B.; Wei, D.; Chen, Y. The challenge of noise pollution in high-density urban areas: Relationship between 2D/3D urban morphology and noise perception. *Build. Environ.* **2024**, *253*, 111313. [CrossRef]
3. Giridharan, R.; Ganesan, S.; Lau, S.S.Y. Daytime urban heat island effect in high-rise and high-density residential developments in Hong Kong. *Energy Build.* **2004**, *36*, 525–534. [CrossRef]
4. Chun, B.; Guldmann, J.-M. Spatial statistical analysis and simulation of the urban heat island in high-density central cities. *Landsc. Urban Plan.* **2014**, *125*, 76–88. [CrossRef]
5. Shen, H.; He, X.; He, J.; Li, D.; Liang, M.; Xie, X. Back to the Village: Assessing the Effects of Naturalness, Landscape Types, and Landscape Elements on the Restorative Potential of Rural Landscapes. *Land* **2024**, *13*, 910. [CrossRef]
6. Hunter, M.R.; Gillespie, B.W.; Chen, S.Y.P. Urban nature experiences reduce stress in the context of daily life based on salivary biomarkers. *Front. Psychol.* **2019**, *10*, 413490. [CrossRef]
7. Kajosaari, A.; Pasanen, T.P. Restorative benefits of everyday green exercise: A spatial approach. *Landsc. Urban Plan.* **2021**, *206*, 10397. [CrossRef]
8. Polajnar Horvat, K.; Ribeiro, D. Urban Public Spaces as Restorative Environments: The Case of Ljubljana. *Int. J. Environ. Res. Public Health* **2023**, *20*, 2159. [CrossRef]
9. Peschardt, K.K.; Stigsdotter, U.K. Associations between park characteristics and perceived restorativeness of small public urban green spaces. *Landsc. Urban Plan.* **2013**, *112*, 26–39. [CrossRef]
10. Lau, K.K.-L.; Yung, C.C.-Y.; Tan, Z. Usage and perception of urban green space of older adults in the high-density city of Hong Kong. *Urban For. Urban Green.* **2021**, *64*, 127251. [CrossRef]
11. Luo, S.; Patuano, A. Multiple ecosystem services of informal green spaces: A literature review. *Urban For. Urban Green.* **2023**, *81*, 127849. [CrossRef]
12. Rupprecht, C.D.; Byrne, J.A. Informal urban green-space: Comparison of quantity and characteristics in Brisbane, Australia and Sapporo, Japan. *PLoS ONE* **2014**, *9*, e99784. [CrossRef] [PubMed]
13. Chen, Z.; Yang, H.; Ye, P.; Zhuang, X.; Zhang, R.; Xie, Y.; Ding, Z. How does the perception of informal green spaces in urban villages influence residents' complaint sentiments? A machine learning analysis of Fuzhou City, China. *Ecol. Indic.* **2024**, *166*, 112376. [CrossRef]
14. Rupprecht, C.D.D.; Byrne, J.A.; Ueda, H.; Lo, A.Y. 'It's real, not fake like a park': Residents' perception and use of informal urban green-space in Brisbane, Australia and Sapporo, Japan. *Landsc. Urban Plan.* **2015**, *143*, 205–218. [CrossRef]
15. Sikorska, D.; Łaszkiewicz, E.; Krauze, K.; Sikorski, P. The role of informal green spaces in reducing inequalities in urban green space availability to children and seniors. *Environ. Sci. Policy* **2020**, *108*, 144–154. [CrossRef]
16. Rupprecht, C.D.D.; Byrne, J.A.; Garden, J.G.; Hero, J.-M. Informal urban green space: A trilingual systematic review of its role for biodiversity and trends in the literature. *Urban For. Urban Green.* **2015**, *14*, 883–908. [CrossRef]
17. Herman, K.; Ciechanowski, L.; Przegalińska, A. Emotional Well-Being in Urban Wilderness: Assessing States of Calmness and Alertness in Informal Green Spaces (IGSs) with Muse—Portable EEG Headband. *Sustainability* **2021**, *13*, 2212. [CrossRef]
18. Włodarczyk-Marciniak, R.; Sikorska, D.; Krauze, K. Residents' awareness of the role of informal green spaces in a post-industrial city, with a focus on regulating services and urban adaptation potential. *Sustain. Cities Soc.* **2020**, *59*, 102236. [CrossRef]
19. Sikorski, P.; Gawryszewska, B.; Sikorska, D.; Chormański, J.; Schwerk, A.; Jojczyk, A.; Ciężkowski, W.; Archiciński, P.; Łepkowski, M.; Dymitryszyn, I.; et al. The value of doing nothing—How informal green spaces can provide comparable ecosystem services to cultivated urban parks. *Ecosyst. Serv.* **2021**, *50*, 101339. [CrossRef]
20. Rupprecht, C.D.D. Informal Urban Green Space: Residents' Perception, Use, and Management Preferences across Four Major Japanese Shrinking Cities. *Land* **2017**, *6*, 59. [CrossRef]
21. Kim, M.; Rupprecht, C.D.D.; Furuya, K. Residents' Perception of Informal Green Space—A Case Study of Ichikawa City, Japan. *Land* **2018**, *7*, 102. [CrossRef]
22. Rupprecht, C.D.D.; Byrne, J.A.; Lo, A.Y. Memories of vacant lots: How and why residents used informal urban green space as children and teenagers in Brisbane, Australia, and Sapporo, Japan. *Child. Geogr.* **2016**, *14*, 340–355. [CrossRef]
23. Rupprecht, C.D.D.; Byrne, J.A. Informal urban greenspace: A typology and trilingual systematic review of its role for urban residents and trends in the literature. *Urban For. Urban Green.* **2014**, *13*, 597–611. [CrossRef]
24. Pietrzyk-Kaszyńska, A.; Czepkiewicz, M.; Kronenberg, J. Eliciting non-monetary values of formal and informal urban green spaces using public participation GIS. *Landsc. Urban Plan.* **2017**, *160*, 85–95. [CrossRef]
25. Biernacka, M.; Kronenberg, J.; Łaszkiewicz, E.; Czembrowski, P.; Amini Parsa, V.; Sikorska, D. Beyond urban parks: Mapping informal green spaces in an urban–peri-urban gradient. *Land Use Policy* **2023**, *131*, 106746. [CrossRef]

26. Stanford, H.R.; Hurley, J.; Garrard, G.E.; Kirk, H. Finding the forgotten spaces: Using a social-ecological framework to map informal green space in Melbourne, Australia. *Land Use Policy* **2024**, *141*, 107114. [CrossRef]
27. Gawryszewska, B.J.; Łepkowski, M.; Pietrych, Ł.; Wilczyńska, A.; Archiciński, P. The Structure of Beauty: Informal Green Spaces in Their Users' Eyes. *Sustainability* **2024**, *16*, 1619. [CrossRef]
28. Anderson, E.C.; Minor, E.S. Vacant lots: An underexplored resource for ecological and social benefits in cities. *Urban For. Urban Green.* **2017**, *21*, 146–152. [CrossRef]
29. Lai, K.Y.; Sarkar, C.; Sun, Z.; Scott, I. Are greenspace attributes associated with perceived restorativeness? A comparative study of urban cemeteries and parks in Edinburgh, Scotland. *Urban For. Urban Green.* **2020**, *53*, 126720. [CrossRef]
30. Malekinezhad, F.; Courtney, P.; bin Lamit, H.; Vigani, M. Investigating the mental health impacts of university campus green space through perceived sensory dimensions and the mediation effects of perceived restorativeness on restoration experience. *Front. Public Health* **2020**, *8*, 578241. [CrossRef]
31. Kaplan, S. The restorative benefits of nature: Toward an integrative framework. *J. Environ. Psychol.* **1995**, *15*, 169–182. [CrossRef]
32. Han, K.T. A review of self-report scales on restoration and/or restorativeness in the natural environment. *J. Leis. Res.* **2018**, *49*, 151–176. [CrossRef]
33. Hartig, T.; Evans, G.W.; Jamner, L.D.; Davis, D.S.; Gärling, T. Tracking restoration in natural and urban field settings. *J. Environ. Psychol.* **2003**, *23*, 109–123. [CrossRef]
34. Berman, M.G.; Jonides, J.; Kaplan, S. The cognitive benefits of interacting with nature. *Psychol. Sci.* **2008**, *19*, 1207–1212. [CrossRef]
35. Van den Berg, A.E.; Jorgensen, A.; Wilson, E.R. Evaluating restoration in urban green spaces: Does setting type make a difference? *Landsc. Urban Plan.* **2014**, *127*, 173–181. [CrossRef]
36. Van den Berg, A.E.; Joye, Y.; Koole, S.L. Why viewing nature is more fascinating and restorative than viewing buildings: A closer look at perceived complexity. *Urban For. Urban Green.* **2016**, *20*, 397–401. [CrossRef]
37. Kang, Y.; Kim, E.J. Differences of Restorative Effects While Viewing Urban Landscapes and Green Landscapes. *Sustainability* **2019**, *11*, 2129. [CrossRef]
38. Brun, M.; Di Pietro, F.; Bonthoux, S. Residents' perceptions and valuations of urban wastelands are influenced by vegetation structure. *Urban For. Urban Green.* **2018**, *29*, 393–403. [CrossRef]
39. Zhang, Z.; Jiang, M.; Zhao, J. The Restorative Effects of Unique Green Space Design: Comparing the Restorative Quality of Classical Chinese Gardens and Modern Urban Parks. *Forests* **2024**, *15*, 1611. [CrossRef]
40. Zhang, T.; Liu, J.; Li, H. Restorative Effects of Multi-Sensory Perception in Urban Green Space: A Case Study of Urban Park in Guangzhou, China. *Int. J. Environ. Res. Public Health* **2019**, *16*, 4943. [CrossRef]
41. Feltynowski, M.; Kronenberg, J.; Bergier, T.; Kabisch, N.; Łaszkiewicz, E.; Strohbach, M.W. Challenges of urban green space management in the face of using inadequate data. *Urban For. Urban Green.* **2018**, *31*, 56–66. [CrossRef]
42. Nguyen, P.-Y.; Astell-Burt, T.; Rahimi-Ardabili, H.; Feng, X. Green Space Quality and Health: A Systematic Review. *Int. J. Environ. Res. Public Health* **2021**, *18*, 11028. [CrossRef]
43. Lin, B.B.; Chang, C.-C.; Andersson, E.; Astell-Burt, T.; Gardner, J.; Feng, X. Visiting Urban Green Space and Orientation to Nature Is Associated with Better Wellbeing during COVID-19. *Int. J. Environ. Res. Public Health* **2023**, *20*, 3559. [CrossRef] [PubMed]
44. Chen, J.; Shaw, S.-L.; Yu, H.; Lu, F.; Chai, Y.; Jia, Q. Exploratory data analysis of activity diary data: A space–time GIS approach. *J. Transp. Geogr.* **2011**, *19*, 394–404. [CrossRef]
45. Wang, H.; Luo, S.; Furuya, K. What attracts tourists to press the shutter in cultural heritage tourism? An analysis of visitor-employed photography and visual attributes: A case study on Japan's Kairakuen Garden. *Tour. Recreat. Res.* **2022**, 1–17. [CrossRef]
46. Korpilo, S.; Virtanen, T.; Lehvävirta, S. Smartphone GPS tracking—Inexpensive and efficient data collection on recreational movement. *Landsc. Urban Plan.* **2017**, *157*, 608–617. [CrossRef]
47. Zhai, Y.; Baran, P.K.; Wu, C. Can trail spatial attributes predict trail use level in urban forest park? An examination integrating GPS data and space syntax theory. *Urban For. Urban Green.* **2018**, *29*, 171–182. [CrossRef]
48. Nitsche, P.; Widhalm, P.; Breuss, S.; Brändle, N.; Maurer, P. Supporting large-scale travel surveys with smartphones—A practical approach. *Transp. Res. Part C Emerg. Technol.* **2014**, *43*, 212–221. [CrossRef]
49. Deng, Y.; Wang, J.; Gao, C.; Li, X.; Wang, Z.; Li, X. Assessing temporal–spatial characteristics of urban travel behaviors from multiday smart-card data. *Phys. A Stat. Mech. Appl.* **2021**, *576*, 126058. [CrossRef]
50. Liu, W.; Wang, B.; Yang, Y.; Mou, N.; Zheng, Y.; Zhang, L.; Yang, T. Cluster Analysis of Microscopic Spatio-Temporal Patterns of Tourists' Movement Behaviors in Mountainous Scenic Areas Using Open GPS-Trajectory Data. *Tour. Manag.* **2022**, *93*, 104614. [CrossRef]
51. Yao, Q.; Shi, Y.; Li, H.; Wen, J.; Xi, J.; Wang, Q. Understanding the Tourists' Spatio-Temporal Behavior Using Open GPS Trajectory Data: A Case Study of Yuanmingyuan Park (Beijing, China). *Sustainability* **2021**, *13*, 94. [CrossRef]
52. Huang, X.; Li, M.; Zhang, J.; Zhang, L.; Zhang, H.; Yan, S. Tourists' spatial-temporal behavior patterns in theme parks: A case study of Ocean Park Hong Kong. *J. Destin. Mark. Manag.* **2020**, *15*, 100411. [CrossRef]
53. Beijing Municipal Bureau of Gardens and Green Spaces. *Code for the Design of Public Parks (GB51192-2016)*; China Architecture & Building Press: Beijing, China, 2016; pp. 6–7.
54. Jiang, J.; Wang, H.; Chen, S.; Xia, Z. Strategy for Optimization and Management of Informal Green Spaces in Shanghai Based on Landscape Perception Evaluation. *Landsc. Archit.* **2024**, *31*, 48–57. [CrossRef]

55. Planning and Construction Committee of the NPC. Master Plan of Suzhou Industrial Park (2012–2030). Available online: https://www.sipac.gov.cn/szghjswyh/ztgh/202003/0c430df700fa46be86677ce30ac899dd.shtml (accessed on 1 March 2024).
56. Liang, H.; Lin, Y.; Chen, Y.; Hao, X.; Gao, D.; Yu, N.; Li, Y.; Qiu, L.; Gao, T. The relationships among biodiversity, perceived biodiversity and recreational preference in urban green spaces—A case study in Xianyang, China. *Ecol. Indic.* **2023**, *146*, 109916. [CrossRef]
57. Fefer, J.P.; Hallo, J.C.; Dvorak, R.G.; Brownlee, M.T.J.; Collins, R.H.; Baldwin, E.D. Pictures of polar bears: Using visitor employed photography to identify experience indicators in the Arctic National Wildlife Refuge. *J. Environ. Manag.* **2020**, *269*, 110779. [CrossRef]
58. Li, Y.; Yang, L.; Shen, H.; Wu, Z. Modeling intra-destination travel behavior of tourists through spatio-temporal analysis. *J. Destin. Mark. Manag.* **2019**, *11*, 260–269. [CrossRef]
59. Jurisic, M.; Plascak, I.; Rendulic, Z.; Radocaj, D. GIS-Based Visitor Count Prediction and Environmental Susceptibility Zoning in Protected Areas: A Case Study in Plitvice Lakes National Park, Croatia. *Sustainability* **2023**, *15*, 1625. [CrossRef]
60. Chang, P.-J.; Ho, L.-C.; Suppakittpaisarn, P. Investigating the interplay between senior-friendly park features, perceived greenness, restorativeness, and well-being in older adults. *Urban For. Urban Green.* **2024**, *94*, 128273. [CrossRef]
61. Fisher, J.C.; Irvine, K.N.; Bicknell, J.E.; Hayes, W.M.; Fernandes, D.; Mistry, J.; Davies, Z.G. Perceived biodiversity, sound, naturalness and safety enhance the restorative quality and wellbeing benefits of green and blue space in a neotropical city. *Sci. Total Environ.* **2021**, *755*, 143095. [CrossRef] [PubMed]
62. Arnberger, A.; Schaper, S.; Eder, R.; White, M.P. Visitor mood, restorativeness and connectedness to nature across four unmanaged urban outdoor swimming sites of varying naturalness. *Urban For. Urban Green.* **2024**, *95*, 128312. [CrossRef]
63. Fischer, L.K.; von der Lippe, M.; Rillig, M.C.; Kowarik, I. Creating Novel Urban Grasslands by Reintroducing Native Species in Wasteland Vegetation. *Biol. Conserv.* **2013**, *159*, 119–126. [CrossRef]

land

MDPI

Article

A Study on the Public Perception of Sports Spaces Under Urban Overpasses from the Perspective of Age Differences

Ziyi Wen, Xiangming Luo, Xin Wang * and Haoran Liu †

School of Civil Engineering, Shandong University, Jinan 250001, China; 201900081027@mail.sdu.edu.cn (Z.W.); 202000201018@mail.sdu.edu.cn (X.L.); 2401213545@stu.pku.edu.cn (H.L.)

* Correspondence: xw@sdu.edu.cn

† Current address: College of Architecture and Landscape, Peking University, Beijing 100086, China.

Abstract: As China's urban development enters the era of stock optimization, the practice of transforming and utilizing spaces under urban overpasses is rapidly gaining momentum. Converting these underpass spaces into sports areas has emerged as a new form of creating public space. Understanding the perceptions of users from different age groups towards these underpass spaces holds significant guiding value for optimizing the design of such areas and improving the quality of service. Taking the Yanshan Interchange Lowline Park in Jinan as an example, this research applied methods of observation, interviews, questionnaires, and importance–satisfaction analysis (ISA) to investigate the activity preferences and the similarities and differences in the perceptions of spatial environment elements in underpass spaces among four age groups: children, youth, middle-aged adults, and the elderly. The findings indicate that different age groups exhibit varying degrees of sensitivity to spatial information, demand levels, and perceptual perspectives in underline parks, which result in distinct spatiotemporal distributions and spatial perception disparities when using the park. All the groups agree that the underpass sports space requires significant improvements in terms of comfort and safety. Based on this, this study proposes age-friendly urban space renewal strategies for spaces under elevated highways, focusing on addressing areas with lower satisfaction across all age groups. These strategies include optimizing the allocation of time, area, and activity types within activity spaces, enhancing the safety and comfort of activity areas, and enriching the cultural connotation and inclusivity of the space. This research provides a theoretical basis for optimizing and creating age-friendly or age-specific urban sports public spaces under elevated highways.

Keywords: sports space under overpasses; public perception; age differences; age-friendly; inclusive design

1. Introduction

As a means to alleviate ground-level traffic congestion and improve transportation efficiency, elevated highways have been widely constructed during China's rapid urbanization phase. However, this has led to the creation of underutilized, negative spaces beneath the bridges. Although these under-bridge spaces are a crucial component of urban public areas, they have often been neglected. Most of these spaces are used in an unplanned manner as parking lots, disorganized green areas, informal structures, or even dumping grounds. By 2020, China's overall urbanization rate exceeded 60%, signaling a transition in China's urban development from an era of growth to an era of optimizing existing urban assets [1]. In 2021, the government released the 14th Five-Year Plan and

2035 Vision for National Economic and Social Development [2], which proposed urban renewal initiatives to optimize urban spatial structures and enhance quality. As part of the focus on existing assets, the potential of under-bridge spaces garnered significant attention, spurring numerous design projects aimed at transforming these areas into recreational and sports spaces for public use. In 2019, the China Planning and Design Commission launched a specialized study on utilizing space under bridges. By analyzing the projected footprint of bridges, it was found that 402 hectares of under-bridge space was available, with approximately 244 hectares already in use and 158 hectares untapped [3]. Given the extensive scale of these spatial resources, cities across China have gradually implemented projects to revitalize and renovate under-bridge spaces. Guidelines and regulations have also been introduced, such as the Beijing Design Guidelines for the Utilization of Space Under Bridges, which was officially issued in January 2023 to promote the structured utilization of these spaces. Jinan, the capital of Shandong Province, boasts an abundance of groundwater resources and relies on its spring water as a primary source for drinking, which has also become a significant cultural emblem of the city. To protect these water resources, Jinan has refrained from developing an extensive subway system to address the growing traffic demand; therefore, elevated highways have become a key element of Jinan's transportation infrastructure. While these highways have accelerated urban development, they have also created large under-bridge spaces that are underutilized. Recognizing the negative impact of these spaces—such as their detrimental effects on the urban environment and community connectivity—the government has begun to transform them into spaces suitable for community activities, enhancing their social, economic, and cultural value and transforming them from negative to positive spaces. In October 2021, Shandong Province's first Lowline Park—the Yanshan Interchange Lowline Park in Jinan—was opened to the public. This park makes full use of the space under the Yanshan interchange, creating a high-quality, multifunctional area under the bridge that integrates cultural, recreational, and fitness functions for city residents. Following its opening, the park received enthusiastic responses from the public. However, research evaluating the use and effectiveness of such sports-oriented repurposed under-bridge spaces remains limited.

A review of the literature reveals that most studies related to elevated highways focus on assessing the potential for the sustainable utilization of under-bridge spaces, analyzing design approaches for these spaces, or examining the impact of elevated highways on the quality of life and spatial usage of nearby residents. For example, Zaki et al. (2023) used a multi-criteria decision-making (MCDM) approach to propose five key criteria—social benefits, economic feasibility, environmental impact, infrastructure enhancement, and urban identity—to systematically evaluate the sustainable use of under-bridge spaces. They applied decision-making trial and evaluation laboratory (DEMATEL) technology to assess the causal relationships between these criteria [4]. Ahmed et al. (2024) discussed various spatial configurations of under-bridge spaces and corresponding approaches to their redevelopment [5]. Lak et al. (2022) conducted surveys to examine residents' perceptions of changes in environmental quality and quality of life before and after elevated highway construction [6]. Meanwhile, a small number of scholars have focused on users of under-bridge spaces as their research subjects and studied user satisfaction with these spaces. For instance, Sarhan et al. (2023) reviewed prior evaluation standards for open spaces and developed a five-level evaluation model encompassing functional performance, accessibility, comfort, safety, pleasure, and meaning [7]. Through surveys, they collected data on satisfaction perceptions from 150 users of three types of under-bridge spaces (vegetation, commercial, and transportation-oriented) in the Heliopolis community in Egypt, examining the attributes that most influenced user satisfaction with these spaces through structural equation modeling. Nunma et al. (2021) used observation, visual questionnaires, and interviews to study children's needs for play spaces under elevated highways, analyzing their satisfaction with these spaces as playgrounds [8]. Although studies like these provide some insights, research on users of elevated bridge spaces remains limited, and there is a lack of studies on public perception of such spaces, especially of the sports-oriented urban under-bridge spaces. Moreover, there is no specialized post-occupancy evaluation system specifically designed to assess the use of these spaces.

Research on urban public space evaluation systems can provide important theoretical support for the establishment of post-occupancy evaluation systems for under-bridge spaces. Zhu (2003) proposed three key types of factors impacting the built environment—human, physical environmental, and social environmental factors—presenting a user-value-oriented subjective evaluation of the built environment from a methodological perspective [9]. Li (2014) suggested that accessibility, functional configuration, environmental quality, and the atmosphere of the space are core factors influencing the vitality of public spaces in residential areas, significantly affecting residents' user experience [10]. Cheng and Liu (2019) focused on the diversity of open spaces and commercial services, finding that increased diversity in public spaces enhances resident usage frequency, indirectly fostering neighborhood interaction and community trust [11]. Zhou and Lin (2020) divided the public space evaluation system into five dimensions—accessibility, safety, ecology, comfort, and culture—and conducted empirical research on public spaces along the banks of the Huangpu River [12]. Yan et al. (2021), based on social ecology theory, proposed that the accessibility, quality, safety, and user characteristics of community public spaces are essential factors in promoting social cohesion [13]. These studies on public space evaluation systems provide a theoretical foundation for establishing evaluation frameworks for spaces beneath elevated highways.

Recent population trends indicate that global cultural diversity is increasing across age groups, which highlights the growing need to address the demands of a multi-generational society. This shift underscores the importance of incorporating age-friendly urban design into planning practices [14,15]. Numerous studies have further emphasized the statistically significant differences among various age groups in their use of urban spaces [16]. These differences arise because each age group has distinct needs and perceptions when interacting with urban environments; however, current urban planning often fails to adequately accommodate vulnerable populations. For instance, due to their limited mobility, children primarily rely on parks within walking or cycling distance as spaces for independent physical activity [17]. Additionally, children tend to favor areas specifically designed for their age group, as such spaces offer both convenience and comfort. These areas also provide opportunities for social interaction with peers, which is essential for their development [18–20]. On the other hand, adults generally prefer visiting parks during quieter periods, such as in the afternoons or evenings, when these spaces are less crowded [21]. Moreover, safety and security play critical roles in influencing adults' choices of activity spaces for their children [22]. When selecting suitable spaces, adults tend to prioritize factors such as the landscape features and the cleanliness of facilities over mere proximity [5,20,23,24]. As individuals age, their needs and preferences for urban spaces evolve. Elderly users, for example, often gravitate towards less physically demanding activities due to declining health [25–29]. Furthermore, noise pollution tends to have a more pronounced negative impact on the elderly, making tranquil and accessible environments particularly important for this group [30]. These findings collectively highlight the pressing need for urban planners to consider the diverse and evolving requirements of different age groups. By fostering inclusive and accessible public spaces, cities can better address the growing societal emphasis on equity and improve the overall quality of life for all generations.

In this context, how different age groups perceive urban public spaces has become an important research topic. Researchers have explored public perceptions of various types of urban public spaces from an age-based perspective, proposing improvement strategies accordingly. For example, Wu (2021) and colleagues evaluated the usage of community public spaces from an age-differentiated perspective, providing optimization strategies for different age groups [31]. Wong (2009) investigated the behavioral patterns of Hong Kong residents visiting urban parks and the perception differences among age groups [32]. Zhai et al. (2021) studied the level of attention given to various elements within 15 urban parks by users of different ages, investigating the correlation between environmental factors in public spaces and the age of respondents [33]. Elin et al. (2020) conducted walking interviews to examine how different age groups used urban parks and their differing management needs [18]. Mak et al. (2019) examined the characteristics, behaviors, and preferences of urban park users and analyzed the influence of various demographic factors on usage patterns by calculating a factor importance index [24]. Li (2019) analyzed the needs of various age groups in urban public spaces from an all-ages perspective, summarizing the theories of inclusive communities and design, and suggested strategies to create age-friendly public spaces [34]. Moore et al. (2015) examined the environmental barriers in children's activity spaces and proposed strategies to enhance their usability and inclusiveness for children with disabilities [35]. Some scholars have also conducted research from the perspective of facility needs among users of different age groups in urban public spaces. For example, Rivera et al. (2021) explored adolescents' perspectives on important park features influencing park visits, park-based physical activities, and social interactions. Through walking interviews, they summarized the characteristics of the "ideal park" [36]. Liao (2022), taking an intergenerational perspective, analyzed the service facility needs among different age groups and proposed an integrated planning approach for community public service facilities. This approach, through policy guidance, spatial planning, functional design, and operational management, aims to integrate multi-generational facilities and promote the development of age-friendly communities [37]. Lindberg et al. (2015) utilized the System for Observing Play and Recreation in Communities (SOPARC) to investigate the usage intensity of different urban green spaces and recreational facilities among age groups, revealing mismatches between the intended target groups and actual users of certain facilities [19]. However, while research on public space perception from an all-age perspective has flourished, most studies focus on community public spaces and urban parks, with relatively limited research on urban spaces under elevated highways. Therefore, there is a pressing need to supplement this area of study.

This study adopts an all-ages perspective, using Jinan's Yanshan Overpass Lowline Park as a case study of a sports-oriented public space under an elevated highway. Through observation, interviews, and importance–satisfaction analysis (ISA), it examines the distribution differences of user groups across age stages and their relationship with spatial elements. The study analyzes variations in the importance and satisfaction ratings of different spatial elements among users of various ages. Based on these findings, it further proposes strategies to enhance the space for each age group, aiming to provide insights for designing age-friendly sports spaces under elevated highways in the future.

2. Materials and Methods

2.1. Research Object

This study focuses on the Yanshan Interchange Lowline Park and its surrounding green spaces. The Yanshan Interchange Lowline Park (hereafter referred to as the "Lowline Park") is located in Jinan, Shandong Province, China (Figure 1), situated in a warm temperate continental monsoon climate zone with hot, rainy summers and cold, dry winters.

The Lowline Park is a sports and recreational area for residents, transformed from an underutilized space beneath an elevated highway. Officially opened on 1 October 2021, it is the first high-quality under-bridge activity space in Shandong Province, initiated by the government, designed by professional architects, and maintained by municipal authorities. The park integrates cultural, recreational, and fitness functions. The surrounding environment of the Lowline Park is complex, including residential, commercial, and office areas. It spans 550 m in length and 50 m in width, with a design area of approximately 28,000 square meters. Divided by the East Second Ring Road, it comprises two distinct functional zones, East and West. The eastern zone includes seven functional areas: a skate park, a multi-sport court (including fields for soccer, badminton, and flexible ball games), a street basketball court, a ping pong court, a 24 h urban library, a performance plaza, and a parking lot. The western zone retains the original parking lot and four gateball courts, with redesigned pedestrian paths, green belts, and parking spaces, along with upgrades to the existing gateball facilities (Figure 2).

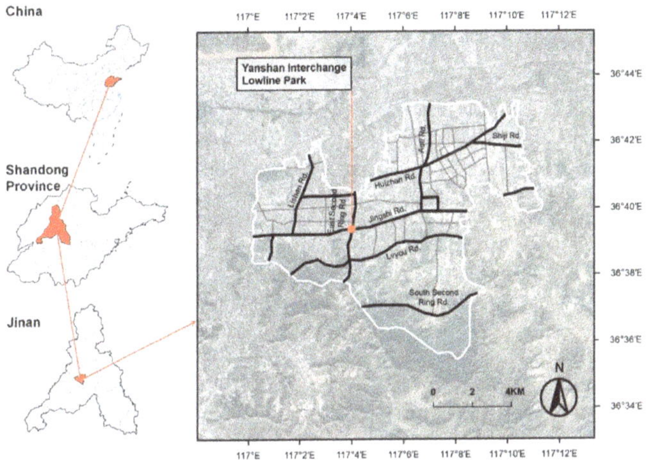

Figure 1. Geographic location of Jinan.

Figure 2. Site plan of the Yanshan Interchange Lowline Park and its functional spaces.

2.2. *Research Methods*

2.2.1. Establishment of an Evaluation System

This study selected evaluation indicators based on theories from behavioral architecture, Maslow's hierarchy of needs [38], *the Chinese Walkable City evaluation standards* [39], the classification of outdoor activities into three types in *Contact and Space* [40], Rancière's theory of aesthetic fairness in "distribution of the sensible" [41], and other urban public space evaluation systems. Five primary criteria were established—"Accessibility", "Safety", "Comfort", "Aesthetic Appeal", and "Cultural Value". These were further refined into 26 secondary evaluation indicators tailored to the specific context of Lowline Park, thereby forming a post-use evaluation system for public spaces under urban elevated highways (Table 1).

Table 1. Post-occupancy evaluation system for public spaces under urban elevated highways and respondents' perceptions of indicator levels.

Criterion Level	Indicator Level	Perception of Indicator Level
Accessibility	1. Convenience of Crossing Roads to Access the Park [42]	The convenience for pedestrians crossing roads from surrounding streets to access the park, including the availability of sufficient safety features for crossing (e.g., crosswalks, traffic lights) and whether the design of crossing paths is reasonable (e.g., appropriate timing of traffic lights, adequate distance, location, and number of crosswalks).
	2. Convenience of Entrance Locations [43]	Whether the geographical distribution of park entrances is reasonable, positioned in high-traffic or easily accessible areas to allow for convenient and quick entry and exit.
	3. Number of Park Entrances [43]	Whether the total number of park entrances is sufficient to minimize detours, allowing visitors to choose the nearest entrance based on their location or preference.
	4. Visual Recognition of Park Entrances [43,44]	Whether entrances have clear signage or directional markers that allow pedestrians to easily identify entrance locations, ensuring they can smoothly find their way into the park.
	5. Connectivity Between Different Functional Areas within the Park [43,45]	The ease of moving between different functional areas within the park, including whether there are convenient pathways and adequate signage that enable people to navigate smoothly between zones.
	6. Number of Nearby Bus Stops [44]	Whether there is a sufficient number of nearby bus stops to ensure high connectivity between the park and public transportation, making it easy for visitors without private vehicles to reach and leave the park by bus.
	7. Location Distribution of Nearby Bus Stops [44]	Whether the locations of bus stops are reasonably distributed in relation to the park entrances. Conveniently located stops can provide easy connections for park visitors, allowing them to quickly walk to the park entrance after disembarking, thus enhancing the park's accessibility for public transportation users.
Safety	8. Security Management in the Park [45]	Whether there are adequate security personnel, patrol frequency, and monitoring measures in place to ensure visitor safety in the park, and if current management practices effectively prevent criminal activity.
	9. Safety of Surrounding Traffic for Pedestrians and Non-Motorized Vehicles [45]	Condition of surrounding roads, including vehicle speed, traffic volume, signage, pedestrian crossings (traffic lights, crosswalks), and designated pathways for pedestrians and non-motorized vehicles to avoid conflicts with motor vehicles.
	10. Lighting Conditions in the Park [44,45]	Adequacy and brightness of lighting installations (especially at night), ensuring visitors can clearly see paths to avoid safety issues from insufficient lighting or glare; assessment of areas lacking lighting or where lighting is overly bright.
	11. Condition of Sports Facilities [44]	Maintenance and usability of sports facilities in the park (e.g., tables and courts); whether they are in good condition, conveniently accessible, and adequately maintained to reduce accident risks for users.
	12. Safety of Internal Facilities [42,44]	Design, structural stability, and regular maintenance of internal facilities, such as benches, railings, and children's play equipment, with warning signs around damaged items to prevent injuries due to facility failure.
	13. Overall Safety of Internal Park Areas [42,44]	General safety conditions of different areas within the park, including pathway smoothness, anti-slip treatment, absence of hazardous terrain, and clear surfaces free from debris or trash that could cause visitors to slip or fall.

Table 1. *Cont.*

Criterion Level	Indicator Level	Perception of Indicator Level
Comfort	14. Sound Environment in the Park [44]	Whether there is noise in the park that disturbs visitors' activities.
	15. Air Quality in the Park [44]	Whether the air quality in the park is good, free from unpleasant odors or pollution, ensuring a healthy and comfortable experience for visitors.
	16. Distribution of Activity Areas [46]	Whether the placement of various activity areas is reasonable, such as quiet areas being separated from lively ones, and whether the layout is convenient for visitors to find desired areas without interference. Presence of clear signage or landmarks.
	17. Number of Activity Areas [46]	Whether the number of activity areas is sufficient to meet the needs of visitors of different ages and interests, providing enough space for people to enjoy activities without overcrowding.
	18. Size of Activity Areas [46]	Whether the size of activity areas is spacious enough to accommodate the expected number of users, avoiding overcrowding and ensuring comfort.
	19. Accessibility Facilities [45,46]	Availability and adequacy of accessibility facilities, such as ramps at entrances, clear signage for accessible paths, tactile paving, non-slip pathways, accessible seating, handrails, accessible restrooms, and activity facilities for those with mobility limitations (e.g., wheelchair users and the elderly).
	20. Placement of Garbage Bins and Seating [46]	Whether garbage bins and seating are sufficient, well located, well maintained, easily accessible, and do not obstruct walking paths or other activities, helping maintain cleanliness and provide resting areas.
	21. Placement of Bookstores and Restrooms [45,46]	Ease of locating bookstores and restrooms within the park, with adequate capacity, good maintenance, and convenient access without interfering with other activity areas.
Aesthetics	22. Aesthetic Quality of Plant Landscape Design [45]	Whether the arrangement of plants in the park is visually appealing and well maintained, including the variety of plant species, seasonal color changes, layering, and overall harmony of plant layouts.
	23. Aesthetic Quality of Pathway Paving [45]	Whether the park's ground surfaces are attractive; whether the paving materials, patterns, and color combinations of park pathways harmonize with the environment, creating a visually appealing appearance.
	24. Aesthetic Quality of Color Coordination in the Park [42,45]	Whether the color coordination of the overall park landscape is harmonious and pleasing, including the colors of the ground, walls, fences, bridge pillars, lighting, plants, facilities, and decorations, enhancing the visitors' visual experience.
	25. Aesthetic Quality of Landscape Features and Structures [42,45]	Whether the appearance of sculptures, lighting fixtures, seating, and other structures in the park is attractive and well-coordinated with the surroundings, including the design aesthetics of landscape elements (e.g., sculptures and water features).
Cultural Relevance	26. Reflection of Jinan's Historical and Cultural Characteristics in the Park [44]	Whether the design and layout of the park incorporate elements of Jinan's history and cultural identity, including traditional architectural styles, statues of historical figures, cultural symbols, and local plants. This also includes decorative items related to historical and cultural characteristics, such as sculptures, slogans, and murals, and whether cultural events are held within the park.

2.2.2. Data Collection Method

This study primarily employed field observations, questionnaire surveys, and interviews to conduct a spatial perception study of Lowline Park users across four age groups: children and adolescents (5–18 years), young adults (19–35 years), middle-aged adults (36–60 years), and seniors (over 60 years) (Table 2).

The study consisted of a total of three rounds of surveys. The first round, a preliminary survey, was conducted in October 2022 with four on-site investigations of the Lowline Park and random interviews with 50 users. Based on the characteristics of the under-bridge space and users' actual experiences, the evaluation system was adjusted accordingly, and a questionnaire and structured interview outline were developed. The second round of surveys took place in November 2022, with six on-site investigations. A questionnaire survey was conducted with 56 users from different age groups (17 children and adolescents, 20 youths, 10 middle-aged users, and 9 elderly users). Based on the responses and feedback, ambiguous or unclear questions in the questionnaire were revised repeatedly. Additionally, structured interviews were conducted with 20 users (5 from each age group: children and adolescents, youth, middle-aged, and elderly) to gather data on user satisfaction, needs, and suggestions for improvement, aiming to investigate the specific needs of different age groups (interview guide is attached as Table S11).

Table 2. Definitions and characteristics of four different age groups.

Population Category	Age Group	Characteristics
Children and adolescents	5–18 years *	This group is in a stage of physical growth, developing cognitive abilities, and learning foundational knowledge. Most individuals in this group are students and have not yet entered society independently.
Youth	19–35 years	This group is in the stage of learning professional skills, building social connections, and integrating into society. Most individuals in this stage complete the transition from family roles to social roles.
Middle-aged	36–60 years	This group is in the stage of building families and advancing careers. Most individuals carry responsibilities for both family and society at this stage, and they pay attention to the education of their children and the care of their parents.
Elderly	60 years and above	This group experiences a decline in physiological functions. Most individuals in this stage focus less on career development and are more involved in family and social activities.

* In this study, only children aged 5 and above with a certain level of comprehension ability were surveyed. Generally, children over the age of 5 can understand the questions and provide responses with the assistance of their guardians.

The third and final round of surveys was conducted from March to April 2023, covering four weekdays and five non-working days. The study began with on-site observations, where the researcher recorded users' activity times, locations, and behavior types in Lowline Park between 7 a.m. and 9 p.m. This information was documented primarily through photography. Through these observations and photographic documentation, the study tracked the usage patterns and user behaviors at different times of the day, while also identifying existing issues within the park. Additionally, a questionnaire survey was conducted, asking users from different age groups to rate the importance and satisfaction of five aspects of Lowline Park: accessibility, safety, aesthetics, cultural value, and comfort. The study first collected data on crowd numbers and spatial distribution for each area. Then, stratified random sampling and questionnaires were administered to users of various age groups at different locations in the park. The first section of the questionnaire collected basic demographic information, including age, gender, occupation, education level, transportation method to the park, distance traveled, frequency of visits, and time spent at the Lowline Park. The second section applied a five-point Likert scale, assigning values from "Very Unimportant" to "Very Important" and "Very Dissatisfied" to "Very Satisfied", with scores ranging from 1 to 5, allowing respondents to rate the importance and satisfaction of the Lowline Park's evaluation indicators. For the final survey, the questionnaire language was adjusted to be comprehensible for the general public, who may not be familiar with academic terminology related to park landscapes and spatial elements. For example, "security management in the park" was rephrased as "Are there enough security personnel and surveillance measures in the park, and do current management practices effectively prevent crime?" (questionnaire is attached as Table S10). For respondents with reading difficulties, such as seniors and children, the answers were obtained through interviews. Structured interviews were conducted for children and seniors who had trouble understanding the questionnaire, with their ratings and other basic information recorded immediately after each interview.

Children aged 5–7 have a short attention span, a limited vocabulary, and restricted comprehension abilities. With the assistance of guardians, researchers use simple and straightforward language to explain the questions in the questionnaire, providing clarification for more abstract questions or unfamiliar vocabulary until the children understand the meaning of the questions and provide an answer.

Informed consent was obtained from all participants, ensuring their confidentiality and the right to withdraw from the study at any time without any consequences. For children, parental consent was obtained, and their participation in interviews and questionnaire completion was supervised by their guardians. All information collected was stored anonymously and used exclusively for research purposes, in strict adherence to ethical guidelines to protect participants' privacy and safety (Table 3).

Table 3. Survey types and number of respondents in three rounds of research.

	Survey Period	Duration	Survey Type	Number of Participants				
				Children	Youth	Middle-Aged	Elderly	Total
First Survey (Pre-survey)	October 2022	4 days	Interview	/	/	/	/	50
Second Survey (Pre-survey)	November 2022	6 days	Questionnaire	16	19	11	6	52
			Interview	5	5	5	5	20
Third Survey (Formal Survey)	March–April 2023	9 days	Questionnaire	38	37	17	17	109

In the formal survey, the sample size was determined using the Slovin's formula. According to statistical data from the management of low-line parks, the average monthly reservation count during fee-charging periods is 1999. In this study, the total number of reservations during both fee-charging and free periods was estimated by multiplying the reservation count by 1.8. This calculation yielded an average daily park visitor count of approximately 131, and subsequently, the total number of park users over 9 days was estimated to be about 1061. Setting the effective population size (N) in Slovin's formula as 1061 and assuming a sampling error (e) of 0.1, this indicates a 90% confidence level that the sample size accurately represents the population. This process determined the minimum required sample size (n) to be 106 respondents.

$$n = \frac{N}{1 + Ne^2} = \frac{1061}{1 + (1061 \times 0.1^2)} = 91$$

Equation: Basic Formula for Ratios [47].

In the third round of the formal survey, a total of 116 questionnaires were distributed, with 109 valid responses collected, resulting in an effective response rate of 93.96%. The sample size was sufficient to evaluate user satisfaction and perceived importance of low-line parks during the 9-day survey period. Stratified sampling was applied to each age group, and snowball sampling was employed at each site to randomly select participants from each age group for the survey. The proportion of respondents from each age group was no less than 10% of their respective population. Ultimately, the collected questionnaires included responses from 38 children and adolescents, 37 young adults, 17 middle-aged adults, and 17 seniors. The statistical analysis software SPSS 26.0 was used for data processing, and the reliability of the questionnaire was tested using Cronbach's alpha, resulting in a reliability coefficient of 0.880, indicating high reliability.

2.2.3. Importance–Satisfaction Analysis (ISA)

Importance–satisfaction analysis (ISA) is a method that has been widely applied in recent years to evaluate urban public spaces. By analyzing the importance and satisfaction ratings users assign to different spatial characteristics, ISA assesses users' spatial perceptions and plots these scores within corresponding quadrants to determine management priorities for improvements across various elements of urban public spaces [48]. The importance–satisfaction method assessed the perceived importance and satisfaction with the spatial attributes of Lowline Park among users from four age groups: children and adolescents, young adults, middle-aged adults, and seniors. By analyzing the differences between these groups, this approach provides valuable insights to planners and designers, helping them understand public perceptions of the park's spatial features. Based on survey questionnaires, users evaluated importance and satisfaction regarding accessibility, safety, and 26 other indicators. According to the evaluation results, a two-dimensional coordinate system was constructed, with the x- and y-axes representing the importance and satisfaction of each evaluation indicator, respectively. The evaluation results for each indicator fall into one of four quadrants, corresponding to different improvement priorities: Quadrant I represents high importance–high satisfaction, indicating that the service level of these indicators should be maintained. Quadrant II represents low importance–high satisfaction, suggesting that the design resources for these indicators can be redirected to improve other services. Quadrant III represents low importance–low satisfaction, indicating that the service of these indicators need not be a design priority. Quadrant IV represents high importance–low satisfaction, suggesting that the service level of these indicators should be prioritized for optimization and improvement.

3. Results

The demographic characteristics of the respondents reflect the diversity of Lowline Park users. Analyzing these characteristics shows that among the 109 respondents, children aged 5–18 comprised the largest proportion (34.9%), followed by young adults aged 19–35 (33.9%). Both middle-aged adults aged 40–46 and seniors over 60 each accounted for 15.5% (Tables S1–S5). In terms of gender distribution, male users (73.4%) outnumbered female users (26.6%), indicating that Lowline Park is more appealing to men, with a lack of amenities specifically catering to female users. Regarding occupation, students made up the largest portion of respondents at 46.7%, followed by retirees, blue-collar workers, and white-collar workers, together accounting for 32%. Homemakers and healthcare professionals were less represented, making up only 4.5% of park visitors. For education, the majority of respondents held a bachelor's or associate degree or lower, comprising 88% of the sample. In terms of usage frequency, 55.9% of respondents visited Lowline Park at least once a week, with 37.6% visiting daily and 18.3% visiting several times a month. Considering the time spent per visit, 66.9% of respondents stayed between 1 and 3 h. Looking at the respondents' residence, 47.7% were local residents from nearby neighborhoods. This suggests that Lowline Park not only serves nearby community residents but also has a broader reach.

The study found significant differences in the spatiotemporal distribution and spatial perceptions of Lowline Park usage across different age groups. In terms of spatiotemporal distribution, each age group displayed unique patterns of temporal and spatial clustering in their use of the park. Regarding spatial perceptions, all age groups placed high importance on safety and comfort, though satisfaction levels for these aspects were low. Additionally, there were notable differences in how each age group perceived secondary spatial environment indicators.

3.1. *Spatial Usage and Differences Among User Groups*

In this study, user activity times were divided into early morning (before 8:00), morning (8:00–11:00), midday (12:00–15:00), afternoon (16:00–19:00), and evening (after 20:00). There are noticeable spatiotemporal differences in the distribution of users from different age groups in Lowline Park. The general age-density ranking of users, from highest to lowest, is as follows: elderly, youth, children, and middle-aged adults. Each age group is present in the park at various times, with elderly and children primarily active in the early morning and morning, youth most densely present at midday and in the afternoon, and middle-aged adults mainly frequenting the park in the afternoon and evening. The activity preferences vary significantly among age groups, with a diverse range of activities. The elderly make versatile use of the gateball courts, engaging in activities like diabolo spinning, gateball, square dancing, and shuttlecock kicking. Additionally, they play ping pong in the designated area and perform music in the performance plaza. Youth and children share similar activity types, mostly utilizing sports facilities for activities such as basketball, soccer, badminton, skateboarding, and reading or studying at the 24 h urban library. However, there are distinctions between the two groups: children might play games on the gateball courts, while youth often prefer performing street dance or engaging in street workouts in those areas. Middle-aged adults mainly visit the gateball courts in the evening. Some join the elderly in square dancing, while others participate in fitness activities with the youth, with many middle-aged users accompanying their children for leisure and recreation (Table 4).

Table 4. Activity records of users in the Yanshan Interchange Lowline Park.

Time	Users	Activities	Location
Before 8:00	Elderly	Morning exercise	gateball courts, performance plaza
8:00–12:00	Elderly	Diabolo, gateball, ping pong, musical performance	gateball courts, performance plaza, ping pong court
	Children	Basketball, reading, skateboarding, games	street basketball court, 24 h urban library, skate park, gateball courts
12:00–15:00	Youth	Basketball, football, badminton, skateboarding, reading	street basketball court, multi-sport court, skate park, 24 h urban library
	Children	Basketball, football, badminton, skateboarding, reading, games	street basketball court, multi-sport court, skate park, 24 h urban library, gateball courts
15:00–20:00	Elderly	Diabolo, gateball, ping pong, musical performance	gateball courts, performance plaza, ping pong court
	Youth	Basketball, football, badminton, skateboarding, reading	street basketball court, multi-sport court, skate park, 24 h urban library
	Children	Basketball, football, badminton, skateboarding, reading, games	street basketball court, multi-sport court, skate park, 24 h urban library, gateball courts
After 20:00	Elderly	Square dancing, shuttlecock kicking	gateball courts
	Middle-aged	Running, square dancing, fitness	gateball courts
	Youth	Running, street dance, fitness, reading	gateball courts, 24 h urban library
	Children	Games	gateball courts

Observations and interviews revealed two key issues with how users utilize the space. First, there is a mismatch between the opening hours of activity areas and peak usage times for different user groups, resulting in a significant supply–demand gap. The usage of activity areas varies considerably over time. In the early morning and morning, the western gateball courts are mainly used by the elderly and children, while the eastern areas, such as the badminton courts, soccer field, and skatepark, have low attendance and utilization. In the afternoon, youth and children mainly engage in activities in the eastern area, with elderly users staying in the west, leading to higher utilization of the eastern spaces during this period. In the evening, activity spaces with facilities in the eastern area close, causing nearly all users to congregate in the western area, where crowding exceeds the area's capacity and occasionally leads to conflicts among groups. Secondly, there is a mismatch between the types of activity spaces and the needs of the user groups. Currently, most spaces are designed for middle-aged and young adults, while the groups who use the park more frequently and for longer periods are children and the elderly. For children, the available activities lack sufficient safety features and appeal. For the elderly, most areas with sports facilities are not suitable for those unable to engage in intense physical activity. The gateball courts, frequently used by the elderly, lack comfort, safety, and accessible design features. As a result, there is a severe shortage of designated spaces suited to the physical and psychological needs of children and elderly users.

3.2. Importance–Satisfaction Evaluation and Differences in Perception of Lowline Park Elements Across Age Groups

3.2.1. Overall Importance–Satisfaction Analysis of Users

The average ratings for the overall importance of and satisfaction with Lowline Park's spatial environment elements reflect users' varying perceptions of accessibility, safety, aesthetics, comfort, and cultural relevance. These differences reveal users' needs to some extent. The average importance ranking of primary indicators is as follows: Safety (4.34) > Comfort (4.18) > Cultural Relevance (4.16) > Accessibility (3.96) > Aesthetics (3.94). The average satisfaction ranking is as follows: Aesthetics (4.16) > Accessibility (4.13) > Cultural Relevance (4.08) > Safety (4.07) > Comfort (3.89). A comparison reveals that the average importance ratings for safety, comfort, and cultural relevance are higher than their satisfaction ratings, indicating that users have higher expectations and needs for these indicators. The largest difference is seen in comfort (difference of 0.29), suggesting that the comfort of the under-bridge space needs improvement. The satisfaction ratings for aesthetics and accessibility are higher than their importance ratings, indicating that users are relatively satisfied with these aspects of the under-bridge space.

According to the matrix analysis, safety and cultural relevance are in Quadrant I, meaning that users value these indicators highly, and they meet public expectations, so these strengths should be maintained. Comfort is in Quadrant IV, indicating that users consider this indicator important, but the actual experience falls short of their expectations, suggesting further optimization is needed. Aesthetics and accessibility fall within Quadrant II, indicating that users find these indicators less important but are highly satisfied with them, implying no immediate need for improvement (Figure 3).

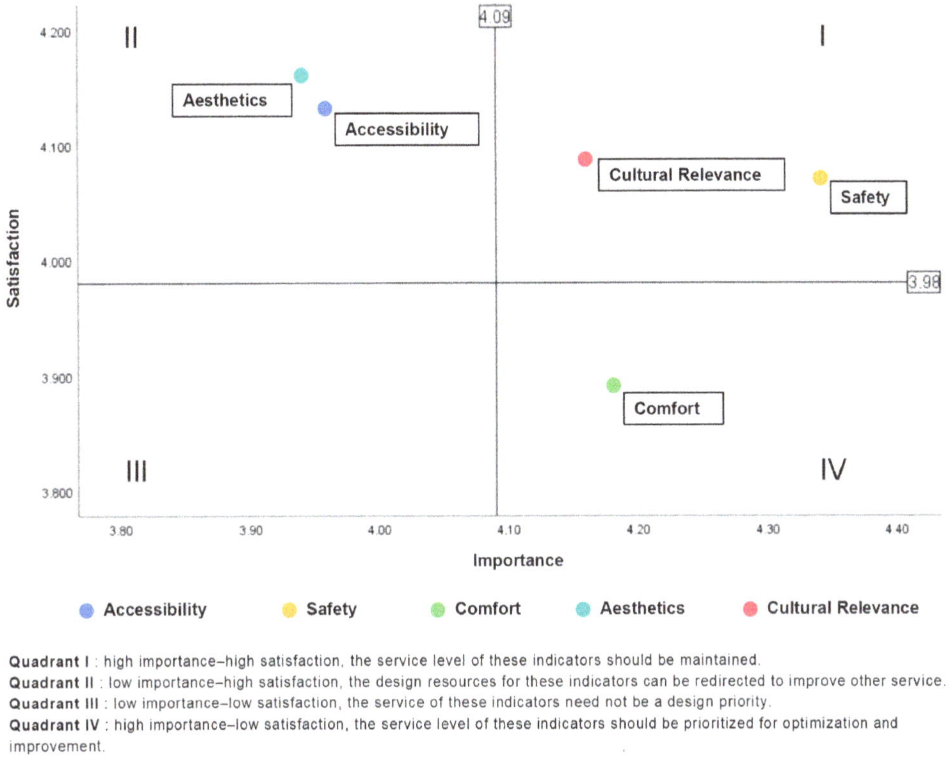

Figure 3. Importance–satisfaction matrix of users' overall perception of Lowline Park.

3.2.2. Importance–Satisfaction Evaluation for the Children's Group

The overall satisfaction mean for the children's group is slightly higher than the importance mean, but the difference is small, indicating that children are generally satisfied with Lowline Park (Figure 3). The importance ranking of the primary indicators is as follows: Comfort (4.14) > Safety (4.10) > Aesthetics (4.04) > Cultural Relevance (4.00) > Accessibility (3.82). The satisfaction ranking is as follows: Cultural Relevance (4.75) > Aesthetics (4.18) ≥ Accessibility (4.18) > Comfort (3.92) > Safety (3.91). A comparison of the differences between importance and satisfaction shows that the satisfaction means for Comfort (difference of 0.23) and Safety (difference of 0.18) are both lower than their respective importance means. The largest gap is in Comfort, indicating an urgent need for improvement in this area, particularly in terms of enhancing comfort in the under-bridge space for children.

The urgency of improvement measures was prioritized through ISA. According to the matrix (Figure 4), the following indicators fall within Quadrant I: convenience of crossing the road to access the park (1), convenience of entrance locations (2), internal safety of the park areas (13), air quality in the park environment (15), location of activity areas within the park (16), number of activity areas (17), size of activity areas (18), and the arrangement of facilities such as trash bins and seating (20). This indicates that the children's group has a high level of recognition and satisfaction with these eight indicators, which should be maintained. The following indicators fall within Quadrant IV: lighting conditions in the park (10), safety of internal facilities (12), sound environment in the park (14), and the arrangement of the book bar and restrooms (21). This suggests that the children's group considers these four indicators to be of high importance, but the actual functionality does not meet their expectations. These should be prioritized for targeted improvement in subsequent design efforts.

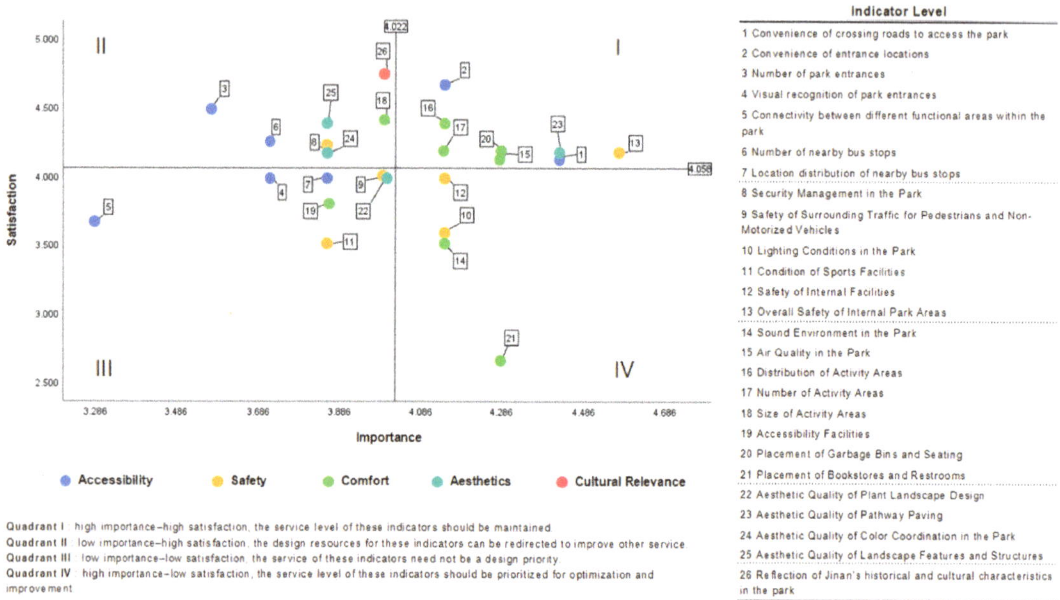

Figure 4. Importance–satisfaction matrix of secondary evaluation indicators for the children's group in Lowline Park.

A paired sample t-test was conducted to assess whether there was a statistically significant difference between the importance and satisfaction ratings among the children's user group for Lowline Park, identifying potential areas for improvement in specific evaluation factors (Table S6). According to the paired sample test results, aside from three aesthetic indicators (park pavement design, landscape color scheme, and landscape features and structures) and two accessibility indicators (convenience of crossing the road and number of entrances), which showed positive directionality (t-value is positive), all the other indicators displayed negative directionality (the t-value is negative). There was a particularly significant difference in the safety and comfort indicators ($p < 0.05$), with the satisfaction ratings being significantly lower than the importance ratings. This indicates that children perceive a need for improvements in comfort and safety, especially regarding the number and size of activity areas, the park's sound and air quality, and the safety of internal spaces.

3.2.3. Importance–Satisfaction Evaluation for the Youth Group

The overall mean satisfaction score for the youth group is significantly lower than the mean importance score, suggesting that several evaluation criteria fail to meet the youth group's needs. The mean importance ranking of the primary indicators is as follows: Cultural Relevance (4.48) > Safety (4.37) > Comfort (4.27) > Accessibility (4.11) > Aesthetics (3.92). For satisfaction, the ranking is: Safety (4.13) > Aesthetics (4.11) > Accessibility (4.09) > Cultural Relevance (4.06) > Comfort (3.93). A comparison of importance and satisfaction scores reveals that satisfaction scores for cultural relevance, comfort, safety, and accessibility all fall below their corresponding importance scores. The largest gaps are observed in cultural relevance (a difference of 0.42) and comfort (a difference of 0.34), indicating an urgent need for improvement in these areas of the under-bridge space for the youth group.

The matrix diagram (Figure 5) shows that the following indicators fall within Quadrant I: pedestrian safety in the surrounding traffic environment (9), lighting conditions in the park (10), current usage of sports tables and courts (11), safety of internal facilities (12), availability of accessible facilities (19), arrangement of trash bins and seating (20), and the reflection of Jinan's historical culture and urban characteristics within the park (26). This placement indicates that the youth group has high recognition and satisfaction with these seven indicators, and their strengths should be maintained. However, the convenience of entrance locations (2), visibility of park entrances (4), safety within park areas (13), air quality in the park (15), number of activity areas (17), size of activity areas (18), and the arrangement of the book bar and restrooms (21) fall within Quadrant IV. This indicates that the youth group considers these seven indicators to be of high importance, but the actual functionality does not meet their expectations, and targeted improvements should be made in future designs.

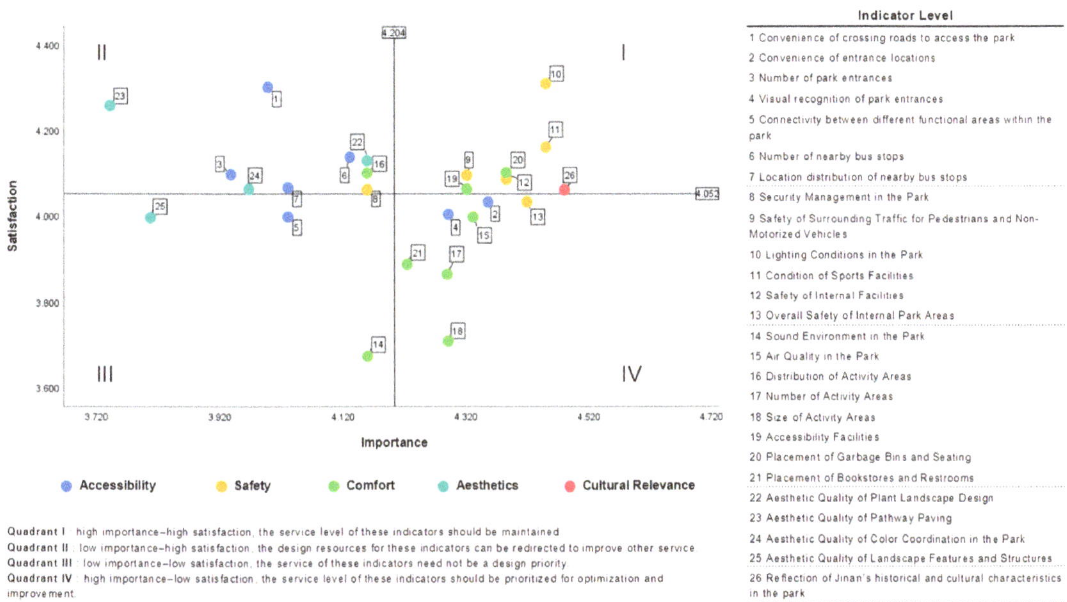

Figure 5. Importance–satisfaction matrix of secondary evaluation indicators for the youth group in the Lowline Park.

A paired-sample *t*-test was conducted to evaluate whether there is a statistically significant difference between the importance and satisfaction ratings for Lowline Park among youth users (Table S7). According to the paired-sample test results, aside from two aesthetic indicators (paving of park pathways and landscape elements/structures) and one accessibility indicator (location of nearby bus stops), which showed positive results (positive *t*-values), all other indicators displayed negative results (negative *t*-values). There is a particularly significant difference in safety and comfort indicators ($p < 0.05$), with satisfaction scores markedly lower than importance scores. This suggests that the youth group perceives a need for improvements in comfort and safety, especially regarding the current condition of sports facilities (such as tables and courts), internal site safety, and the park's sound environment and air quality.

3.2.4. Importance–Satisfaction Evaluation for the Middle-Aged Group

The overall mean satisfaction score for the middle-aged group is slightly lower than the mean importance score, suggesting that some evaluation criteria do not fully meet this group's needs. The mean importance ranking for the primary indicators is as follows: Safety (4.69) > Comfort (4.22) > Cultural Relevance (4.17) > Accessibility (4.05) > Aesthetics (3.92). In terms of satisfaction, the ranking is: Aesthetics (4.21) > Accessibility (4.19) > Safety (4.18) > Comfort (3.83) > Cultural Relevance (3.82). Comparing the differences between importance and satisfaction, it is evident that satisfaction scores for safety, comfort, and cultural relevance fall below their importance scores, with relatively large gaps. This indicates an urgent need for improvements in these aspects of the under-bridge space to better serve the middle-aged group's expectations.

The matrix diagram (Figure 6) shows that the convenience of entrance locations (2), security management within the park (8), current usage of sports tables and courts (11), internal facility safety (12), safety within park areas (13), and accessibility features (friendliness toward people with limited mobility) (19) fall within Quadrant I. This indicates that the middle-aged group has high recognition and satisfaction with these six indicators, and their strengths should be maintained. Indicators in Quadrant IV include pedestrian and non-motorized vehicle safety in the surrounding traffic environment (9), lighting conditions in the park (10), air quality in the park environment (15), the number of activity areas in the park (17), the size of activity areas (18), and the arrangement of the book bar and restrooms (21). These placements suggest that the middle-aged group considers these six indicators highly important, but their actual functionality does not meet expectations. Therefore, targeted improvements should be made in future designs.

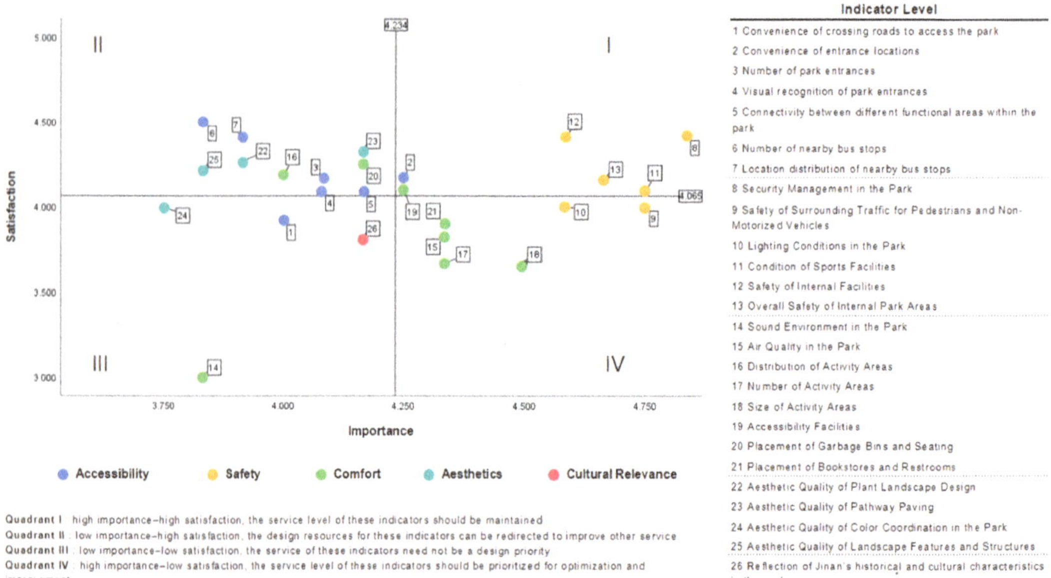

Figure 6. Importance–satisfaction matrix of secondary evaluation indicators for the middle-aged group in the Lowline Park.

A paired-sample *t*-test was used to assess whether there is a statistically significant difference between the importance and satisfaction ratings of middle-aged users for Lowline Park, identifying potential areas for improvement in specific evaluation elements (Table S8). According to the paired-sample test results, aside from four aesthetic indicators (park paving, landscape colors, landscape elements, and built structures) and two accessibility indicators (ease of crossing streets and number of entrances) showing positive responses (with positive t-values), all other indicators exhibited negative responses (with negative *t*-values). Notably, there was a significant difference in safety and comfort indicators ($p < 0.05$), with satisfaction ratings significantly lower than importance ratings. This indicates that middle-aged users perceive room for improvement in aspects such as comfort and safety, especially regarding the number and size of activity areas, sound and air quality, and safety within internal park spaces.

3.2.5. Importance–Satisfaction Evaluation for the Elderly Group

The overall mean satisfaction score for the elderly group is slightly lower than the mean importance score, though the difference is minimal, indicating that the elderly group is generally satisfied with Lowline Park. Nonetheless, a few evaluation criteria still fall short of fully meeting their needs. The importance ranking for the primary indicators is as follows: Safety (4.45) > Comfort (4.02) > Accessibility (3.86) > Cultural Relevance (3.82) > Aesthetics (3.77). For satisfaction, the ranking is: Safety (4.21) > Aesthetics (4.15) > Accessibility (4.02) > Cultural Relevance (3.81) > Comfort (3.78).

A comparison of the importance and satisfaction scores reveals that satisfaction levels for safety and comfort are notably lower than their importance scores, with relatively large gaps. This suggests a pressing need for improvements in these areas of the under-bridge space to better address the elderly group's expectations.

According to the matrix diagram (Figure 7), the following indicators fall within Quadrant I, meaning that the elderly group has high recognition and satisfaction with these seven factors: security management of the park (8), safety of pedestrians and non-motorized vehicles crossing nearby roads (9), lighting conditions in the park (10), safety of internal facilities (12), safety of internal park areas (13), number of activity areas in the park (17), and the arrangement of trash bins and seating facilities (20). These strengths should be maintained. The following indicators fall within Quadrant IV, indicating that the elderly group considers them important, but the actual functionality does not meet their expectations: convenience of the location of park entrances (2), current usage of sports tables and courts (11), air quality in the park environment (15), availability of accessible facilities (friendly for people with limited mobility) (19), and the arrangement of the book bar and restrooms (21). These aspects should be improved in future designs. The following indicators fall within Quadrant III, meaning both importance and satisfaction are relatively low: number of park entrances (3), visibility of park entrances (4), ease of moving between different areas in the park (5), size of activity areas in the park (18), aesthetic quality of landscape features and structures in the park (25), and reflection of Jinan's historical culture and urban characteristics in the park (26). These factors do not need to be prioritized in the future development of under-bridge spaces.

A paired-sample *t*-test was conducted to assess whether there is a statistically significant difference between importance and satisfaction ratings for Lowline Park among elderly users (Table S9). According to the paired-sample test results, aside from accessibility indicators (excluding "convenience of entrance locations") and aesthetic indicators, which showed positive results (positive t-values), all other indicators displayed negative results (negative t-values). There is a particularly significant difference in comfort indicators ($p < 0.05$), with satisfaction scores significantly lower than importance scores. This suggests that the elderly group perceives a need for improvement in comfort, particularly regarding air quality in the park environment, accessibility features, and the arrangement of the book bar and restrooms.

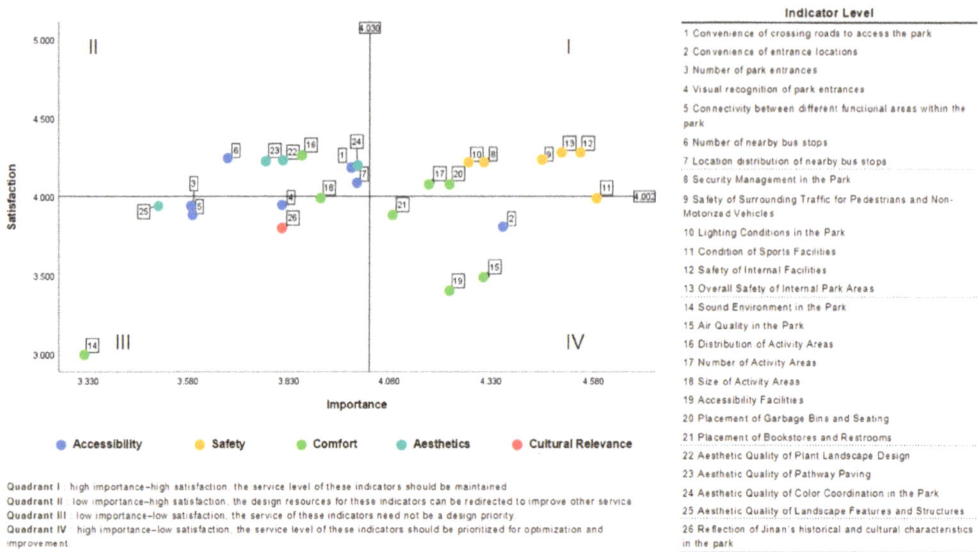

Figure 7. Importance–satisfaction matrix of secondary evaluation indicators for the elderly group in the Lowline Park.

3.2.6. Comparative Analysis of Importance–Satisfaction Evaluations Across Age Groups

Based on the differences in the mean importance–satisfaction values, there are both similarities and differences in how the four groups perceive the evaluation indicators. For children, the importance ratings for safety and comfort are higher than the satisfaction ratings. Youth rate accessibility, safety, comfort, and cultural relevance as more important than satisfactory. The middle-aged and elderly groups also rate safety, comfort, and cultural relevance as more important than satisfactory. This further analysis shows that all the age groups have high expectations for improvements in safety and comfort. Youth have additional expectations for improvements in accessibility, while youth and middle-aged and elderly groups express expectations for improvements in cultural relevance, which are particularly strong for youth and middle-aged users. No group shows a notable demand for improvements in aesthetics (Table 5).

Table 5. Primary indicators with importance scores higher than satisfaction scores in spatial perception for different groups and the differences.

Category of User Groups	Difference Between Importance and Satisfaction Scores for Primary Indicators				
	Accessibility	Safety	Comfort	Aesthetics	Cultural Relevance
Overall	/	0.27	0.29	/	0.08
Children	/	0.18	0.23	/	
Youth	0.02	0.24	0.34	/	0.42
Middle-aged	/	0.51	0.39	/	0.35
Elderly	/	0.24	0.24	/	0.01

The quadrant placement of the 26 secondary indicators also shows both similarities and differences among the four groups. All the groups agree on the importance of the arrangement of the book bar and restrooms (21) and express dissatisfaction with them (Table 6).

Table 6. Importance–satisfaction comparative analysis for children, youth, middle-aged, and elderly groups (Quadrants I, II, III, and IV represent the importance–satisfaction quadrants).

Criteria Level	Indicator Level	Category of User Groups				Inconsistency between Importance and Satisfaction
		Children	Youth	Middle-Aged	Elderly	
Accessibility	1. Convenience of crossing the road to access the park	I	II	III	II	Important—Children; Not important—Youth, Middle-aged, Elderly Satisfied—Children, Youth, Elderly; Not satisfied—Middle-aged
	2. Convenience of the location of park entrances	I	**IV** *	I	**IV**	Important—All; Satisfied—Children, Middle-aged; Not satisfied—Youth, Elderly
	3. Number of park entrances	II	II	II	III	/
	4. Visibility of park entrances	III	**IV**	II	III	Important—Youth; Not important—Children, Middle-aged, Elderly Satisfied—Middle-aged; Not satisfied—Children, Youth, Elderly
	5. Ease of moving between different areas within the park	II	III	II	III	/
	6. Number of nearby public transportation stops	II	II	II	II	/
	7. Location of nearby public transportation stops	III	II	II	II	/
Safety	8. Security management of the park	II	II	I	I	Important—Middle-aged, Elderly; Not important—Children, Youth Satisfied—All
	9. Safety of pedestrians and non-motorized vehicles in the surrounding traffic environment	III	I	**IV**	I	Important—Youth, Middle-aged, Elderly; Not important—Children Satisfied—Youth, Elderly; Not satisfied—Children, Middle-aged
	10. Lighting conditions in the park	**IV**	I	**IV**	I	Important—All; Satisfied—Youth, Elderly; Not satisfied—Children, Middle-aged
	11. Current usage of sports tables and courts in the park	III	I	I	**IV**	Important—Youth, Middle-aged, Elderly; Not important—Children Satisfied—Youth, Middle-aged; Not satisfied—Elderly, Children
	12. Safety of internal park facilities	**IV**	I	I	I	Important—All Satisfied—Youth, Middle-aged, Elderly; Not satisfied—Children
	13. Safety of internal park areas	I	**IV**	I	I	Important—All; Satisfied—Children, Middle-aged, Elderly; Not satisfied—Youth

Table 6. *Cont.*

Criteria Level	Indicator Level	Category of User Groups				Inconsistency between Importance and Satisfaction
		Children	Youth	Middle-Aged	Elderly	
Comfort	14. Noise levels in the park environment (whether there is noise disturbance)	IV	III	III	III	Important—Children; Not important—Youth, Middle-aged, Elderly Not satisfied—All
	15. Air quality in the park environment	I	**IV**	**IV**	**IV**	Important—All; Satisfied—Children; Not satisfied—Youth, Middle-aged, Elderly
	16. Location of activity areas within the park	I	II	II	II	Important—Children; Not important—Youth, Middle-aged, Elderly Satisfied—All
	17. Number of activity areas in the park	I	**IV**	**IV**	I	Important—All; Satisfied—Children, Elderly; Not satisfied—Youth, Middle-aged
	18. Size of activity areas in the park	II	IV	IV	III	Important—Youth, Middle-aged; Not important—Children, Elderly Satisfied—Children; Not satisfied—Youth, Middle-aged, Elderly
	19. Availability of accessible facilities in the park (friendliness towards people with limited mobility)	III	I	I	**IV**	Important—Youth, Middle-aged, Elderly; Not important—Children Satisfied—Youth, Middle-aged; Not satisfied—Children, Elderly
	20. Arrangement of trash bins, seating, and other facilities in the park	I	I	II	I	Important—Children, Youth, Elderly; Not important—Middle-aged Satisfied—All
	21. Arrangement of the book bar and restrooms in the park	**IV**	**IV**	**IV**	**IV**	Important—All; Not satisfied—All
Aesthetics	22. Aesthetic quality of the park's plant landscape arrangement	III	II	II	II	/
	23. Aesthetic quality of the park's pavement design	I	II	II	II	Important—Children, Not important—Youth, Middle-aged, Elderly; Satisfied—All
	24. Aesthetic quality of the park's landscape color combinations	II	II	III	II	/
	25. Aesthetic quality of landscape features, structures, and other facilities in the park	II	III	II	III	/
Cultural Relevance	26. Reflection of Jinan's historical culture and urban characteristics in the park	II	I	III	III	Important—Youth; Not important—Children, Middle-aged, Elderly Satisfied—Children, Youth; Not satisfied—Middle-aged, Elderly

* Quadrant IV in this table has been bolded. It represents high importance and low satisfaction, suggesting that the service level of these indicators should be prioritized for optimization and improvement.

Children consider the lighting conditions (10), safety of internal facilities (12), and sound environment (14) important but unsatisfactory. Compared to other groups, they are less sensitive to air quality (15) but place more importance on the convenience of crossing the road to the park (2), sound environment (14), location of activity areas (16), and aesthetic quality of pavement design (23).

Youth find the convenience of entrances (2), visibility of park entrances (4), safety of internal areas (13), air quality (15), number of activity areas (17), and size of activity areas (18) important but unsatisfactory. Compared to other groups, they place more importance on the visibility of park entrances (4) and the representation of Jinan's history and culture (26).

Middle-aged users consider pedestrian and non-motor vehicle safety in the surrounding traffic (9), lighting conditions (10), air quality (15), and the number (17) and size (18) of activity areas important but unsatisfactory. Compared to the other groups, they are less sensitive to the visibility of park entrances (4), express lower satisfaction with the convenience of crossing the road to the park (1), and view facilities like trash bins and seating (20) as less important.

Elderly users consider the convenience of entrance locations (2), current condition of sports facilities like tables and courts (11), air quality (15), and accessibility facilities (19) important but unsatisfactory. Compared to the other groups, they have the highest demand for improvements in accessibility facilities (19).

3.3. *Usage Needs and Differences Across Age Groups*

Using a structured interview guide, this study conducted interviews with 20 users (5 from each age group: children, youth, middle-aged, and elderly), gathering evaluations of Lowline Park and more specific suggestions for future improvements from different age groups. Overall, the main improvement requests from park users focused on enhancing windbreak facilities, removing entry fees, increasing child-friendly areas, and adjusting the timing of nighttime lighting. Adult users expressed clear concerns about the fee system and lack of windbreak facilities, highlighting that Lowline Park's comfort level in winter is inadequate and that the fairness of public amenities needs improvement. Both middle-aged and elderly groups emphasized the importance of making the park more child-friendly, indicating a strong public demand for family-friendly public spaces.

Specific to different age groups, people of different ages have varying needs for Lowline Parks: children hope for more activity spaces and facilities tailored to them, such as additional play areas, cooling facilities, and convenient commercial services to meet their needs for social interaction and activities. The youth group is concerned with the park's comfort, convenience, and affordability, suggesting longer lighting hours, a simplified reservation process, and more windbreak facilities to improve the climate adaptability of sports areas. Middle-aged users, in addition to free space use and more windbreak and rest facilities, particularly requested more child-friendly activity areas to create a family-friendly experience. The elderly group emphasized simplifying management, reducing fees, increasing restroom facilities, and using greenery and color improvements to reduce noise and enhance aesthetic appeal (Table 7).

Table 7. Suggestions for Lowline Park from different age groups in interviews.

Age Group	Age	Gender	Suggestions for Lowline Park
Children	7	Male	In summer, it's hot; hope there are more areas for children to play and run, and also a playground.
	5	Female	Hope there are more places for children to play to attract more children to come, as there are no friends in the park now.
	11	Female	Hope there are more places for children to play.
	5	Male	Hope a shop will open.
	11	Female	Turn on the lights earlier in winter.
Youth	20	Female	Turn on the lights earlier.
	22	Male	Turn on the lights earlier.
	22	Male	1. Complex reservation system, low accessibility for elderly. 2. The badminton court is greatly affected by windy weather.
	31	Female	Mop more frequently; there is a lot of dust, and the ground is very slippery.
	34	Male	Issue with charging fees.
Middle-aged	54	Male	Issue with charging fees.
	47	Female	1. Wind-blocking facilities needed. 2. Issue with charging fees.
	39	Male	Increase wind-blocking resting areas.
	37	Male	Expand the area of the basketball court.
	52	Female	Add wind-blocking facilities.
Elderly	64	Female	1. Complex management. 2. Issue with charging fees. 3. Restrooms are too far away.
	69	Male	Add wind-blocking facilities to courts for light sports like badminton and table tennis.
	62	Male	1. Issue with charging fees. 2. Wind-blocking facilities and ball containment enclosures are essential for table tennis courts. 3. Add channels for supervision and feedback.
	65	Female	Noise is high; should increase tall, dense trees on the park edges near roads to reduce noise and decrease traffic distraction for those exercising.
	64	Male	Road paving should be more vibrant and colorful, suitable for children.

In summary, feedback from all age groups reflects a comprehensive demand for the multifunctionality, convenience, and comfort of Lowline Park, indicating that park planning needs to balance the specific needs of different age groups to enhance overall satisfaction and inclusivity in urban public services.

4. Discussion

4.1. Analysis of the Reasons for Perceptual Differences Among the Four Groups

Our observations reveal that the activity spaces in Lowline Park are primarily used by the intended target age groups, and no specific area is used by all age groups in similar proportions. The Theory of Planned Behavior suggests that physiological needs, psychological motivations, social environment, and facility suitability influence behavioral choices [49], while an individual's attitude toward a behavior, subjective norms (perceived social pressure from significant others), and perceived behavioral control (such as barriers and facilitators) determine their intent to use a space [50]. Regarding the design and planning of Lowline Park, there is a degree of inclusive, activity-oriented intent, though it remains at a superficial level without deep analysis of the complex needs of each age group. For example, a skatepark for children, soccer and basketball courts for young adults, a ping pong court for middle-aged users, and a gateball court for seniors were included, resulting in clustering by age group. The spatiotemporal differences in Lowline Park usage among age groups are mainly influenced by physical and psychological factors, as well as available leisure time. Elderly and children tend to visit in the morning, largely due to their flexible schedules, early bedtime habits, and the belief among seniors that morning exercise benefits health [40]. The accompanying travel patterns of elderly and children also contribute to higher numbers of child users in the morning. Young adults and middle-aged users, on the other hand, prefer visiting the park in the afternoon or evening when they are less busy. This trend is also supported by the study of Reichert et al. (2007) [21].

The results show that child users have high expectations for improved safety and are dissatisfied with the lighting conditions, the safety of internal facilities, and the noise levels. Safety is a fundamental internal driving force for children's development, playing a crucial role in their cognitive, social, emotional, and self-evaluation growth [51]. In urban public spaces, a sense of security is the foundation for children to engage in social activities. Therefore, in urban public spaces under elevated highways, child users primarily seek safety features such as sufficient lighting and security facilities. In addition to "safety and security", the "basic services" of a site are also important indicators for child users in perceiving an environment as friendly [23]. Therefore, child users tend to prefer convenient and comfortable facilities and desire designated areas suited to their age group, where they can interact with peers. This conclusion is also supported by other studies [18–20]. This also explains why child users have higher expectations for improved comfort. The study results further indicate that child users value the ease of crossing to the park, the acoustic environment, the distribution of activity areas, and the aesthetic quality of the pathways more than the other age groups. This can be explained by the fact that children are in a stage of physical development, where they are generally more sensitive to noise [52], which may account for their prioritization of the comfort of the acoustic environment. Additionally, due to their relatively lower physical strength and lower endurance, having to cross multiple roads with traffic lights before entering the park may make children feel both exhausted and unsafe. As a result, the road accessibility of the park is highly valued by child users. When children participate in activities, a safe, convenient, and aesthetically pleasing environment enhances their experience and sense of engagement, making the location of activity areas particularly important to them. For children, simple and refreshing colors directly stimulate healthy psychological development [53]. As the park under the bridge is covered by an elevated highway, with a predominantly gray roof, children—due to their limited life experience and unique perspective—are more easily attracted by the park's facilities and vivid visual colors than by elements like landscape features or urban culture.

In addition to having high expectations for improvements in safety and comfort, the youth users also have expectations for enhanced accessibility and cultural relevance. They express dissatisfaction with the convenience and visibility of park entrances, the safety of the internal areas, the air quality, and the quantity and size of the activity areas. They place greater importance on the visibility of entrances and elements within the park that reflect Jinan's historical and cultural characteristics. According to data from the National Bureau of Statistics, young people are generally optimistic and cheerful, and their environmental needs are primarily driven by self-fulfillment and relaxation [54]. Therefore, compared to the aesthetic quality of the environment, young people are more concerned with the quality of the activity spaces and the physical and mental enjoyment these spaces provide [36]. In addition, they have a stronger preference for sports that are competitive and cathartic [55].

They place greater emphasis on the accessibility of spaces, such as the visibility of entrances, as well as the safety and comfort of sports fields, plazas, and leisure facilities. Additionally, compared to other age groups, youth users are often not satisfied with just the basic functionality of activity spaces, they also value the cultural atmosphere created in public spaces [56]. As a result, they have a stronger desire for urban public spaces under elevated highways to convey cultural meaning.

In addition to having high expectations for improvements in safety and comfort, middle-aged users also have higher expectations for cultural relevance. They express dissatisfaction with the pedestrian and non-motorized vehicle safety in terms of the surrounding traffic, the lighting conditions, the air quality, and the quantity and size of the activity areas. Compared to other groups, they are less satisfied with the ease of crossing roads. Middle-aged users often face the greatest life and work pressures, have limited leisure time, and have a strong sense of family responsibility [57]. Previous studies have shown that more than one-third of middle-aged park users are married with children, exhibiting a significant pattern of family-accompanied visits [23,24,58]. When visiting parks under elevated highways, they typically bring their families along, with some even needing to use strollers [59]. Therefore, their primary concern is the safety of the space, followed by the comfort that the facilities provide for them and their families, with particular attention to the park's child-friendliness. This also explains why middle-aged users have high expectations for improvements in safety and comfort, as well as their lower satisfaction with road accessibility. The importance and expectations that middle-aged users place on the cultural aspects of urban public spaces are not only a reflection of their personal psychological needs but are also closely related to their roles in family and society. They hope to find a sense of belonging through these spaces, fulfill the mission of cultural heritage, and enhance their quality of life.

The elderly group has high expectations for improvements in safety and comfort. They express dissatisfaction with the convenience of park entrances, the condition of exercise facilities, air quality, acoustic environment, and the availability of accessible facilities. Compared to other groups, they have a higher demand for improvements in accessibility. As elderly users age, they face challenges such as declining physical health, significantly increased economic vulnerability, transportation difficulties, reduced spatial navigation ability, and a shrinking social network due to factors like death, relationship breakdowns, health issues, retirement, and accessibility challenges [25–29]. From a physical needs perspective, the deterioration of physical functions, decreased mobility, and reduced stamina result in a diminished ability for elderly users to respond to environmental factors [58,60–62]. Therefore, elderly users prioritize the safety and comfort of spaces and closely monitor the availability of facilities. They place high value on the convenience of entrance locations, the current condition of sports facilities like tables and courts, air quality in the park, and the presence of accessible facilities. From a psychological needs perspective, as elderly individuals' social networks decrease, they become particularly susceptible to social isolation [63]. Consequently, they seek to participate in social and physical activities in Lowline Park to strengthen their connections with others. Additionally, economic vulnerability has a greater impact on elderly users' willingness to participate in park activities than accessibility does; if there are fees associated with using the park facilities, their willingness to use the space would be significantly reduced [64–66]. As a result, elderly users have expressed strong dissatisfaction with the fee policy in Lowline Park, and compared to other age groups, they have a higher demand for affordability in activity spaces.

4.2. Optimization Strategies for Under-Bridge Parks Based on Age-Specific Needs

In summary, sports-oriented public spaces under urban elevated highways, such as Jinan's Yanshan Interchange Lowline Park—government-led, professionally designed, and municipally managed—should focus on addressing two key issues: the mismatch between activity times and peak usage hours, and the inconsistency between facility provision and the needs of different user groups. Additionally, improvements should be prioritized for the primary and secondary indicators in Quadrant IV of the importance–satisfaction matrix, which are high in importance but low in satisfaction. It is essential to fully consider the unique needs of each user group to create an under-bridge public space that accommodates all age groups.

The first recommendation is to optimize the time, area, and activity-type configuration of the activity space. To address the issues of the uneven spatial area and distribution identified by the elderly, youth, and child user groups, the size and function of the spaces could be flexibly adjusted according to the numbers of users and activity types at different times. For example, fixed partitions could be replaced with movable adjustable ones, allowing spaces to be reconfigured based on user needs. In the morning, a larger fitness area could be allocated for elderly users, while in the afternoon, the same area could be transformed into a soccer field for children, and in the evening, it could serve as an entertainment zone for youth. This approach would allow for the efficient and quality use of low-traffic spaces through quick adjustments to meet demand. To ensure that vulnerable groups can conduct activities safely, comfortably, and conveniently in the park, exclusive activity venues can be set up to meet the actual needs of the child and elderly users. For child users, the most important thing is to set up a dedicated children's activity area, including play structures, interactive game zones, a plant garden, and small sports fields. Nearby, a designated parent supervision area should be provided for safety. Parent–child interaction facilities, such as playgrounds and science-based interactive experience areas, can be enhanced. These spaces could include parent–child swings, climbing structures, and interactive educational exhibits to promote parent–child relationships. For elderly users, an area with barrier-free activity facilities can be set up and kept away from basketball courts and football fields, to avoid mutual interference.

The second recommendation is to enhance the safety and comfort of the activity spaces. An intelligent lighting system would improve users' experience. An intelligent lighting system should automatically adjust the brightness based on the time, weather, and activity type, including nighttime lighting to avoid overly bright or dark areas, thereby enhancing safety. Installing wider ticket gates at park entrances and additional ramps or replacing stairs with ramps could improve the accessibility of families with strollers and balance the experience of people with mobility difficulties and the elderly. Furthermore, pedestrian entrances should be separated from non-motor vehicle parking areas to prevent entrance blockages and reduce risks for pedestrians. To reduce air pollution, additional greenery and air-purification facilities are recommended. For instance, trees, shrubs, grass, and ivy could be planted along the park side near roads to form a multi-layered green barrier. Dust-removing and air-purification devices could also be installed in gathering areas to mitigate the effects of vehicle emissions, ensuring comfort and health for all age groups. Additionally, a feedback channel for users should be established, enabling a sustainable and effective optimization mechanism over time. In addition, around sports areas such as badminton courts, basketball courts, and ping pong tables, flexible windbreak facilities should be installed. During cooler seasons—autumn, winter, and early spring—wind panels can be closed to reduce wind exposure, while in summer, they can be opened to allow airflow and natural cooling. Auxiliary facilities, such as storage and changing rooms, should be added around sports areas, providing indoor areas with suitable temperatures during extreme seasons and offering temporary rest areas for users. Stores selling drinking water and snacks and offering the rental or sale of sports equipment could also enhance user convenience. In particular, for elderly users, comfortable and easy-to-clean seating should be provided along the main routes and scenic areas to facilitate rest. In the design, anti-slip flooring, protective barriers, and accessible restrooms should be installed to make the space more elderly friendly. Restrooms should be located near activity areas, equipped with clear and prominent directional signage; users should not intersect with motor vehicles along traffic routes, ensuring that people can access the facilities safely, conveniently, and comfortably.

The third recommendation is to improve the spatial inclusion to ensure that the needs of all age groups are met. For child users, a child-friendly wayfinding system that matches their height could be added along pathways or on either side, using colors or patterns to mark routes to children's activity areas. For youth and middle-aged users, a cultural and arts exhibition zone can be created to showcase Jinan's historical culture. Regular cultural events, such as traditional festival celebrations and art exhibitions, could enrich the cultural content of the under-bridge space. For elderly users, considering the complex traffic environment around the bridge space, an intelligent traffic signal control system can be implemented according to the real-time traffic flow, pedestrian crossing demand, and parking lot operating time, automatically adjusting the traffic-light waiting time, reducing the drivers and pedestrians, and setting an emergency control button, which would enhance the crossing safety and improve the traffic efficiency. Furthermore, a balanced approach to management revenue and citizens' rights should be considered when setting usage rules and fee standards. This could include reducing fees to accommodate lower-income individuals and offering discounts for vulnerable groups like the elderly and children.

5. Conclusions

This study examined user behavior and spatial perceptions in the Yanshan Interchange Lowline Park in Jinan, Shandong Province, China. Using observation and the importance–satisfaction analysis (ISA) method, combined with questionnaires and interviews, this study focused on analyzing the activity preferences and spatial environment perceptions of four age groups: child, youth, middle-aged, and elderly users. It explored the differences in perceptions among these groups and investigated the reasons behind these differences, providing a basis for prioritizing the renovation and optimization of spaces under overpasses.

This study found that the low-line park beneath the overpass provides a rich public space for exercise, significantly enhancing the quality of daily life for nearby residents. However, there are notable differences in the spatial and temporal usage patterns and perceptions among different age groups. The four age groups (child, youth, middle-aged, and elderly) exhibited distinct patterns in how they use the low-line park in time and space. A mismatch exists between the park's opening hours and the peak usage times for different user groups, as well as between the activity areas and user needs. The importance–satisfaction analysis indicates that, overall, users place a higher importance on safety, comfort, and cultural features than their satisfaction levels suggest, highlighting greater expectations in these areas—particularly comfort, which requires urgent improvement. In contrast, users are relatively satisfied with the park's aesthetics and accessibility. All age groups express high expectations for improvements in safety and comfort. Youth users also desire better accessibility, while youth and middle-aged users have higher expectations for cultural relevance. None of the groups demonstrated significant expectations for improvements in aesthetics, suggesting that the current redesign of spaces under overpasses is generally well-received in this regard. However, meeting the needs of all age groups will require further improvements in safety, comfort, and cultural relevance. In addition, differences were observed in the perceptions of the 26 secondary indicators across the four age groups. Child users are dissatisfied with the lighting conditions, the safety of the internal facilities, and the noise levels, and they value the ease of crossing to the park, the distribution of the activity areas, and the aesthetic quality of the pathways more than other groups. Youth users express dissatisfaction with the convenience and visibility of park entrances, the internal area's safety, the air quality, and the quantity and size of the activity areas. They also place greater importance on entrance visibility and elements within the park that reflect Jinan's historical and cultural features. Middle-aged users are dissatisfied with the pedestrian and non-motorized vehicle safety in terms of the surrounding traffic, the lighting conditions, the air quality, and the quantity and size of the activity areas. Compared to other groups, they are less satisfied with the ease of crossing roads. Elderly users are dissatisfied with the convenience of park entrances, the condition of the exercise facilities, the air quality, and the availability of accessible facilities, showing a higher demand for accessibility improvements compared to other groups. The differing physical and psychological characteristics of each age group result in varying sensitivities to spatial information and distinct needs within the low-line park.

Based on the above analysis, this study proposes a series of optimization strategies, including adjusting the allocation of the time, area, and activity types within the activity spaces, enhancing the safety and comfort of the activity areas, and enriching the cultural connotation and inclusivity of the space. To address the specific needs of each age group, this study recommends tailoring space functionalities to different activity periods, for example, dedicated play areas for child users, additional rest areas and accessible facilities for elderly users, and enhanced cultural facilities to meet the cultural atmosphere expectations of youth and middle-aged users. Furthermore, this study highlights that the transformation of spaces under overpasses requires not only considerations of physical layout but also a focus on refining the management and operational mechanisms. This ensures that the accessibility, safety, and comfort of spatial facilities align with users' needs, particularly in balancing the requirements across different time periods and age groups.

This study provides practical theoretical support for the design and optimization of public spaces under overpasses, offering significant reference value for creating age-friendly spaces. However, this study has certain limitations, such as a relatively small sample size, which may affect the generalizability and accuracy of the results. Additionally, the timeliness of the findings may be impacted as park management continues to improve. Future research could expand the sample size and conduct ongoing tracking of changes in park management and user perceptions.

Supplementary Materials: The following supporting information can be downloaded at: https://www.mdpi.com/article/10.3390/10.3390/land14020293/s1, Table S1: User Demographic Characteristics; Table S2: Demographic Characteristics of Child Users; Table S3: Demographic Characteristics of Young Users; Table S4: Demographic Characteristics of Middle-aged Users; Table S5: Demographic Characteristics of Elderly Users; Table S6: Paired Sample *t*-test of Importance and Satisfaction of Children; Table S7: Paired Sample *t*-test of Importance and Satisfaction of Youth; Table S8: Paired Sample *t*-test of Importance and Satisfaction of Middle-Aged Adults; Table S9: Paired Sample *t*-test of Importance and Satisfaction of Elderly; Table S10: Post-Use Evaluation Questionnaire for the Lowline Park under Elevated Bridges; Table S11: Post-Use Evaluation Structured Interview Guide for the Lowline Park under Elevated Bridges.

Author Contributions: Conceptualization, X.W.; Methodology, Z.W., X.L. and X.W.; Software, Z.W.; Formal analysis, Z.W.; Investigation, Z.W., X.L. and H.L.; Writing—original draft, Z.W. and X.L.; Writing—review and editing, X.W.; Visualization, X.L. and H.L.; Project administration, H.L.;

Funding acquisition, X.W. and H.L. All authors have read and agreed to the published version of the manuscript.

Funding: This research was funded by Ministry of Education of the People's Republic of China, grant number 202210422099, and Department of Education of Shandong Province of China, grant number M2023303.

Data Availability Statement: Data are contained within the article.

Acknowledgments: We sincerely thank the respondents who generously contributed their time to participate in the questionnaire and interview process. We also extend our gratitude to the Yanshan Interchange Bridge Management Authority for their invaluable cooperation. Furthermore, we would like to express our appreciation to Li Te, Li Luhan, and Guo Qiuhua for their support and assistance in this study.

Conflicts of Interest: The authors declare no conflicts of interest.

References

1. Beijing Municipal Commission of Planning and Natural Resources. *Design Guidelines for Space Utilization Under Bridges in Beijing*; Beijing Municipal Commission of Planning and Natural Resources: Beijing, China, 2022.
2. People's Publishing House. *The 14th Five-Year Plan for National Economic and Social Development of the People's Republic of China and the Long-Range Objectives Through the Year 2035*; People's Publishing House: Beijing, China, 2021; ISBN 978-7-01-023255-3.
3. Wu, S.; Wang, J.; Zhao, D.; Rong, J. Research on Under-Bridge and Its Public Demand Evaluation in Mega City Based on Multi-Source Data: A Case Study of Beijing. *J. Hum. Settl. West. China* **2023**, *38*, 87–93. [CrossRef]
4. Zaki, S.H.; Haggag, A.G.; Selim, A.M. Novel Criteria for Sustainable Use of Urban Spaces under Bridges in Cities by Applying DEMATEL Technique. *Int. J. Urban Sustain. Dev.* **2023**, *15*, 299–320. [CrossRef]
5. Adams, A.K.; Harvey, H.; Brown, D. Constructs of Health and Environment Inform Child Obesity Prevention in American Indian Communities. *Obesity* **2008**, *16*, 311–317. [CrossRef] [PubMed]
6. Lak, A.; Amiri, N.; Aghamolaei, R. Urban Elevated Highways in Residential Districts: New Developed Elevated Highways from Residents' Perspectives. *J. Urban Plan. Dev.* **2022**, *148*, 05022006. [CrossRef]
7. Sarhan, Y.; Elfouly, H. Constructing an Analytical-Based Model for User Satisfaction towards under-Bridge Spaces within the Egyptian Urban Fabric. *Civ. Eng. Arch.* **2023**, *11*, 948–985. [CrossRef]
8. Nunma, P.; Kanki, K. Playing under the Flyover in Bangkok from the Children's Point of View. *J. Asian Arch. Build. Eng.* **2021**, *21*, 865–883. [CrossRef]
9. Zhu, X. *Research on Subjective Evaluation Methods of the Built Environment*; Southeast University Press: Nanjing, China, 2005.
10. Li, Q. A Study on the Properties of Public Space with the Orientation of Social Cohesion in Residential Area: Two Cases of Chuangzhifang and Caoyangyicun in Shanghai. *Urban. Plan. Forum* **2014**, *04*, 88–97.
11. Cheng, X.; Liu, Z.; Wang, X. Effects of Neighborhood-Scale Public Spaces on Social Cohesion in the Context of Home-Work Separation: The Case of Beijing. *Urban. Dev. Stud.* **2019**, *26*, 28–36, 45.
12. Zhou, G.; Lin, Y. Research on Comprehensive Assessment of Public Space in Huangpu River Waterfront Area. *City Plan. Rev.* **2020**, *44*, 62–73.
13. Yan, S.; Zheng, X. Promotes Physical and Mental Health: The Impact of Community Public Space on Social Cohesion. *Urban Dev. Stud.* **2021**, *28*, 117–124.
14. Abdulmunem, S.A.; Shok, M.E.; Al-Hussaini, Z.I.H.; Alkinani, A.S. Multigenerational Urban Design: Creating Urban Spaces That Support Active Aging and Intergenerational Interaction. *Planning* **2024**, *19*, 3737–3747. [CrossRef]
15. Zhong, S.; Lee, C.; Foster, M.J.; Bian, J. Intergenerational Communities: A Systematic Literature Review of Intergenerational Interactions and Older Adults' Health-Related Outcomes. *Soc. Sci. Med.* **2020**, *264*, 113374. [CrossRef] [PubMed]
16. Pazit, L.; Bronwyn, L.; Dreher, S.; Sze-Ee, S.; Dow, B.; Batchelor, F.; Hill, K.D. Multigenerational Spaces: The Usage and Activities of Different Age Groups within Six Community Parks in Victoria, Australia. *J. Public Health* **2024**, *46*, 403–411. [CrossRef]
17. Veitch, J.; Salmon, J.; Ball, K. Children's Active Free Play in Local Neighborhoods: A Behavioral Mapping Study. *Health Educ. Res.* **2008**, *23*, 870–879. [CrossRef] [PubMed]
18. Sundevall, E.P.; Jansson, M. Inclusive Parks across Ages: Multifunction and Urban Open Space Management for Children, Adolescents, and the Elderly. *Int. J. Environ. Res. Public Health* **2020**, *17*, 9357. [CrossRef]
19. Lindberg, M.; Schipperijn, J. Active Use of Urban Park Facilities–Expectations versus Reality. *Urban For. Urban Green.* **2015**, *14*, 909–918. [CrossRef]

20. Veitch, J.; Bagley, S.; Ball, K.; Salmon, J. Where Do Children Usually Play? A Qualitative Study of Parents' Perceptions of Influences on Children's Active Free-Play. *Health Place* **2006**, *12*, 383–393. [CrossRef]
21. Reichert, F.F.; Barros, A.J.D.; Domingues, M.R.; Hallal, P.C. The Role of Perceived Personal Barriers to Engagement in Leisure-Time Physical Activity. *Am. J. Public Health* **2007**, *97*, 515–519. [CrossRef]
22. Tucker, P.; Gilliland, J.; Irwin, J.D. Splashpads, Swings, and Shade: Parents' Preferences for Neighbourhood Parks. *Can. J. Public Health* **2007**, *98*, 198–202. [CrossRef]
23. Nordström, M. Children's Views on Child-Friendly Environments in Different Geographical, Cultural and Social Neighbourhoods. *Urban Stud.* **2010**, *47*, 514–528. [CrossRef]
24. Mak, B.K.L.; Jim, C.Y. Linking Park Users' Socio-Demographic Characteristics and Visit-Related Preferences to Improve Urban Parks. *Cities* **2019**, *92*, 97–111. [CrossRef]
25. Jessica, F.; Thea, F.; Heather, M.; Sims-Gould, J. Therapeutic Landscapes and Wellbeing in Later Life: Impacts of Blue and Green Spaces for Older Adults. *Health Place* **2015**, *34*, 97–106. [CrossRef]
26. Bath, P.A.; Deeg, D. Social Engagement and Health Outcomes among Older People: Introduction to a Special Section. *Eur. J. Ageing* **2005**, *2*, 24–30. [CrossRef] [PubMed]
27. Vaughan, M.; Lavalley, M.P.; Alheresh, R.; Keysor, J.J. Which Features of the Environment Impact Community Participation of Older Adults? A Systematic Review and Meta-Analysis. *J. Aging Health* **2016**, *28*, 957–978. [CrossRef]
28. Clarke, P.; Gallagher, N.A. Optimizing Mobility in Later Life: The Role of the Urban Built Environment for Older Adults Aging in Place. *J. Urban Health* **2013**, *90*, 997. [CrossRef]
29. Hemingway, A.; Jack, E. Reducing Social Isolation and Promoting Well Being in Older People. *Qual. Ageing Older Adults* **2013**, *14*, 25–35. [CrossRef]
30. Day, R. Local Environments and Older People's Health: Dimensions from a Comparative Qualitative Study in Scotland. *Health Place* **2008**, *14*, 299–312. [CrossRef]
31. Wu, S.; Xia, X.; Wu, W.; Zhou, J. Research on the Leisure Environment of Resettlement Community from the Perspective of Age Difference—A Case of Ningbo Jiuqu Community. *Urban. Archit.* **2021**, *18*, 59–63.
32. Wong, K.K. Urban Park Visiting Habits and Leisure Activities of Residents in Hong Kong, China. *Manag. Leis.* **2009**, *14*, 125–140. [CrossRef]
33. Zhai, Y.; Wu, C. Identification of Major Factors Influencing Overall Satisfaction in Urban Parks—Analysis Based on Users from Different Age Groups. *Landsc. Archit.* **2021**, *28*, 57–62.
34. Li, X. Inclusive Design: Public Space Renewal Strategy for the Targets of Community for All Ages. *Urban Dev. Stud.* **2019**, *27*, 27–31.
35. Moore, A.; Lynch, H. Accessibility and Usability of Playground Environments for Children under 12: A Scoping Review. *Scand. J. Occup. Ther.* **2015**, *22*, 331–344. [CrossRef] [PubMed]
36. Rivera, E.; Timperio, A.; Loh, V.H.Y.; Deforche, B.; Veitch, J. Critical Factors Influencing Adolescents' Active and Social Park Use: A Qualitative Study Using Walk-along Interviews. *Urban For. Urban Green* **2021**, *58*, 126948. [CrossRef]
37. Liao, Y. *Research on Integrated Planning Path of Community Public Service Facilities from the Perspective of Intergenerational Integration*; Shenzhen University: Shenzhen, China, 2022.
38. Maslow, A.; Lewis, K.J. *Maslow's Hierarchy of Needs*; Salenger Incorporated: New York, NY, USA, 1987; Volume 14.17.
39. Yao, Y. Evaluation of Chinese Cities' Walkability. *Urban Transp. China* **2019**, *17*, 85–88, 126.
40. Gehl, J. *Livet Mellem Husene, Udeaktiviteter Og Udemiljøer*; China Architecture & Building Press: Beijing, China, 2002.
41. Rancière, J. The Aesthetic Dimension: Aesthetics, Politics, Knowledge. *Crit. Inq.* **2009**, *36*, 1–19. [CrossRef]
42. Yan, Y. *Research on the Evaluation of Outdoor Activity Spaces for Community Children from a Health Perspective*; Qingdao University of Technology: Qingdao, China, 2022.
43. Yang, G. *Research on Micro Renewal Strategy of Public Space in Wuhan Old Residential Area Based on Resilience Evaluation*; Huazhong Agricultural University: Wuhan, China, 2022.
44. Wang, H.; Jiang, D. Evaluation System Research on Vitality of Urban Public Space. *J. Railw. Sci. Eng.* **2012**, *9*, 56–60.
45. Zou, X. *Research on the Construction and Optimal Design of the Aging Evaluation System of Community Parks*; Chengdu University of Technology: Chengdu, China, 2021.
46. Dou, Y. Research on Sustainable Aging-Friendly Renewal Methods for Public Spaces in Existing Residential Areas. Master's Thesis, Southeast University, Nanjing, China, 2018.
47. Lamola, A.A.; Yamane, T. Sensitized Photodimerization of Thymine in DNA. *Proc. Natl. Acad. Sci. USA* **1967**, *58*, 443–446. [CrossRef]
48. Martilla, J.; James, J. Importance-Performance Analysis. *J. Mark.* **1977**, *41*, 77–79. [CrossRef]
49. Buchan, D.S.; Ollis, S.; Thomas, N.E.; Baker, J.S. Physical Activity Behaviour: An Overview of Current and Emergent Theoretical Practices. *J. Obes.* **2012**, *2012*, 546459. [CrossRef]
50. Ajzen, I. The Theory of Planned Behavior. *Organ. Behav. Hum. Decis. Process.* **1991**, *50*, 179–211. [CrossRef]

51. Liu, T. Sense of Security: Its Educational Implication and Implementation. *Educ. Res.* **2016**, *37*, 50–56, 77.
52. van Kamp, I.; Davies, H. Noise and Health in Vulnerable Groups: A Review. *Noise Health* **2013**, *15*, 153–159. [CrossRef] [PubMed]
53. Huang, Z.; Yang, L. Discussion on The Application of the Design of Environmental Colour in the Children Activity Area. *J. Henan For. Sci. Technol.* **2010**, *30*, 21–23+33.
54. Li, X.; Tan, S.; Dai, Y. Research on Urban Public Space and Planning Strategy for Promoting the Communication Between People. *Archit. Cult.* **2017**, *06*, 230–231.
55. Cabello-Manrique, D.; Lorente, J.A.; Padial-Ruz, R.; Puga-González, E. Play Badminton Forever: A Systematic Review of Health Benefits. *Int. J. Environ. Res. Public Health* **2022**, *19*, 9077. [CrossRef]
56. Pugliese, J.; Tinsley, B. Parental Socialization of Child and Adolescent Physical Activity: A Meta-Analysis. *J. Fam. Psychol.* **2007**, *21*, 331. [CrossRef]
57. Shaw, S.M.; Dawson, D. Purposive Leisure: Examining Parental Discourses on Family Activities. *Leis. Sci.* **2001**, *23*, 37–41. [CrossRef]
58. Head, D.; Isom, M. Age Effects on Wayfinding and Route Learning Skills. *Behav. Brain Res.* **2010**, *209*, 49–58. [CrossRef]
59. Veitch, J.; Salmon, J.; Crawford, D.; Abbott, G.; Giles-Corti, B.; Carver, A.; Timperio, A. The REVAMP Natural Experiment Study: The Impact of a Play-Scape Installation on Park Visitation and Park-Based Physical Activity. *Int. J. Behav. Nutr. Phys. Act.* **2018**, *15*, 10. [CrossRef]
60. Kavanagh, J.J.; Barrett, R.S.; Morrison, S. Age-Related Differences in Head and Trunk Coordination during Walking. *Hum. Mov. Sci.* **2005**, *24*, 574–587. [CrossRef]
61. Imms, F.J.; Edholm, O.G. Studies of Gait and Mobility in the Elderly. *Age Ageing* **1981**, *10*, 147–156. [CrossRef] [PubMed]
62. Wright, R.E. Aging, Divided Attention, and Processing Capacity. *J. Gerontol.* **1981**, *36*, 605–614. [CrossRef] [PubMed]
63. Nicholson, N.R. A Review of Social Isolation: An Important but Underassessed Condition in Older Adults. *J. Prim. Prev.* **2012**, *33*, 137–152. [CrossRef] [PubMed]
64. Townsend, B.G.; Chen, J.T.H.; Wuthrich, V.M. Barriers and Facilitators to Social Participation in Older Adults: A Systematic Literature Review. *Clin. Gerontol.* **2021**, *44*, 359–380. [CrossRef] [PubMed]
65. Fardanesh, A.; Zeraatkish, Y. An Investigation on the Promenade Value of Javanmardan Park in Tehran, Using Contingent Valuation Method (CVM). *Int. Bus. Manag.* **2016**, *10*, 1639–1641.
66. Membrebe, Z.O.; Santos, A.J.G.; Valeroso, J.C.C.; Ancheta, A.A. Urban Forest Park as Eco-Space for Liveable City: Arroceros Forest Park, Manila, Philippines. *Int. J. Real Estate Stud.* **2017**, *11*, 23–34.

land

MDPI

Article

Resident-Centered Narrative Mapping for Micro-Morphological Analysis: Case of a Marginalized Lilong Compound in Downtown Shanghai

Yuqi Zhai

College of Art and Media, Tongji University, No. 1239 Siping Road, Shanghai 200092, China; zhaiyuqi@tongji.edu.cn; Tel.: +86-15900552720

Abstract: While informal settlements have been extensively studied in the Global South, their counterparts in the Global North remain under-researched, despite their critical role in shaping urban morphology. This paper introduces "Resident-Centered Narrative Mapping", a framework designed to uncover micro-morphological knowledge through the lived spatial experiences of marginalized residents. By examining the epistemological question "whose morphology?", this study critiques conventional urban morphological methods, which often disregard spatial practices embedded in the everyday lives of marginalized communities. Focusing on a marginalized lilong settlement in downtown Shanghai, this research work integrates critical cartography with ethnographic fieldwork to develop a micro-morphological mapping process centered on resident narratives. This process, structured around the phases of finding, inscription, and simplification, demonstrates how residents' daily practices actively shape and reconfigure their built environment. This study offers an alternative perspective to understand the dynamic processes of urban renewal in informal settlements and emphasizes the dialectical relationship between resident-driven spatial practices and the transformation of the urban form. By broadening urban morphology's methodological framework, this research provides insights into how resident-driven mapping can inform localized regeneration strategies. The findings highlight the potential for marginalized communities to shape urban regeneration policies, advocating for inclusive, resident-centered development.

Keywords: micro-urban morphology; cartography; everyday practices; lilong; marginalized residential community

1. Introduction

In the spring of 2016, as part of my doctoral research, my supervisor guided me to an atypical lilong building near People's Square in downtown Shanghai. While the exterior of the building appeared indistinguishable from the surrounding modern structures, its interior unveiled a complex spatial configuration that highlighted unique socio-spatial relationships. Windows acted as gateways to terraces, residents' reclining chairs stretched across corridors, and at the end of the hallway, a bathroom was separated only by a curtain. Traditional urban analyses, which often rely on urban planning and regulation, fail to capture the complex and nuanced realities of this Shanghai lilong community's living conditions. Instead, this lilong community fluctuates at multiple intermediate and extreme positions that defy the "status quo of urban studies [1]".

This seemingly ordinary building exemplifies the lilong housing form, constructed between the late 19th and early 20th centuries in Shanghai's foreign concessions, born from the collision of Chinese vernacular traditions and colonial modernity, reflecting hybrid spatial practices common to cities of the Global South [2,3]. Like informal settlements in Lagos or Mumbai, lilongs emerged as adaptive responses to socio-economic marginalization and transnational power asymmetries, yet they also subverted Eurocentric urban norms by embedding communal lifeworlds into rigid spatial frameworks [4]. This tension between colonial imposition and local agency aligns with the Global South's broader struggles to negotiate modernity on its own terms [5].

The Global South is a concept that highlights urban practices and conditions historically marginalized in mainstream urban theory. These conditions include extreme informality [6], distinct social relations [7,8], the absence of institutional governance [9], and lingering post-colonial legacies [10]. While these conditions are often associated with cities in the Global South, they can also be observed in cities of the Global North [11]. For example, hybrid spaces shaped by colonial encounters and grassroots adaptations appear in Shanghai's lilongs, emerging forms of civic participation and activism are evident in Seoul [12], and informal street markets thrive in London [13]. In this sense, the Global South signals new and dissident forms of urban governance, manifesting through both formal and informal actions and structures [14]. This opens up the possibility to explore how findings, theories, and practices from specific case studies can be employed, expanded, or evaluated across diverse urban contexts, regardless of geographic location [15].

Morphological research can be viewed as a tool—a "prosthesis"—that extends human vision and memory [16], enabling the detection, naming, and description of spatial patterns and forms. It also serves to correct the simplifications introduced by unilateral globalization which have been disseminated from colonizers to the colonized [17].

Typomorphologists such as Muratori, Caniggia, and Conzen have studied the role of building types, plots, streets, and blocks in the evolution of the urban form through successive iterations [18–20] Within this tradition, buildings, plots, streets, and blocks are identified as core morphological elements that combine to produce the urban fabric. While this approach has advanced our understanding of formal urban structures, it struggles to address the dynamic, informal practices prevalent in Global South cities, where socio-political precarity necessitates fluid, adaptive spatial uses.

Scholarly efforts in urban morphology have increasingly sought to address the discipline's historical blind spots, particularly its limited engagement with the dynamic socio-spatial practices of the Global South. Dovey and King introduced the concept of "informal typology" in Dharavi [21], but their approach frames informality as a "problem", overlooking the residents' dynamic use of materials and spaces for flexible, adaptive purposes. Building on Italian School insights, Oliveira emphasizes residents' incremental adaptations that gradually transform the urban fabric [22]; however, these Eurocentric frameworks struggle to account for contexts like Shanghai's lilongs, where socio-spatial fluidity—such as curtains demarcating bathrooms or corridors being repurposed as communal terraces—defies formal typological categories. Recent efforts to provincialize these paradigms show promise: Duarte Cardoso, for example, adapts morphological analysis to Amazonian socio-environmental repair [17], though her regional-scale focus leaves micro-scale improvisations understudied. Meanwhile, Bianchi employs ethnography to document informal practices in European historic neighborhoods [23], yet her rich empirical work stops short of theorizing how such agency operates in Southern contexts marked by structural precarity.

These studies collectively reveal a critical gap: while urban morphology has broadened its empirical scope, its epistemological roots in Northern planning traditions limit its ability to interpret Southern urbanisms as constitutive logics rather than deviations [4]. The lilong serves as a clear example of this challenge, with its socio-spatial fluidity—such as the blurring of private/public boundaries and the incremental repurposing of corridors—calling for methodologies that move beyond Conzenian typology and the formal/informal binary. To address this, urban morphology could incorporate temporal analysis and ethnographic mapping [8], while also provincializing Northern theoretical assumptions. Such an approach would not negate existing frameworks but rather extend their analytical reach, enabling the discipline to engage more fully with the micro-political realities of Southern cities.

In this context, this paper focuses on the enduring ontological question of "whose morphology?" and seeks to re-examine the epistemological and methodological foundations of urban morphology through the lens of a "Resident-Centered Narrative Mapping" approach. Since the late 20th century, Western cartography has turned its attention to the "lived spaces" of "the other", aiming to transcend traditional cartographic conventions by representing spaces shaped by lived experiences [24–26]. This emerging methodology not only strives to articulate local needs but also transforms cartography into a strategic tool for local actors to express and resolve spatial relationships. Margaret's work on embedding narrative techniques directly into the visual variables of cartographic language [27] demonstrated significant success in depicting experiential spaces. However, it has yet to be widely applied or integrated into the production of morphological knowledge.

To effectively map resident-centered micro-morphologies and the micro-governance mechanisms they engender, traditional morphological approaches must undergo critical revision. Informed by Henri Lefebvre's critique of everyday life, there exists a significant opportunity to integrate and advance both critical cartography and micro-urban morphology, particularly in relation to the everyday spatial practices of marginalized communities. For instance, in Shanghai's lilong communities, residents frequently repurpose corridors and balconies for informal activities, such as creating shared spaces for relaxation or communal cooking. While these adaptations are not formally recognized in urban planning, they exemplify a form of micro-governance, where residents actively negotiate space usage through everyday interactions. In doing so, they establish flexible spatial boundaries that challenge the formal regulations governing public housing usage. Two key pathways for innovation in this domain are (1) narrative homology, which emphasizes "place" narratives centered on residents as active agents, and (2) knowledge interconnectivity, which connects local morphological knowledge embedded in daily practices. This paper proposes a resident-focused micro-morphological analytical framework, employing narrative-based analysis of spatial discourses from individual interviews to construct cartographies that represent local spatial realities.

This paper introduces an innovative resident-centered micro-morphological research methodology and demonstrates its relevance and applicability through a case study of Shanghai's lilong neighborhoods. By utilizing narrative mapping, this study uncovers residents' everyday practices and the spatial morphologies embedded within these practices, offering new methodological pathways for generating localized morphological knowledge. The second section reviews the existing literature on everyday practices, critical cartography, and micro-morphological studies, focusing on the core questions of "whose morphology?" and "how should morphological analysis be conducted?". It emphasizes the need to integrate residents' lived experiences and subjective perspectives into urban morphological research, advocating for a resident-centered analytical framework. The third section outlines the methodology itself, detailing the processes of finding, inscription, and simplification that structure the research framework. This section also demonstrates the applicability of this approach through a case study of Shanghai's lilongs, exploring how the spatial characteristics of these historically marginalized housing forms are shaped and reshaped by everyday practices. It examines how individual residents influence the dynamic construction of lilong micro-morphologies, either facilitating or constraining these transformations through their daily interactions. The Xiangsheng Building, a lilong apartment complex in central Shanghai, is used as the core case study to illustrate the practical application of the methodology. The fourth section presents the findings from the core case study, focusing on how the processes of finding, inscription, and simplification inform the analysis of everyday practices and their spatial manifestations. The fifth section offers a conclusion, summarizing the three key steps in resident-centered morphological research (finding, inscription, and simplification) and emphasizing the role of critical cartographic narrative as a vital analytical tool. It reinterprets residents' everyday practices, shedding light on the micro-morphological knowledge and mechanisms of urban renewal embedded in these practices. Finally, the discussion section addresses the advantages and limitations of this approach, reflecting on its broader contributions to the field of urban studies, particularly in the context of informal and adaptive urban environments.

2. Theoretical Perspectives

2.1. Everyday Practices in Shaping Urban Morphology

Henri Lefebvre, Guy Debord, and Michel de Certeau identified everyday life as a crucial arena of modern culture and society [28]. In Lefebvre's three-volume *Critique of Everyday Life*, he argues that capitalism dominates the cultural and social world, necessitating an examination of alienation that views everyday life as a significant terrain of struggle [29]. Research into new or dissident forms of urban governance, through formal and informal actions and structures, highlights aspects historically neglected in urban theory and practice [14]. This focus is particularly relevant in areas characterized by extreme informality [6] and distinct social relations [8,30] in both the Global South and North.

McFarlane argue that everyday life constitutes a "set of processes with distinct geographies and rhythms [31]", often characterized by their informality and invisibility in traditional urban studies, which shape the urban landscape and offer insights into the social dynamics and rhythms that govern urban life. Research within the emergent strand of Southern urbanism temporally emphasizes "the mundane" [32] and "ordinary/ordinariness" [33] in contrast to the spectacular, and spatially emphasizes "street-level" and "close-focus explorations", focusing on micro-scale experiences, offering spatial and temporal approaches in research of everyday practices. Research emphasizes the importance of acknowledging daily routines and mundane concerns in urban environments [28]. These "everyday spaces" emerge naturally from daily practices, rather than being designed by architects or planners. They represent zones of social transition and offer potential for new social arrangements and forms of imagination. This approach also emphasizes everyday methods through bottom-up, process-oriented research which has a specific focus on the micro scale.

Related detailed examinations capture the lived experiences of residents and reject the given properties or meanings of materiality (such as pipes or grates), instead emphasizing "material configurations" that drive daily life [34,35]. Reflecting on subjectivity and constructing strategies that defy the official establishment, through incorporating various agencies and the importance of materiality into analyses [36], scholars address the complex "lifeworlds" and "lived vitalities" of urban spaces, resulting in a new type of urban morphology that can inform urban policy and city planning [37].

Although some [8,11,38] express concern that reducing studies of everyday life to local, micro, and descriptive methods risks falling into the trap of infinite particularism and increasing obscurity [37], Lefebvre provides a "third way" by examining the details of everyday life and their interrelation with structural dynamics [39] and through thinking about how everyday practices in specific places have significant scalar reverberations and theoretical significance. This leads to the introduction of multiple or alternative narratives of cities and city-making [38], offering insights that move seamlessly between the particular and the generalizable [37], avoiding both particularism and structural abstraction.

This article follows the third way of Lefebvre by focusing on everyday research and engaging with theories and practices of urban morphology to raise the question of resident-centered morphological research, aiming to balance ethnographic narratives of urban residents with macro-urban morphological processes.

2.2. *Mapping the Residents' Everyday Practices*

Everyday space is not pre-given; it is formed by bodies, relationships, affects, forces, and conflicts—elements that are often non-represented, silenced, or oppressed [40]. These aspects remain hidden behind the shapes, esthetics, scales, or programs of traditional maps. Critical cartography seeks to reveal these overlooked elements and bring visibility to the displacements and affections that are typically suppressed in conventional mapping practices [41].

A major critique of traditional Western cartography, originating in the 1600s in Europe, is its anonymous generation and guise of scientific objectivity, despite being deeply political [42–45]. Critical cartography, emphasizing that all maps are authored and interpreted differently, reflects subjective perspectives [41] and is often supported by community residents, other development activists, artists, and new media innovators [46], thus challenging Western cartographic conventions. It reshapes maps using alternative methods—such as color, perspective, and audio—to highlight the subjectivity inherent in map-making [24]. Some cartographers enhance map symbols through techniques like color coding for emotional structure and tilted angles for viewpoint representation. Scholars categorize mapping strategies into drift [47], layering [48], game-board [49], and rhizome [50] strategies. Multi-layered structures in maps, including sound, text, and sketch layers, encode narratives and perceptions [51]. Additionally, emerging technologies enable the mapping of personal experience trajectories on digital maps using GPS and video. For example, Marie Cieri uses GPS technology to map personal travel trajectories on digital maps, overlaying these paths with historical texts, memories, and visual data to showcase an individual's experiences and identity in specific spaces [52].

These evolving approaches to cartography challenge traditional notions of space by emphasizing subjectivity and layered narratives. Henri Lefebvre's concept of "representations of space" provides a theoretical foundation for understanding how places, people, actions, and things intersect in the production of cartographies [15,53]. This involves processes of gathering, reworking, assembling, relating, revealing, sifting, and speculating [54] to uncover realities that would otherwise remain unseen. Margaret proposed narrative-focused experiential mapping, viewing place as a "narrative-like synthesis", existing in a tension between being both centered and de-centered [55], encoding each day with its own narrative palette to express a sense of place [27]. Víctor focuses on mapping dissident spatial practices, using bodies, occupations, and affections to uncover usually invisible relationships between informal micro-activities and formal public spaces. These cartographic practices are linked to "ethnographic drawing" [13], as observing and drawing urban spaces from users' perspectives rather than those of architects or planners. This approach empowers marginalized groups by giving them a voice in shaping the built environment and challenging the dominance of elitist structures of power [15].

Emphasizing subjective perspectives, combined with ethnographic drawing strategies and various visualization techniques, provides opportunities to recognize the significance of local knowledge and everyday spatial practices, far beyond relying solely on official data and physical housing surveys.

2.3. The Micro-Morphological Research on Everyday Control, Use, and Activity Practices

Research on urban morphology increasingly focuses on micro-urban morphology, broadening the scope to encompass daily practices and spatial redevelopment at the micro scale [56–61]. A fundamental role of urban morphology is to identify the repeating patterns in urban structures [20,62,63]. In this context, the term "room" is broadly interpreted as "any void formed by structures", thereby conceptualizing a building as a specific arrangement of various room types [64]. Consequently, a building type is defined as the class of all buildings that share the same room types, quantities, and spatial arrangements.

Alterations in everyday practices intrinsically change the "typological diversity" of buildings, affecting room usage patterns and spatial arrangements through both formal and informal processes [65–68]. These changes significantly influence the formation and transformation of urban morphology. Such changes may occur through individual residents' daily practices or through large-scale, collective adjustments by state or city institutions. The latter can transform urban models by altering everyday life within buildings rather than through physical deconstruction and reconstruction; notable examples include the communal reconstruction of Bologna and Chinese cities in the 1960s, as well as the socialist housing transformations in China during the 1950s.

Everyday practices, often more economical and less durable, lead to both formal and informal changes in property rights, building layouts, usage patterns, and resident relations, interacting with dominant modes of production, legal frameworks, and guidance systems. Formal elements of urban space are those that conform to legally recognized and institutionalized norms, including building layouts, land use patterns, and property rights, which are regulated by urban planning policies and official legal frameworks. On the other hand, informal elements refer to the adaptive, everyday practices of residents that often fall outside the scope of formal regulations. Consequently, property is not univocal; urban morphology should be identified as a physical form, a unit of land use, and a unit of control [69,70]. Several authors have highlighted the ambiguity and complex relationships between property, control, use, and form in their studies of urban form [66,71–73]. Alexander argues that the built environment is better understood as a network of overlapping sets, where "sets" refer to categories or systems describing different dimensions of urban space, such as formal property rights, informal spatial uses, and social control mechanisms [74]. In overlapping sets, a given element can be a member of two or more sets simultaneously, and identifying aspects does not deny overlaps but rather articulates them.

Building on Kropf's concepts of control, use, and activity, this paper identifies two types of relationships between humans and physical forms, extending the notion of buildings beyond the plot-scale typically examined in existing studies [70,72,75,76]. While foundational studies by Muratori, Caniggia, and Conzen established rigorous methodologies for analyzing the urban fabric through buildings, plots, and streets [8,62], their analytical frameworks—even when incorporating bottom-up processes as in Oliveira's adaptation of Italian School insights—remain constrained by Northern epistemological lenses [22].This limitation manifests in two key aspects: (1) an overreliance on technical documentation (cadastral maps, typological classifications, etc.) that privileges researcher-defined categories over voices and spatial practices of residents [31,35,77,78] and (2) the persistent formal/informal binary that misrepresents adaptive reuse as deviations rather than constitutive logics [4].

Addressing this gap, this paper introduces the concept of resident-centered morphology, exploring ethnographic approaches to capture the characteristics of everyday practices of control, use, and activity. Recognizing residents' everyday practices as authentic data sources, this dynamic process involves multiple residents' perspectives and encompasses interactions among hierarchical practices, from individual neighbors to the community of a single building. This highlights the necessity for urban planning policies that are adaptable and sensitive to the evolving needs and practices of urban inhabitants.

3. Materials and Methods

3.1. Methodology

To address the complex and often obscured socio-spatial relationships and regeneration challenges inherent in lilong housing, this study established a Resident-Centered Narrative Mapping methodology, utilizing narrative as a foundational structure. An ethnographic research approach was employed, integrating critical cartography with micro-morphological analysis to delineate a methodological framework for the finding, inscription, and simplification of everyday practices within marginalized communities. These stages—finding, inscribing, and simplifying—are inherently iterative, characterized by overlap, recursion, and mutual reinforcement.

Step 1: Finding. The finding process was approached as a collaborative effort between residents and the author, using various interview methods to explore key topics such as residents' identities, their knowledge of the building's history, and their everyday practices. Iterative cooperation between residents and researchers allowed for the continuous refinement of data, uncovering narratives that closely reflect the residents' lived experiences, thereby providing critical insights into their morphological perspectives.

Step 2: Inscription. Narrative structures were inscribed onto maps within a critical cartography framework, consistent with Latour's concept of "inscription" [79]. This process traced individual everyday practices, mapping experiential spaces that transcend purely physical spatial representation. Key decisions included whether to map individuals or groups, choose single maps or atlases, and use traditional or innovative formats, all while resolving contradictions between multiple discourses and the mismatch between experiential and physical spaces.

Step 3: Simplification. The simplification of narrative structures into morphological analysis enhanced "readability" [80] and proposed a resident-centered micro-urban morphology. Applying Kropf's "control-use" morphological framework [57] enabled the recognition of spaces governed by residents, the identification of overlapping control and usage units, and the re-examination of the complex spatial rights and relationships embedded in everyday practices. The challenge lay in creating morphological analysis maps that were both detailed and coherent, effectively synthesizing localized knowledge of spatial rights shaped by daily practices of control and use. By better understanding the narrative-like qualities that capture specific relational structures between people and places [55], a deeper understanding of place was achieved.

3.2. The Lilong Housing Morphology in Downtown Shanghai

3.2.1. Marginalized Lilong Housing Residential Dwellings in the Center of Shanghai

Emerging as a defining residential typology, lilong houses originated from late 19th-century colonial encounters, constructed through both formal concession planning and informal rural encroachments. These neighborhoods arose to house displaced populations—foreign settlers, Taiping Rebellion refugees, and economic migrants drawn to Shanghai's burgeoning capitalism [81]. Architecturally, they materialized a trans-cultural synthesis: British rowhouse layouts were reimagined through Jiangnan courtyard craftsmanship, creating a hybrid form that spatially encoded colonial power dynamics and local adaptation strategies [81].

Initially peripheral settlements, lilong districts gained urban centrality through Shanghai's 20th-century spatial expansion [58]. This geographical paradox intensified their socio-economic liminality. During the concession era, lilong compounds epitomized Shanghai's identity as a "migrant city" [82], transitioning from single-family dwellings to high-density tenements where individual rooms housed entire families. Post-1990s economic reforms accelerated this transformation: original residents relocated to modern apartments, while vacated units became subdivided rentals for low-income migrants. By 2017, central lilong neighborhoods accommodated over 800,000 residents at extreme densities (avg. 9 m^2/person), sustaining their characterization as "villages within the city" [83,84].

3.2.2. Formal and Informal Everyday Practices Change Lilong Housing Morphology

The historical formalization and transformation of lilong houses have resulted in a complex socio-spatial morphology characterized by layered, overlapping, and conflicting everyday practices. During the concession period, lilong houses were initially owned by single, above-middle-class families. However, the influx of rural immigrants and declining owner incomes during wartime led to room-by-room sales, resulting in multiple owners, fragmented management, and the informal construction and use of public spaces of lilong houses [85]. During the planned economy period, the state's expropriation and redistribution of lilong properties, granting each family a room along with shared kitchen and bathroom facilities, transformed lilong houses into organized communal living spaces through everyday control, usage, and activities. In the socialist market economy period, the State-owned housing system allowed residents to sublet, transfer, and lease their rooms indefinitely. Some residents purchased adjacent rooms, disrupting the one-room-per-family arrangement and informally occupying public spaces, thereby breaking previously homogeneous control and usage patterns.

The development and evolution of lilong compounds demonstrate that everyday practices have consistently played a crucial role in shaping these spaces, resulting in significant micro-morphological transformations over time and continuing to influence the micro-urban morphology of lilong houses today. The socio-spatial relationships within lilong buildings are complex, chaotic, and fragile, yet they are an essential part of Shanghai's urban culture. As Amos Rapoport stated, "A house is a human fact… clearly showing the relationship between form and lifestyle" [86]. Scholars have long recognized that the everyday "marketplace" (市井) form of Shanghai's lilongs represents their fundamental characteristics, exploring their societal form through narratives of residents' daily lives. Existing research primarily focuses on well-known heritage buildings and significant individuals, while studies on non-heritage-listed lilong buildings and the everyday practices of their residents are scarce.

3.2.3. Micro-Morphology of Lilong Residents and Urban Regeneration Challenges

The complex control and usage rights associated with lilong buildings pose significant challenges for Shanghai's urban regeneration efforts. Traditional strategies, such as large-scale demolitions and comprehensive preservation, often necessitate the complete relocation of residents. This process not only risks the destruction of lilongs' living heritage but also exacerbates urban social segregation and displaces low-income rural migrants [87]. Public housing residents, who receive government compensation rather than profits, often face protracted and costly redevelopment negotiations. These negotiations are further complicated by rising land value in central Shanghai. Additionally, the high-density occupancy and informal practices of residents accelerate the deterioration of both public facilities and lilong buildings. This situation places a significant maintenance burden on the government, which is further complicated by an emphasis on social benefits and low rental income, obstructing daily maintenance, preservation, and regeneration efforts. Since 2015, Shanghai has adopted micro-renewal strategies focused on residents' daily needs and reorganizing resources at the micro scale [87,88]. However, these studies remain predominantly case-based and lack a comprehensive methodology recognizing the micro-morphology of residents' practices, limiting their effectiveness.

3.3. Application of the Resident-Centered Morphology: A Case Study of the Xiangsheng Lilong Apartment Complex

3.3.1. Case Selection

This paper introduces an innovative resident-centered micro-morphological research methodology, applied to the Xiangsheng Lilong Apartment Building in Shanghai. Located in the West Guizhou Community of the East Nanjing Street subdistrict in Huangpu District, the Xiangsheng Building represents an atypical example of the lilong typology (Figure 1). Situated less than a kilometer from the commercial center at People's Square, the West Guizhou Community stands out for its stark differences in population density, per capita residential area, income levels, and rent compared to neighboring areas, underscoring its marginal status within the city.

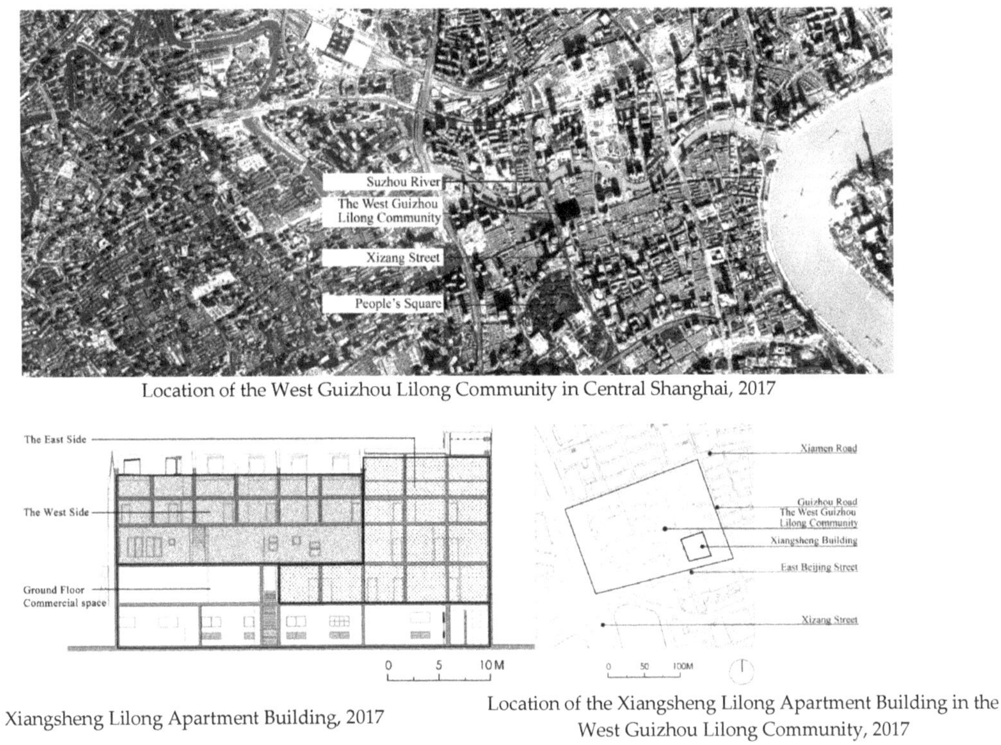

Figure 1. Location of the Xiangsheng Lilong Apartment Building in West Guizhou Lilong Community in Central Shanghai, 2017. Source: Authors.

What makes the Xiangsheng Building particularly significant is its intergenerational mix of long-term residents whose everyday practices intersect with those of newer, often lower-income inhabitants. This creates a complex social fabric that reflects broader trends found in other lilong communities. Given these characteristics, the Xiangsheng Building provides a particularly rich context for exploring the relationship between residents' spatial practices and the physical environment (Figure 2). Its ongoing transformation, both physically and socially, allows for an in-depth investigation of how marginal residents adapt to and shape their living spaces.

Street-facing south façade (Left) and rear north façade (Right) of the Xiangsheng Building in the West Guizhou Community during 2017

The building features a variety of residential unit types

Shared kitchen and bathroom facilities and terraces, with significant privatization of public spaces

Figure 2. Façade and interior of the marginal Xiangsheng Lilong Apartment Building in Central Shanghai. Source: Authors.

Unlike traditional lilong housing, the building was originally designed in the concession period as a hybrid between an office building and residential spaces, with a floorplan modeled after commercial buildings. Despite its initial design, over time, it has undergone a transformation similar to that of traditional lilong buildings, evolving into a residential space where each room functions as a self-contained unit, with shared kitchen and bathroom facilities. This historical adaptation mirrors the transformation of other lilong buildings, making it a relevant case study despite its atypical origins.

3.3.2. Method

This study explores the resident-centered morphology of the Xiangsheng Lilong Apartment Building through three core processes: "finding", "inscription", and "simplification". Conducted between 2017 and 2018, the research involved in-depth interviews with all 34 residents, resulting in 34 maps documenting their everyday spatial practices. These maps illustrate how different social groups perceive and inhabit space, offering insights into the morphology and dynamics of marginalized apartment forms.

The research team consisted of my Ph.D. supervisor (Professor Ming Tong), myself (a Ph.D. student during the research period), five Master's students, and two professional architects from the TM Studio architecture firm.

During the fieldwork, we adopted an open and flexible interview approach, treating the study as a collaborative effort between residents and researchers. Through iterative data collection and verification, we uncovered contradictions and ambiguities, gradually revealing the distinct aspects of each resident's life and addressing the central question: "Whose Xiangsheng Building?". Initial background information was gathered through a basic questionnaire.

Data collection involved note-taking, with occasional audio recordings. Interviews were conducted in teams of two to three, allowing for role adjustments as needed. After each interview, the team promptly reviewed the notes for accuracy, and weekly meetings with supervisors were held to assess progress and refine methods, ensuring a dynamic and methodologically robust process.

The interviews focused on themes such as residents' identity, subjective perceptions, and daily practices. This flexible approach shaped the format, location, and content of each session. As the interviews progressed, a variety of dynamic conversational techniques were employed. In-depth Interviews: A total of 34 semi-structured interviews, covering 100% of the households, were conducted in residents' homes, each lasting 60–90 min. Neighborhood Discussions: These discussions were usually initiated by the residents themselves during interviews with long-term residents. Two or three familiar neighbors would often join the conversation, contributing additional insights into daily life. Design Charrettes: My mentor team, which also leads micro-renewal practices in the community, held workshops with the neighborhood committee nearly every month. Each workshop included 3–5 resident representatives, some from the Xiangsheng Building, providing information about the building's history and development.

In addition to interviews, we utilized several archival materials, including historical aerial photographs (1937, 1948, 1976, 2015, 2017) to track spatial transformations and incremental encroachments, architectural surveys documenting informal modifications to the building, and policy archives (2017) that documented residential usage rights and room allocations.

This combination of fieldwork and archival research provided a comprehensive understanding of the spatial dynamics and the evolving social fabric of the Xiangsheng Lilong Apartment Building.

The research team's dual role as interviewers and participants in the community's redevelopment project brought both advantages and challenges. Their involvement made residents more willing to participate, as many saw the interviews as part of the redevelopment process's preliminary research. However, this also led to distortions, with some residents fabricating historical narratives about the building's original design, claiming it should be restored to a "historic" state, a claim later disproven through digital reconstructions. This dual role also encouraged residents to share their concerns, offering valuable insights into their views on urban changes and spatial agency. For example, second-generation residents and tenants expressed anxiety about losing ownership. One resident noted, "My siblings and I are co-owners of this public housing unit. Since my situation is worse, I've been allowed to live here. If the renovation improves the space, my siblings will return to claim it, and I don't want that".

4. Results

4.1. *Results of the Finding Stage: Everyday Narratives Regarding "Finding" Through the Collaboration of Cartographers and Residents*

The micro-morphology of Xiangsheng Building is shaped by two main groups of residents: public housing residents and tenants. Public housing residents are divided into first-generation residents, who moved in before 1949 during the concession period through purchase or lease, and second-generation residents, who were allocated housing between 1950 and 1994 under the public housing distribution system. Tenants who moved in after 1994, when lilong public housing was permitted to be rented out by public housing residents, can be categorized into three groups: transient tenants with short-term stays, kinship tenants renting through family connections, and subletting tenants who function as sublandlords. The choice of interview locations highlights differing perceptions of public and private spaces (Table 1): first-generation residents are comfortable in both, second-generation residents balance public and private spaces, and tenants rely more on the security of their units, reflecting a weaker connection to communal areas.

Table 1. Interview formats preferred by different resident groups.

Resident Groups	Primary Interview Format
First-generation	Typically conducted in communal areas or their units, with little engagement from others.
Second-generation	More often conducted in shared spaces like laundry rooms or kitchens, where neighbors frequently join; private interviews occur only after trust is built.
Tenants	Usually conducted inside their units, as they feel less comfortable in public spaces.

The everyday practices of public housing residents within Xiangsheng Building exhibit a pattern governed by relatively stable rules: dwelling units are fixed as communal facilities belonging to individuals, and daily pathways are predetermined. First-generation residents, having the longest tenure and the most extensive architectural knowledge, generally hold a strong sense of ownership. However, interviews reveal distinctions between former owners and tenants despite their equal public housing rights after the socialist housing reform. For instance, Resident A, whose family historically owned both interior and exterior units, continues to occupy the side corridor as family property, disregarding the needs of the tenant of a former exterior unit, Resident B, illustrating the persistence of pre-revolution landlord–tenant relationships in shaping current spatial practice. Second-generation residents' everyday practices are shaped by assumed usage rights within communal living spaces and a tradition of sharing public areas. Gradually, the communal stoves and sinks in the shared kitchen were increasingly treated as personal property, fragmenting once-collective spaces into private zones. In contrast to the first generation of residents, second-generation residents engage with each other on more equal terms, treating historical information as casual conversation during leisure moments without sparking conflict or cognitive dissonance among neighbors.

The everyday practices of tenants within the Xiangsheng Building are typically restricted to their residences, stairwells, and corridors within the building, primarily conforming to external demands from other public housing residents. Due to the transient nature of their living situation and frequent mobility, most tenants perceive their residence as temporary, lacking a sense of community and belonging. They show minimal interest in the building's history and prioritize concerns over potential rent increases tied to future renovation plans. Tenants generally exhibit limited ownership or engagement with public spaces, viewing their apartments as the primary secure and stable areas. This behavior is reinforced by public housing residents who have installed private kitchen and bathroom facilities within their units to increase rental prices, reducing tenants' need to use communal facilities. Tenants who lack these amenities often rely on their landlord's private communal kitchen and bathroom, sometimes facing verbal restrictions from public housing residents. For example, a second-generation resident was observed telling tenants "Could you please not cook at the same time as us? You could cook earlier or later; it's very crowded. I'll have to discuss this with your landlord. It's inconvenient for us". Interactions between tenants and public housing residents occur more frequently and amicably on the east side of the third floor of the Xiangsheng Building, while communication in other areas remains limited.

4.2. Results of the Inscription Stage: Simultaneous Mapping of Individual Experiential Narratives and Comprehensive Non-Experiential Space

Despite the continuous nature of everyday life, individual experiences collected from interviews present a fragmented narrative, predominantly centered on a few repetitive, fixed activities. A three-dimensional cartographic approach was used to visualize these daily practices, highlighting activities such as cooking and laundry as key components in shaping residents' cognitive mapping of their immediate environment and informing their sense of place (Figure 3). Spaces that residents identify as central to their daily routines are mapped with distinct colors and detailed features, while areas of less personal significance are left unshaded, blank, or outlined with minimal linework. The contrast between spaces shaped by residents' daily routines and those left unformed provides key insights into how individual practices shape spatial perceptions, while the cartographic approach illustrates how these personal activities influence the configuration of residential units and the building's form, enhancing the understanding of lived experience within a localized context.

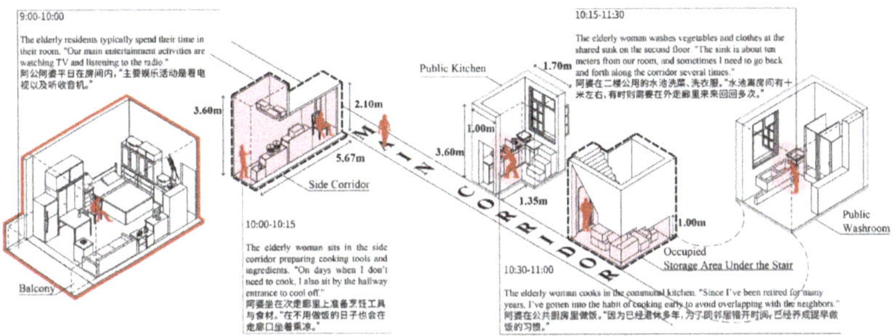

Figure 3. Narrative mapping of the everyday practices of the "Grandpa" who placed his reclining chair across the corridor, as mentioned in the Introduction, within the Xiangsheng Building. Source: Authors.

In the Xiangsheng Building, the daily trajectories of both first- and second-generation public housing residents are marked by stability, involving fixed living spaces spanning one or more public housing rooms, shared facilities, varying degrees of control over communal areas, and consistent daily pathways. Tenants' daily trajectories are marked by fluidity and transience, typically limited to a single room (excluding subletting tenants who may control multiple rooms) and focused on entry and exit pathways, with varying usage of communal kitchens and bathrooms, and occasional restrictions on public space access imposed by public housing residents.

The detailed examination of residents' daily lives and neighborhood dynamics revealed significant discrepancies in how individuals perceive their relationships with one another. A resident who holds a prominent position in one person's cognitive map may be of little importance in another's. To address this complexity, a narrative layer was introduced, incorporating first-person accounts through textual symbols, thus capturing conflicting information between residents and providing everyone with a distinct narrative voice. The integration of direct statements and experiential details enriched the maps, shifting from a detached "he is there" perspective to a more intimate "I am here" view, fostering a deeper personal engagement with the residents through the cartographic representation.

4.3. Results of the Simplification Stage: Microscopic Urban Morphological Analysis from the Perspective of Resident Dynamics

This study simplifies the cartographic representation to highlight residents' everyday control, usage patterns, and spatial interactions within the lilong building's floor plan. The author mapped these patterns under the category of "Residents' Everyday Control and Usage Patterns" (Figure 4).

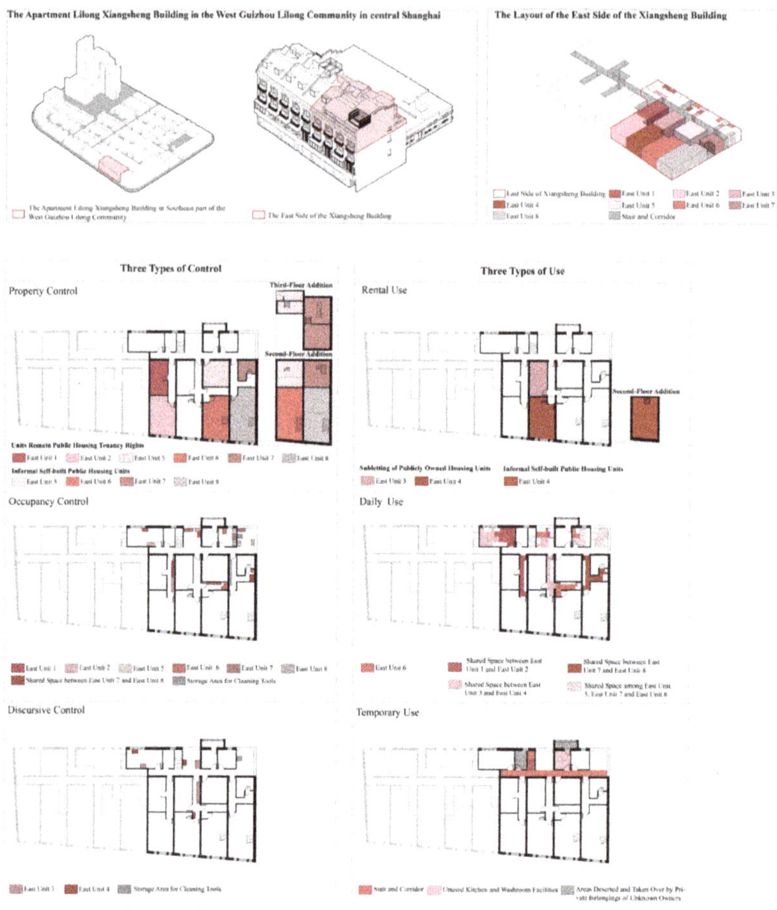

Figure 4. Residents' everyday control and usage patterns: example of the east part of the fourth floor of the Xiangsheng Building. Source: Authors.

"Control" in this context includes property control, occupancy control, and discursive control (Figure 4). Property control refers to ownership rights, such as selling, leasing, modifying, and using space. Occupancy control involves the use of non-owned spaces, such as public areas like kitchens or corridors, limiting access for others. Discursive control arises in public spaces where informal rules are set by residents, often targeting tenants. These forms of control frequently coexist, as seen with an elderly resident who holds property control over their room while asserting occupancy control over a corridor with a reclining chair. "Usage" refers to human activities and spatial functions, identified in three types in the Xiangsheng Building: rental use, daily use, and temporary use (Figure 4). Rental use is governed by formal contracts, giving tenants rights to lilong residential units and shared facilities, often restricted by owners or neighbors. Daily use refers to informal, non-contractual activities in units or public areas, tolerated by neighbors but subject to withdrawal. Temporary use includes non-obstructive, irregular activities like resting on the terrace or chatting with neighbors.

Through a micro-morphological analysis of residents' everyday control and usage patterns, a resident-centered unit morphology is developed, grounded in these daily practices. The author developed a series of maps to represent individual resident units, beginning with data collection and information gathering, progressing to vivid narratives of daily life, and culminating in standardized 3D architectural drawings, all presented in Figure 5 as an example. The resident-centered unit morphology plans are shown separately in Figure 6 as an example. The use of standardized 3D drawings is essential because many residents exhibit highly segmented control over public spaces, such as shared corridors. For instance, two households might share a storage space on the same wall, with one occupying the lower shelves and the other using the upper shelves. Therefore, mapping individual resident units requires integrating both spatial data and three-dimensional representations.

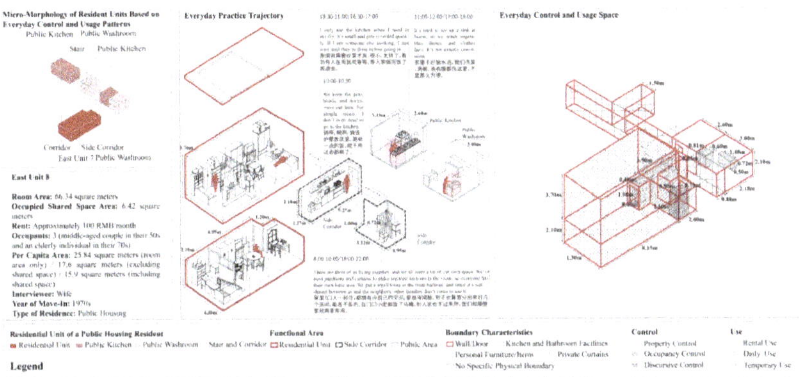

Figure 5. Micro-morphological analysis of resident units based on everyday control and usage patterns: example from the fourth floor, east unit 8 of the Xiangsheng Building. Source: Authors.

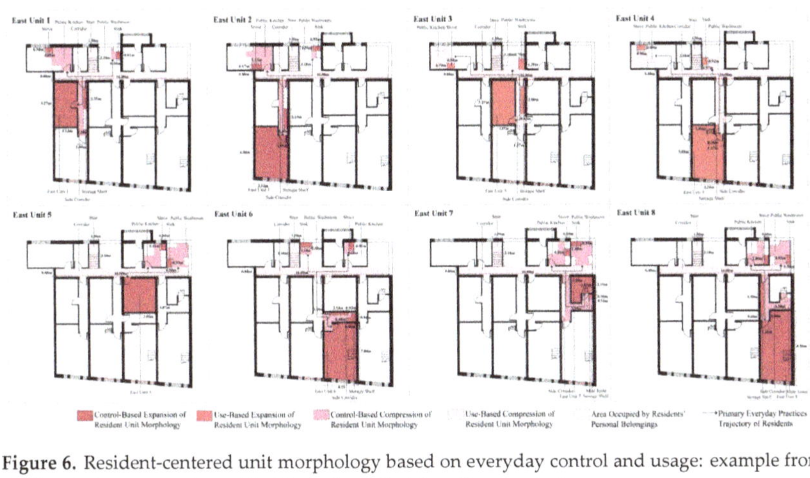

Figure 6. Resident-centered unit morphology based on everyday control and usage: example from the fourth floor, east wing of the Xiangsheng Building. Source: Authors.

By further examining the overlapping relationships of control and usage within this lilong building, zones of rational rights distribution, overlapping rights, and ambiguous rights can be identified, illustrating the complex spatial and social dynamics embedded in these units, as shown in Figure 7 as an example. Zones of rational rights distribution include residential units and public spaces without disputes. Overlapping rights zones are either internal—where multiple property owners, such as siblings, share ownership of a room, complicating regeneration due to differing opinions—or external, where residents share spaces like corridors, enforce access through verbal rules, and install shared facilities for exclusive use, occasionally permitting close neighbors. Zones of ambiguous rights typically appear in public areas used for furniture storage or underutilized spaces like kitchens and bathrooms. These areas result from misaligned information, as seen in conflicting accounts about a stone-walled area on the fourth floor, unclear ownership of items in cluttered spaces, and the absence of responsible residents, leading to neglected areas that no one manages.

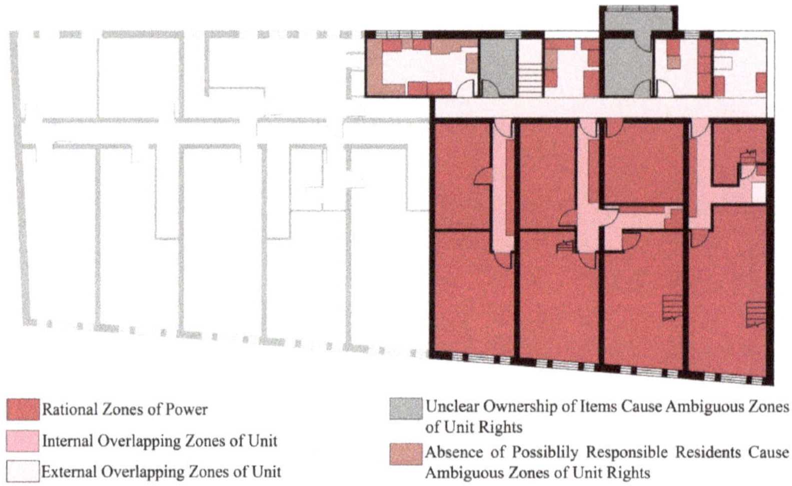

Figure 7. Rational zones of power distribution, overlapping zones of unit rights, and ambiguous zones of unit rights in the Xiangsheng Building. Source: Authors.

5. Conclusions

The socio-spatial fluidity of marginalized urban neighborhoods demands methodologies that bridge lived experiences with morphological analysis. Through a case study of a peripheral lilong house in central Shanghai, this paper foregrounds everyday spaces—domains shaped organically by residents' practices rather than formal design—as critical sites for resisting established spatial orders and generating micro-urban landscapes. This paper demonstrates how residents' everyday practices reshape urban form and governance. By synthesizing Lefebvrian epistemology, narrative cartography, and rights-based typologies, this study advances three core contributions to urban morphology:

Adoption of Lefebvre's Third Way: By emphasizing everyday research and engaging with urban morphology theories, this paper provokes an epistemological reflection on "whose morphology", shifting the focus from perspectives centered on researchers and technologists to prioritizing residents' everyday practices as legitimate morphological knowledge and cartographic context.

Introduction of Ethnographic Methods: By employing an ethnographic approach, this paper introduces the Resident-Centered Narrative Mapping method as a critical cartographic technique that reconfigures the methodology of morphological analysis. This shift moves from conventional Western cartographic language—defined by visual variables and grammar that convey homogeneous and modern spaces—to an alternative methodological framework informed by residents' experiential knowledge, thereby establishing the resident-centered unit morphology based on everyday control and usage.

Investigation of the Resident-Centered Narrative Mapping Method: Central to this investigation is the development and application of the Resident-Centered Narrative Mapping method, a three-stage methodology (finding, inscription, and simplification) that integrates ethnographic research, critical cartography, and micro-morphological analysis to decode marginalized communities' spatial practices.

Rights-Based Micro-Urban Morphology: Through examining the overlapping relationships of control and usage, this paper identifies zones of rational rights distribution, overlapping rights, and ambiguous rights, investigates the transformative potential of integrating everyday practices with the Resident-Centered Narrative Mapping method to enhance the revitalization and governance of historical neighborhoods, and emphasizes the dynamic interplay between residents' everyday experiences and urban regeneration efforts to foster participatory governance frameworks that promote inclusive, context-sensitive strategies honoring the historical and cultural fabric of these communities.

The Resident-Centered Narrative Mapping method not only uncovers hidden socio-spatial dynamics but also repositions urban morphology as a discipline capable of engaging with Southern urbanisms on their own terms. Its application in a Shanghai lilong demonstrates how marginalized communities reshape cities through micro-practices, offering a model to integrate grassroots agency into global urban theory and inclusive governance frameworks.

6. Discussion

Since our initial in-depth investigation of the lilong compound in central Shanghai uncovered a marked divergence between residents' property rights and their actual usage patterns, this paper stresses the need to critique and expand traditional urban morphology—typically focused on property rights and physical form—to encompass a more comprehensive understanding of the spatial realities shaped by residents' everyday "control" and "use", with the aim of revealing the informal and often invisible micro-morphologies that define living spaces within lilong communities. Our fieldwork conceptualizes the everyday practices of residents as a form of morphological agency, emphasizing that in the continuous reconstruction of public and private boundaries, individuals do not entirely dismiss the physical forms of urban structures or the normative usage orders established by state entities. Instead, residents' everyday practices, shaped by institutional changes, neighborhood agreements, and individual actions, engage in the ongoing subdivision and reconfiguration of existing forms, resulting in a micro-institutional framework of morphological operations.

Utilizing the Resident-Centered Narrative Mapping method, we spatialize these data, using them as a tool to find, inscribe, and simplify the everyday practices of residents. This advancement in micro-morphological research enriches our theoretical frameworks and discussions. Furthermore, these micro-institutions of morphological operations diverge from formal regulations; rather, they represent a fabric of novel spaces defined by everyday practices. Although often fragmented and challenging to systematize, they can also serve as entry points and foundational micro-institutions for resident engagement in urban regeneration and governance.

A significant challenge encountered in this study was conducting in-depth interviews with all residents while maintaining a consistent level of detail. Although our research on Shanghai's lilong neighborhoods benefited from the enthusiastic participation of residents—who often perceived our interviews as part of a pre-renovation assessment led by my supervisor's team—discrepancies remained in both the quantity and quality of the information gathered. Certain interviews resulted in vague, contradictory, or exaggerated responses. Consequently, this research should not be interpreted solely as an ethnographic narrative or mapping of local residents, but rather as a collaborative "finding" process involving residents, the author, and the reader.

Undoubtedly, the transformation from physical landscapes to map representations involves a process of reduction, selection, consolidation, distortion, and exaggeration [79,89]. This process inevitably imprints the map with the intentions and knowledge of both the cartographer and the narrator. The map presented in this study features two perspectives: a close-up view, where residents detail their daily practices, resulting in small-scale, three-dimensional spatial mappings, and a broader analytical perspective from my objective cartographic approach, which encompasses large-scale representations of both three-dimensional and two-dimensional spaces. The maps produced for this project are based not on verbatim recordings of all resident interviews but rather on selectively chosen phrases that capture the essence of their everyday practices, along with the differentiated spaces I chose to highlight. The spatial morphologies and narratives are only broadly outlined by discourse and remain incomplete. Readers are required to draw upon their own experiences and interpretations to contextualize and understand the everyday practices within these homes.

Moreover, compared to studies of marginalized neighborhoods in Chinese and other cities, research on "the everyday" has become a central focus in Southern urban studies, which, rather than being defined by latitude, emphasizes areas marked by extreme informality [80] and distinctive social relations [8,30]. The everyday practice approach offers a vital framework for analyzing new or dissident forms of urban governance, expressed through both formal and informal actions and structures [14]. Extensive studies of marginalized neighborhoods within Global North cities that exhibit Southern characteristics—such as diachronic socio-spatial structures shaped by institutional changes or rapidly evolving socio-spatial structures resulting from specific industrial developments—offer valuable case studies. Employing micro-morphological approaches, such as Resident-Centered Narrative Mapping, within these studies enables a deeper understanding of the spatial dynamics shaped by everyday practices, facilitating the development of interventions that are both adapted to local conditions and responsive to the intricacies of urban socio-spatial regeneration and governance.

Future research should broaden its diachronic exploration and analysis to encompass not only individual practices but also factors such as neighborhood collaboration, community governance, and urban institutional frameworks. This expansion would enhance the development of Resident-Centered Narrative Mapping and micro-morphological studies across diverse urban scales and temporal contexts. Such research may include an investigation of how multi-layered everyday practices contribute to the continuous transformation of urban morphology.

To conclude, we argue that the details of everyday life evolve and endure with a seemingly consistent and unremarkable rhythm, and while this daily existence appears both lasting and quietly impactful, its sustained continuity holds the potential to trigger swift, widespread, and significant urban transformations. As the renowned Shanghai writer Wang Anyi captures in her reflection on the century-long transformation of the city, "Each day is filled with the simple essentials of life, lived with quiet diligence and without grand ambitions, yet upon looking back, it has quietly transformed into a legend" [90]. However, the individual everyday practices are often limited and fragile, resulting in a micro, fragmented, and vulnerable spatial realm of everyday life that can be irreversibly erased, highlighting the necessity for such spaces to be "discovered", thoughtfully preserved, and adapted within the context of urban development.

Funding: The project is supported by the National Key Research and Development Program of China (2023YFC3805503).

Data Availability Statement: The original contributions presented in this study are included in the article. Further inquiries can be directed to the corresponding author.

Acknowledgments: The author sincerely appreciates the invaluable guidance and insightful comments provided by doctoral supervisor Ming Tong and D.G. Shane. Special thanks also to T. Hou and Y. Cao for their support in preparing the figures.

Conflicts of Interest: The author declares no conflicts of interest.

References

1. Oldfield, S. Critical Urbanism. In *The Routledge Handbook on Cities of the Global South*; Routledge: Oxfordshire, UK, 2014; p. 8.
2. Escobar, A. *Designs for the Pluriverse: Radical Interdependence, Autonomy, and the Making of Worlds*; Duke University Press: Durham, NC, USA, 2018.
3. Robinson, J. *Ordinary Cities: Between Modernity and Development*; Routledge: Oxfordshire, UK, 2006.
4. Roy, A. Slumdog cities: Rethinking subaltern urbanism. *Int. J. Urban Reg. Res.* **2011**, *35*, 223–238. [CrossRef] [PubMed]
5. Miraftab, F. Insurgent planning: Situating radical planning in the Global South. *Plan. Theory* **2009**, *8*, 32–50. [CrossRef]
6. Campos, M.J.Z.; Kain, J.H.; Oloko, M.; Scheinsohn, M.; Stenberg, J.; Zapata, P. Residents' collective strategies of resistance in Global South cities' informal settlements: Space, scale and knowledge. *Cities* **2022**, *125*, 103663. [CrossRef]
7. Simone, A.; Rao, V. Counting the uncountable: Revisiting urban majorities. *Public Cult.* **2021**, *33*, 151–162. [CrossRef]
8. Simone, A.M. People as infrastructure: Intersecting fragments in Johannesburg. *Public Cult.* **2004**, *16*, 407–429. [CrossRef]
9. Mabin, A. Grounding Southern City Theory in Time and Place. In *The Routledge Handbook on Cities of the Global South*; Routledge: Oxfordshire, UK, 2014; pp. 21–36.
10. Lara, F.L.; Hernández, F. (Eds.) *Spatial Concepts for Decolonizing the Americas*; Cambridge Scholars Publishing: Newcastle, UK, 2021.
11. Roy, A. Wordling the South: Toward a Post-Colonial Urban Theory. In *The Routledge Handbook on Cities of the Global South*; Routledge: Oxfordshire, UK, 2014; pp. 9–20.
12. Seo, U. Emerging Post-Political City in Seoul. In *Post-Politics and Civil Society in Asian Cities*; Routledge: Oxfordshire, UK, 2019; pp. 54–71.
13. Kelley, V. London's street markets: The shifting interiors of informal architecture. *Lond. J.* **2020**, *45*, 189–210. [CrossRef]
14. Swyngedouw, E. The Post-Political City. In *Urban Politics Now: Re-Imagining Democracy in the Neoliberal City*; BAVO, Ed.; NAI Publishers: Rotterdam, The Netherlands, 2007; pp. 58–76.
15. Cano-Ciborro, V.; Medina, A. Invisible networks: Counter-cartographies of dissident spatial practices in La Jota Street, Quito. *Cities* **2023**, *140*, 104435. [CrossRef]
16. Haraway, D. Situated knowledges: The science question in feminism and the privilege of partial perspective. *Fem. Stud.* **1988**, *14*, 575–599. [CrossRef]
17. Duarte Cardoso, A.C. Morphological analysis as a tool for socio-environmental reparation: Contributions from the Amazon context. *Urban Morphol.* **2024**, *28*, 117–131.
18. Cataldi, G.; Maffei, G.; Vaccaro, P. Saverio Muratori and the Italian school of planning typology. *Urban Morphol.* **2002**, *6*, 3–14. [CrossRef]
19. Caniggia, G.; Maffei, G. *Architectural Composition and Building Typology*; Alinea Editrice: Florence, Italy, 2001.
20. Conzen, M.R.G. *Alnwick, Northumberland: A Study in Town-Plan Analysis*; Institute of British Geographers: London, UK, 1960.
21. Dovey, K.; King, R. Forms of informality: Morphology and visibility of informal settlements. *Built Environ.* **2011**, *37*, 11–29. [CrossRef]
22. Oliveira, V. *Urban Morphology: An Introduction to the Study of the Physical Form of Cities*; Springer: Berlin/Heidelberg, Germany, 2016.
23. Bianchi, A. Beyond the plan: Informal urbanism and everyday spatial practices in historic neighborhoods. *J. Urban Des.* **2020**, *25*, 589–607.
24. Corner, J. The Agency of Mapping: Speculation, Critique, Invention. In *Mappings*; Cosgrove, D., Ed.; Reaktion: London, UK, 1999.
25. Kanarinka, F.A. Art-machines, body-ovens and maprecipes: Entries for a psychogeographic dictionary. *Cartogr. Perspect.* **2006**, *53*, 24–40. [CrossRef]
26. Lilley, K.D. Landscape Mapping and Symbolic Form: Drawing as a Creative Medium in Cultural Geography. In *Cultural Turns/Geographical Turns: Perspectives on Cultural Geography*; Naylor, S., Ryan, J., Cook, I., Crouch, D., Eds.; Routledge: London, UK, 2000.
27. Margaret, W.P. Framing the Days: Place and Narrative in Cartography. *Cartogr. Geogr. Inf. Sci.* **2008**, *35*, 17–32.

28. Cobarrubias, S.; Pickles, J. Spacing Movements: The Turn to Cartographies and Mapping Practices in Contemporary Social Movements. In *The Spatial Turn: Interdisciplinary Perspectives*; Warf, F.B., Arias, S., Eds.; Routledge: London, UK, 2009; pp. 36–58.
29. Holston, J. Insurgent citizenship in an era of global urban peripheries. *City Soc.* **2009**, *21*, 245–267. [CrossRef]
30. Simone, A.M. The Missing People. In *The Routledge Handbook on Cities of the Global South*; Routledge: London, UK, 2014.
31. McFarlane, C.; Desai, R.; Graham, S. Informal urban sanitation: Everyday life, poverty, and comparison. *Ann. Assoc. Am. Geogr.* **2014**, *104*, 989–1011. [CrossRef]
32. Barac, M. Place resists: Grounding African urban order in an age of global change. *Soc. Dyn.* **2011**, *37*, 24–42. [CrossRef]
33. Lombard, M. Constructing ordinary places: Place-making in urban informal settlements in Mexico. *Prog. Plan.* **2014**, *94*, 1–53. [CrossRef]
34. Silver, J. Incremental infrastructures: Material improvisation and social collaboration across post-colonial Accra. *Urban Geogr.* **2014**, *35*, 788–804. [CrossRef]
35. Cirolia, L.R.; Suraya, S. Towards a Multi-Scalar Reading of Informality in Delft, South Africa: Weaving the 'Everyday' with Wider Structural Tracings. *Urban Stud.* **2019**, *56*, 594–611. [CrossRef]
36. Amin, A.; Cirolia, L.R. Politics/matter: Governing Cape Town's informal settlements. *Urban Stud.* **2018**, *55*, 274–295.
37. Pieterse, E. Grasping the unknowable: Coming to terms with African urbanisms. *Soc. Dyn.* **2011**, *37*, 5–23. [CrossRef]
38. Peck, J. Cities Beyond Compare? *Reg. Stud.* **2015**, *49*, 160–182. [CrossRef]
39. Lefebvre, H. *Critique of Everday Life*; Verso: London, UK, 1991; p. 95.
40. Thrift, N.; Dewsbury, J.D. Dead geographies—And how to make them live. *Environ. Plan. D Soc. Space* **2000**, *18*, 411–432. [CrossRef]
41. Medina, A.; Cano-Ciborro, V. Cartographies of everyday conflicts in public spaces. Informal micro-activities on formal infrastructure. Carapungo Entry Park, Quito. *Rev. INVI* **2022**, *37*, 149–176.
42. Kim, A. Critical cartography 2.0: From 'participatory mapping' to authored visualizations of power and people. *Landsc. Urban Plan.* **2015**, *142*, 215–225. [CrossRef]
43. Krygier, J.; Crampton, J.W. An introduction to critical cartography. *ACME Int. E J. Crit. Geogr.* **2006**, *4*, 11–33.
44. Pickles, J. *A History of Spaces: Cartographic Reason, Mapping and the Geo-Coded World*; Routledge: London, UK, 2004.
45. Wood, D. *The Power of Maps*; Guilford Press: New York, NY, USA, 1992.
46. Tang, M.; d'Auria, V. Popular cartography: Collaboratively mapping the territorial practices of/with the urban margin in Mumbai. *City* **2023**, *27*, 321–346. [CrossRef]
47. Debord, G. Discourse on the Passions of Love: Psychogeographic Descents of Drifting and Localisation of Ambient Unities. In *Psychogeographic Guide of Paris*; Bauhaus Imaginiste, Ed.; Permild & Rosengreen: Copenhagen, Denmark, 1957.
48. Koolhaas, R.; Man, B. *S,M,L,XL*; Monacelli Press: New York, NY, USA, 1995.
49. Bunschoten, R. Proto-Urban Conditions and Urban Change. In *Beyond the Revolution: The Architecture of Eastern Europe: Architectural Design Profile 119*; Toy, T., Ed.; London Architectural Design: London, UK, 1996; pp. 17–21.
50. Deleuze, G.; Guattari, F. *A Thousand Plateaus: Capitalism and Schizophrenia*; Massumi, B., Translator; University of Minnesota Press: Minneapolis, MN, USA, 1987.
51. Cieri, M. Between being and looking: Queer tourism promotion and lesbian social space in Greater Philadelphia. *ACME* **2003**, *2*, 147–166.
52. Kwan, M.P. Affecting geospatial technologies: Toward a feminist politics of emotion. *Prof. Geogr.* **2007**, *59*, 22–34. [CrossRef]
53. Lefebvre, H.; Enders, M.J. Reflections on the politics of space. *Antipode* **1976**, *8*, 30–37. [CrossRef]
54. Corner, J. The agency of mapping: Speculation, critique and invention. In *The Map Reader*; John Wiley & Sons: New York, NY, USA, 2011; pp. 89–101.
55. Entrikin, J.N. *The Betweenness of Place: Towards a Geography of Modernity*; Macmillan: Basingstoke, UK, 1991.
56. Whitehand, J.W.R.; Larkham, P.J.; Jones, A.N. The Changing Suburban Landscape in Post-War England. In *Urban Landscapes: International Perspectives*; Whitehand, J.W.R., Larkham, P.J., Eds.; Roulicdge: London, UK, 1992; pp. 227–265.
57. Whitehand, J.W.R.; Morton, N.J.; Carr, C.M.H. Urban Morphogenesis at the Microscale: How Houses Change. *Environ. Plan. B Plan. Des.* **1999**, *26*, 503–515. [CrossRef]
58. Whitehand, J.W.R.; Gu, K.; Conzen, M.P.; Whitehand, S.M. The typological process and the morphological period: A cross–cultural assessment. *Environ. Plan. B Plan. Des.* **2014**, *41*, 512–533. [CrossRef]
59. Zhai, Y.; Tong, M. The transformation of the lilong form: A morphological study on changing boundaries. *Urban Des Int* **2024**, *29*, 153–170. [CrossRef]
60. Eisa, E. The longevity of Persian urban form: Maywood from late antiquity to the fifteenth century. *Urban Morphol.* **2015**, *9*, 57–71.
61. Stefano, B. Morphology as the Study of City Form and Layering. In *Reconnecting the City: The Historic Urban Landscape Approach and the Future of Urban Heritage*; Bandarin, F., van Oers, R., Eds.; Wiley-Blackwell: Chichester, UK, 2014; pp. 85–104.
62. Moudon, A.V. Urban Morphology as an emerging interdisciplinary field. *Urban Morphol.* **1997**, *1*, 3–10. [CrossRef]

63. Scheer, B.C. The epistemology of urban morphology. *Urban Morphol.* **2015**, *20*, 5–17. [CrossRef]
64. Kropf, K. Ambiguity in the definition of built form. *Urban Morphol.* **2013**, *18*, 41–57. [CrossRef]
65. Matthijs, V.O. Urbanizing villages: Informal morphologies in Shenzhen's urban periphery. *J. Urban Des.* **2018**, *23*, 732–748.
66. Dovey, K.; van Oostrum, M.; Chatterjee, I.; Shafique, T. Towards a morphogenesis of informal settlements. *Habitat Int.* **2020**, *104*, 102240. [CrossRef]
67. Matthijs, V.O. Appropriating public space: Transformations of public life and loose parts in urban villages. *J. Urban. Int. Res. Placemak. Urban Sustain.* **2022**, *15*, 84–105.
68. Çalışkan, F.; Mashhoodi, R. Typological diversity and morphological continuity in the modern residential fabric: The case of Ankara, Turkey. *Habitat Int.* **2023**, *104*, 102240. [CrossRef]
69. Kropf, K. Aspects of urban form. *Urban Morphol.* **2009**, *13*, 105–120. [CrossRef]
70. Kropf, K.S. Plots, property and behaviour. *Urban Morphol.* **2018**, *22*, 5–14. [CrossRef]
71. Lynch, K. *Good City Form*; MIT Press: Cambridge, MA, USA, 1984.
72. Rossi, A. *The Architecture of the City*; Ockman, J., Ghirardo, D., Eds.; Walden, B.V.d.G., Translator; MIT Press: Cambridge, MA, USA, 1982.
73. Scheer, B. *The Evolution of Urban Form: Typology for Planners and Architects*; American Planning Association: Chicago, IL, USA, 2010.
74. Alexander, C. A city is not a tree. *Archit. Forum* **1965**, *122*, 58–62 (Part I), 58–62 (Part II).
75. Scheer, B. Toward a minimalist definition of the plot. *Urban Morphol.* **2018**, *22*, 162–163.
76. Scardigno, N. *Landscape as Forma Mentis: Interpreting the Integral Dimension of the Anthropic Space. Mongolia*; Franco Agneli Edizioni: Milano, Italy, 2018.
77. Lemanski, C. Everyday human (in) security: Rescaling for the Southern city. *Secur. Dialogue* **2012**, *43*, 61–78. [CrossRef]
78. Sletto, B.; Palmer, J. The liminality of open space and rhythms of the everyday in Jallah Town, Monrovia, Liberia. *Urban Stud.* **2017**, *54*, 2360–2375. [CrossRef]
79. Latour, B. Visualization and Cognition: Drawing Things Together. In *Knowledge and Society: Studies in the Sociology of Culture Past and Present*; Kuklick, H., Ed.; JAI Press: Greenwich, UK, 1986; Part 6; pp. 1–40.
80. James, C.S. *Seeing Like a State: How Certain Schemes to Improve the Human Condition Have Failed*; Yale University Press: New Haven, CT, USA, 1998.
81. Wang, S.; Chen, Z. *Lilong Architecture*; Shanghai Social Science and Technology Literature Publishing House: Shanghai, China, 1987; p. 52.
82. Li, J. The Historical Transformation of Shikumen and Shanghai Identity in Mass Media. *J. Univ.* **2004**, *4*, 32–39.
83. Lu, H. *Beyond the Neon Lights—Everyday Shanghai in the Early Twentieth Century*; Shanxi People's Publishing House: Taiyuan, China, 2018; p. 13.
84. Zhai, Y.; Zhang, J. Back and Forth: Ontological Analysis on the Regeneration of Lilong in Shanghai. *Hous. Sci. Technol.* **2024**, *44*, 1–7.
85. Fan, W. *The Conservation and Regeneration of Shanghai Lilong*; Shanghai Scientific and Technical Publishers: Shanghai, China, 2004.
86. Rapoport, A. *House Form and Culture*; Yang, S., Ed.; Tianjin University Press: Tianjin, China, 2020.
87. Zhang, S. The Idea and Path of Establishing the Mechanism of Preserving the Urban Living Heritage: Shanghai's Experience and Challenges in Historic Townscape Preservation. *Urban Plan. Forum* **2021**, *06*, 100–108.
88. Dong, C.; Li, K.; Wang, H. Spatial Revolution: Reflections on the Renovation of 0.8 Shikumen Residence. *Archit. J.* **2014**, *2*, 101–105.
89. Bai, X.; Tong, M. Urban Micro-Renewal: From Network to Nodes, and From Nodes to Network. *Archit. J.* **2020**, *10*, 8–14.
90. Wang, A. *The Song of Everlasting Sorrow: A Novel of Shanghai*; Columbia University Press: New York, NY, USA, 2008.

land

MDPI

Article

The Role of Urban Landscape on Land Surface Temperature: The Case of Muratpaşa, Antalya

Mehmet Tahsin Şahin [1,*], Halil Hadimli [1], Çağlar Çakır [1], Üzeyir Yasak [2] and Furkan Genişyürek [1]

[1] Department of Geography, Akdeniz University, Dumlupinar Boulevard Campus, Antalya 07058, Türkiye; halilhadimli@akdeniz.edu.tr (H.H.); caglarcakir@akdeniz.edu.tr (Ç.Ç.); furkangyurek@gmail.com (F.G.)

[2] Department of Geography, Marmara University, Göztepe Campus, Kadıköy, İstanbul 34722, Türkiye; uzeyir.yasak@marmara.edu.tr

* Correspondence: tahsinsahin@akdeniz.edu.tr

Abstract: The role of landscape configuration in urban heat island effects is crucial for sustainable urban planning. This study examines the impact of land-use changes on land surface temperature (LST) in the Muratpaşa District of Antalya from 1984 to 2024. Data from 1984, 1989, 1994, 1999, 2004, 2009, 2014, 2019, and 2024 were analyzed at five-year intervals. Land-use maps and LST data were derived from the thermal infrared bands of Landsat-5 TM and Landsat-8 OLI-TIRS. LST values, categorized into seven groups, were calculated by converting radiance values into spectral radiation and Kelvin temperatures. Land-use classes, including green land, agricultural land, constructive land, water land, and bare land, were identified using interactive supervised classification. Landscape patterns were analyzed using ten indices within the framework of landscape ecology. ArcGIS 10.8.1 and Fragstats 4.2 software were used for analyses. Findings reveal a significant increase in surface temperatures over four decades, driven by urban expansion. Increased impervious surfaces created more high temperature zones, while reduced green spaces intensified the urban heat island effect. A strong correlation between LST and land-use patterns was identified, providing insights for urban heat management and climate change adaptation.

Keywords: urban landscape; land surface temperature; Antalya

1. Introduction

The rapid growth of the global population and economy has led to an accelerated increase in both urban populations and the spatial extent of cities. Daily rural-to-urban migration continues to fuel urban population growth. Currently, about 56% of the global population, equivalent to 4.4 billion people, resides in urban areas. This proportion is expected to rise to 70% by 2050, with an estimated addition of 1.2 million square kilometers of new urban areas being added globally by 2030 [1–3]. As of 2023, the urban population ratio in Türkiye stands at 77%, which exceeds the global average [4]. Urban areas, due to their dense populations, are significant drivers of environmental change. The increase in rural-to-urban migration leads to rising demands for essential services, employment challenges, a surge in housing needs, and inadequacies in infrastructure systems such as transportation, communication, and sanitation, all of which force cities to expand. The expansion of urban areas and associated infrastructure often comes at the expense of green spaces and agricultural lands, which are destroyed to meet these needs. Consequently, the land-use pattern changes, with green spaces and agricultural lands replaced by impervious surfaces. The proliferation of impervious surfaces alters heat capacity, albedo, and evaporation rates, leading to higher land surface temperature (LST) levels and intensifying the urban heat island (UHI) effect [5–7].

Land surface temperature is a critical parameter related to surface energy and water balance at both local and global scales [8,9]. LST serves as a key climatic indicator, reflecting the surface temperature of the Earth, which is influenced by radiative transfer processes [10,11]. The variability of LST is shaped by multiple factors, including landscape texture, vegetation cover, wetlands, soil moisture, slope, elevation, diurnal cycles, urban structural characteristics, building density and height, roof types and colors, tree heights, cloud cover, and albedo [12–19]. LST measurements can be obtained through two primary methods: in situ measurements recorded at meteorological stations or satellite-based data, which have become increasingly reliable in recent years, providing datasets at both local and global scales (e.g., ASTER, Landsat, and MODIS). However, meteorological station data are limited to the station's immediate area and do not represent the entire landscape, making them less practical. Despite being affected by cloud cover, satellites equipped with various sensors have emerged as the most preferred data sources in recent research [20–22].

The concept of the urban heat island (UHI) was first defined by Oak in 1982. UHIs are phenomena resulting from human-induced alterations to natural environments, where impervious surfaces replace natural landscapes. This concept describes the condition whereby surface temperatures in urban areas, characterized by poor vegetation and a high density of impervious surfaces, exceed those in surrounding natural environments [13,23,24]. Changes in land cover lead to varying heating processes, amplifying the UHI effect [5,25]. UHIs contribute to extreme temperatures, increased rates of disease and mortality, air pollution, and increased water and energy consumption [26–31].

In recent years, the impact of land-use characteristics on LST has become a focal point for numerous studies. Specifically, the effects of land-use changes in urban areas on UHI and LST are of critical importance in the context of climate change, public health, and sustainable urban planning. For example, Weng (2009) examined the effects of land-use changes on LST, evaluating the influence of various land-use types using Landsat TM data. Their study concluded that intense urbanization significantly increases surface temperatures, providing a fundamental understanding of the urban heat island phenomenon temperature. Further investigated the relationship between urban growth and UHI, analyzing how this relationship influences LST. Using satellite data, Weng demonstrated that dense urbanization and the reduction of green spaces considerably elevate LST in urban areas [32]. Similarly, Zhang, Odeh, and Han (2009) analyzed the thermal characteristics of urban areas in Fuzhou, Fujian Province, using Landsat Thematic Mapper (TM) and Enhanced Thematic Mapper Plus (ETM+) imagery from 1989 and 2001. They explored relationships among LST, the impervious surface area percentage, the Normalized Difference Vegetation Index (NDVI), and the Normalized Difference Built-up Index (NDBI). Their results revealed weak correlations between NDVI and LST but strong positive correlations between impervious surface area, NDBI, and LST [33]. Li et al. investigated how landscape composition and configuration influence UHI in Shanghai's metropolitan area by analyzing LST in relation to the Normalized Difference Vegetation Index (NDVI), vegetation fraction (Fv), and impervious surface area (ISA) percentage. Using Landsat ETM+ imagery from 13 March and 2 July 2001, they identified significant variations in LST per pixel based on specific vegetation or impervious surface ratios. Although NDVI, Fv, and ISA percentages were good predictors of LST on a regional scale, a strong negative linear relationship was found between LST and NDVI, as well as an even stronger negative linear relationship between LST and Fv. The study concluded that urban vegetation better mitigates surface UHI effects during summer than in early spring, while a strong positive relationship exists between the mean LST and ISA percentages. Furthermore, residential areas were identified as the largest contributors to UHI, followed by industrial zones [23]. Zhou et al. analyzed daily and seasonal surface UHI intensity in 32 major cities in China between 2003 and 2011 using MODIS data. Their study examined the relationship between land use and LST, finding that the annual average urban-to-rural temperature difference ranged from 0.01 to 1.87 °C during the day and 0.35 to 1.95 °C at night, with significant spatial heterogeneity. The study highlighted that artificial surfaces increase LST in densely urbanized regions, while green areas mitigate this effect [34]. Naserikia et al. characterized land cover and its effects on LST for 54 highly populated cities using Landsat-8 imagery. They observed that urban surface features and thermal responses differ significantly across climate regimes, with the largest discrepancies occurring in arid climate cities. While cold cities showed the greatest seasonal variability, tropical and arid cities exhibited minimal seasonality. In tropical, temperate, and cold climates, the Normalized Difference Built-up Index (NDBI) was the strongest contributor to LST variability during warmer months, followed by NDVI. In arid climates, the Normalized Difference Bare Soil Index (NDBaI) was the most influential factor. These findings provide a climate-sensitive foundation for future land-cover planning to mitigate local surface warming [35]. Hasan et al. investigated changes in land cover, LST, soil moisture, and evapotranspiration over the Kutupalong Rohingya Refugee Camp in Bangladesh. Utilizing index- and spatial algorithm-based remote sensing methodologies applied to Landsat-8 satellite imagery, they analyzed the impact of rapid population growth and land-use changes on LST. The study concluded that, across the entire Rohingya camp area, the land experienced a shift from moderately moist soil conditions to predominantly dry soil conditions due to increases in LST [36]. Shu et al. employed land-use and land-cover data, surface temperatures, the Normalized Difference Vegetation Index (NDVI), the Normalized Difference Built-up Index (NDBI), and the Modified Normalized Difference Water Index (MNDWI) obtained from Landsat-5 TM and Landsat-8 OLI for four-year intervals between 1992 and 2021. Their analysis revealed a positive correlation between urban expansion, the spatial extent of impervious surfaces, and surface temperatures, alongside climate change [37]. Mathew et al. demonstrated significant relationships between LST and various factors in their study spanning the summer and winter seasons from 2001 to 2021. They employed a wide range of input variables, including spectral indices, land characteristics, land-use/land-cover (LULC) data, atmospheric parameters, and spatial datasets, to perform their analyses [6].

Numerous studies emphasize the link between land surface temperature (LST) and landscape characteristics. This connection plays a crucial role in understanding ecosystem functions and shaping urban planning strategies. The aforementioned studies emphasize that LST varies significantly depending on factors such as land-cover types, landscape structure, and human activities. For instance, areas with dense vegetation typically exhibit lower LST values due to high evapotranspiration rates and shading effects, whereas impervious surfaces such as asphalt and concrete contribute to higher LST values through the urban heat island effect. Moreover, the spatial configuration of the landscape (e.g., fragmentation, connectivity, and landscape diversity) has been identified as a determinant of the spatial distribution of LST. In this context, sustainable landscape design is highlighted as an effective strategy for controlling LST and adapting to climate change [5,15,28,37].

Land surface temperature has a broad impact range, extending from atmospheric processes to ecological dynamics, from human activities to energy balances. In the physical dimension, surface temperature plays a decisive role in soil moisture, evaporation, precipitation patterns, wind dynamics, and microclimatic conditions [15,28]. Its effects on moisture and energy balance are particularly critical for the hydrological cycle and ecosystem services. From a human perspective, changes in land surface temperature intensify the urban heat island effect, directly influencing air quality, energy consumption, human health, and biodiversity. Rising temperature levels exert pressure on agricultural productivity, the sustainability of water resources, and infrastructure systems, thereby transforming social and economic structures. Therefore, a comprehensive analysis of the multidimensional effects of land surface temperature is crucial for developing climate change adaptation strategies and establishing sustainable environmental planning frameworks [2,23,28,38,39].

This study aims to analyze the effects of landscape on land surface temperature (LST) and land-use classification in the Muratpaşa District of Antalya by utilizing bands derived from Landsat satellite images for the years 1984, 1989, 1994, 1999, 2004, 2009, 2014, 2019, and 2024. The specific objectives of the research are as follows: to investigate the role of the landscape in illustrating LST, to measure the relationship between LST and different land-use types in the landscape, to examine and evaluate the temporal changes in these dynamics, and to provide concrete data for district-scale spatial planning and climate-change adaptation projects by correlating landscape metrics with LST analyses using Fragstats 4.2 software. Considering the lack of similar district-scale research in the national literature, this study aims to fill this gap and serve as a guide for comparable research. While global studies on LST and landscape relationships typically focus on broad geographic scales (e.g., national or regional levels), this research contributes detailed micro-scale (district-based) analyses to the international literature. Additionally, by addressing the relationship between landscape metrics and LST within a long-term change perspective, the study adopts a less common approach in the international literature. In this regard, it aims to provide a novel methodological and scale-based reference point for international research.

2. Data and Methods

Antalya, located in the Mediterranean region of Türkiye, is a city distinguished by its physical and human geographical characteristics and which occupies a strategic position. Known for its natural beauty, climate, and historical heritage, Antalya holds significant importance on both regional and national scales. While the city contributes substantially to the national economy through agricultural, tourism, trade, and industrial activities, rapid urbanization and population growth have brought about challenges in environmental and spatial planning. Issues such as unplanned construction, the reduction of agricultural areas, water pollution, and the destruction of natural vegetation are becoming increasingly prominent. Muratpaşa, one of Antalya's major districts, was granted its district status in 2008. Covering an area of 92 km^2 and consisting of 55 neighborhoods, it had a population of 512,700 as of 2023. The district experiences a typical Mediterranean climate, characterized by hot and dry summers and mild, rainy winters. In Antalya's city center, the long-term average temperature is recorded at 18.4 °C, with annual precipitation of approximately 1000 mm. Most of the precipitation occurs during the winter months, underscoring the region's importance for agricultural and water resources [40–43].

This study employs ArcGIS 10.8.1 and Fragstats 4.2.1, two widely used tools for spatial data analysis and landscape metrics evaluation. ArcGIS facilitates geographic data visualization and mapping, while Fragstats is specifically designed for quantifying landscape patterns. The integration of these tools allows for a comprehensive assessment of how landscape changes influence land surface temperature (LST) over time. By analyzing both spatial and temporal variations, this approach provides insights into how different land classifications contribute to temperature shifts. [5,15,44,45].

2.1. Land Surface Temperature (LST)

The baseline year for the analysis was selected as 1984, as it provided the most suitable historical data for conducting the study. To effectively capture changes over time, five-year intervals were chosen, and Landsat satellite imagery with low cloud cover, available from the United States Geological Survey (USGS), was utilized. The study analyzed imagery from Landsat-5 TM (Thematic Mapper), manufactured by Raytheon Santa Barbara Remote Sensing, Santa Barbara, CA, USA, for the years 1984, 1989, 1994, 1999, 2004, and 2009, and Landsat-8 OLI-TIRS (Operational Land Imager and Thermal Infrared Sensor), with the OLI instrument developed by Ball Aerospace & Technologies Corp., Boulder, CO, USA and the TIRS instrument built by NASA Goddard Space Flight Center, Greenbelt, MD, USA, for the years 2014, 2019, and 2024. As this study is based on a retrospective time series, the data were obtained from the USGS. These images formed the foundational dataset for the research and strengthened the analytical processes. The technical specifications of the imagery are presented in detail in Table 1. Landsat-5 TM captures data in visible, shortwave infrared (SWIR), near-infrared (NIR), and thermal infrared (TIR) ranges. These bands offer a spatial resolution of 30 m for bands 1–5 and 7 and 120 m for band 6. Landsat-8 OLI-TIRS captures images across multiple spectral and thermal ranges, with a spatial resolution of 30 m for bands 1–7 and 9, 15 m for band 8, and 100 m for bands 10 and 11. This satellite was integral to the spectral analyses carried out in this study [44,46].

Table 1. Metadata details of Landsat-5 TM (1984–2009) and Landsat-8 OLI-TIRS (2014–2024) satellite imagery.

Satellite Image	Image Number	Shooting Date	Path/Row	Format
LANDSAT-5 TM	LT51780341984198AAA05	16 July 1984	178-034	GeoTiff
LANDSAT-5 TM	LT51780341989211FUI00	30 July 1989	178-034	GeoTiff
LANDSAT-5 TM	LT51780341994193FUI00	12 July 1994	178-034	GeoTiff
LANDSAT-5 TM	LT51780341999207FUI00	26 July 1999	178-034	GeoTiff
LANDSAT-5 TM	LT51780342004205MTI00	23 July 2004	178-034	GeoTiff
LANDSAT-5 TM	LT51780342009202MOR01	21 July 2009	178-034	GeoTiff
LANDSAT-8 OLI-TIRS	LC81780342014184LGN01	3 July 2014	178-034	GeoTiff
LANDSAT-8 OLI-TIRS	LC81780342019182LGN00	1 July 2019	178-034	GeoTiff
LANDSAT-8 OLI-TIRS	LC81780342024196LGN00	14 July 2024	178-034	GeoTiff

To determine land surface temperature (LST) values, metadata from Landsat-5 TM (1984–2009) and Landsat-8 OLI-TIRS (2014–2024) satellite imagery were obtained from the USGS service (Table 1) [47].

The thermal infrared (TIR) bands of the satellite images were assessed for their suitability and pixel accuracy for the study area using ArcGIS 10.8.1. After verifying the data, the digital number (DN) values from the thermal infrared (TIR) bands were first converted to spectral radiance values using Formula 1 for land surface temperature (LST) calculation [5,47,48].

$$L_\lambda = \frac{(LMax_\lambda - LMin_\lambda)}{(QCalMax_\lambda - QCalMin)} \; X \; (DN - QCalMin) + LMin_\lambda \tag{1}$$

Formula (1) was used to calculate radiance values in LST.

In Formula (1), L_λ represents the spectral radiance value, DN is the cell value, Lmin and Lmax are the minimum and maximum spectral radiance values obtained from the metadata, and QCalMin and QCalMax are the calibrated minimum and maximum cell values. The calculated spectral radiance value was then converted to a Kelvin temperature value using Formula (2) [5,47,48].

$$T = \frac{K_2}{ln\left(\frac{K_1}{L_\lambda}\right) + 1} \tag{2}$$

Formula (2) was used to convert radiance value to Kelvin value.

In Formula (2), T represents the temperature in Kelvin, while K1 and K2 are calibration constants for the specific bands of the Landsat satellite imagery used in the study (Table 2). The obtained Kelvin temperature values were subsequently converted to degrees Celsius using Formula (3), enabling the calculation of LST values [5,47,48].

$$T(°C) = T - 273(K) \quad (3)$$

Table 2. Calibration constants for Landsat satellite used in LST calculation.

	Landsat-5 TM	Landsat-8 OLI TIRS	
		Band 10	Band 11
K1	607.76	774.89	1321.08
K2	1260.56	480.89	1201.14

Formula (3) was used to convert Kelvin value to degrees.

After conducting the LST analyses for the study area, the standard deviation classification method was employed to categorize the LST values into seven classes. This categorization of LST values into seven classes was chosen to facilitate a more detailed analysis of the spatial and temporal variations in temperature changes. Such classification helps identify extreme heat or cold regions and assess environmental risks. Additionally, the use of the standard deviation method allows for a more precise definition of temperature zones by considering the natural distribution of the data. This enables more accurate identification of relationships between temperature, land-use, or ecological processes. The equation required for determining the upper and lower temperature limits in the classification is provided in Formula (4) [5,38,45].

$$D = X \pm a \times S \quad (4)$$

Formula (4) was used to determine upper and lower temperature limits.

In Formula 4, D represents the boundaries of each category, aa denotes multiples of 0.5 ranging from 0 to 2.5, X refers to the mean LST of the image, and S represents the standard deviation. For the nine selected years in the study, X and S values were calculated separately. The categories defined are as follows: Extremely High Temperature Zone (EHTZ), High Temperature Zone (HTZ), Slightly High Temperature Zone (SHTZ), Medium Temperature Zone (MTZ), Slightly Low Temperature Zone (SLTZ), Low Temperature Zone (LTZ), and Extremely Low Temperature Zone (ELTZ). The formula for temperature thresholds used for each category is presented in Table 3 [5,49,50].

Table 3. Classification of the land surface temperature.

Temperature Grade	The Temperature Threshold
Extremely Low Temperature Zone (ELTZ)	$D \leq x - 2.5$ s
Low Temperature Zone (LTZ)	$X - 2.5$ s $< D \leq x - 1.5$ s
Slightly Low Temperature Zone (SLTZ)	$X - 1.5$ s $< D \leq x - 0.5$ s
Medium Temperature Zone (MTZ)	$X - 0.5$ s $< D \leq x + 0.5$ s
Slightly High Temperature Zone (SHTZ)	$X + 0.5$ s $< D \leq x + 1.5$ s
High Temperature Zone (HTZ)	$X + 1.5$ s $< D \leq x + 2.5$ s
Extremely High Temperature Zone (EHTZ)	$D > x + 2.5$ s

2.2. Land-Use Classification

Land use refers to the ways humans modify or utilize the natural or physical environment for various purposes. As part of this study, Landsat-5 and Landsat-8 OLI-TIRS (Operational Land Imager and Thermal Infrared Sensor) satellite images with low cloud cover (Table 1) for the years 1984–2024 were selected. These images were selected due to their high spatial resolution, which provides clearer visuals for their respective periods. To highlight changes between 1984 and 2024, a five-year time interval was chosen, and the best Landsat satellite image of each year was used to ensure high-quality analysis. The best satellite image for each year was selected based on the absence of cloud cover and the clear visibility of the district. To ensure consistency and optimal image quality, all images were chosen from July, during the summer season, when atmospheric conditions were most favorable for clear visibility. To obtain the natural appearance of Muratpaşa, a combination of red, green, and blue bands was created using ArcGIS 10.8.1 [51,52]. A sample group consisting of five land-use categories was created using the classification tool. These categories are green land (areas covered with forests, grasslands, parks, and natural vegetation), agricultural land (cultivated areas used for farming activities), constructive land (areas covered with human-made buildings and other infrastructure), water land (areas containing water bodies such as rivers and lakes) and bare land (areas without vegetation, generally covered with soil or rock). For each category, 50 samples were collected from the study area, and detailed land-use maps for the years 1984–2024 were created using the interactive supervised classification analysis tool within classification [5,20,39].

2.3. Landscape Metric Analysis

In another phase of the study, 10 indices were selected and evaluated within the framework of landscape ecology to analyze the landscape pattern characteristics of land cover in detail [5]. These indices were chosen to gain a deeper understanding of the structural and functional characteristics of land cover in the context of landscape ecology.

Patch-level indices examine smaller, granular units of land use and assess the numerical, morphological, and structural characteristics of individual patches in the landscape. These indices are used to understand the diversity and complexity in the internal structure of the landscape. For example, the number of patches (NP) measures the number of patches in the landscape, indicating the level of fragmentation of the landscape. Percentage of landscape (PLAND) determines the proportion of area of each land class in the landscape, which helps to understand the level of homogeneity or heterogeneity of the landscape. The largest patch index (LPI) expresses the proportion of the area of the largest patch in the landscape, revealing how large areas or large patches occupy a place in the landscape. The landscape shape index (LSI) measures the complexity of the landscape shape, helping to understand whether the landscape is flat or complex. The interspersion and juxtaposition index (IJI) determines how different classes in the landscape interact with each other and how close they are. Landscape-level indices analyze the overall structure and layout of the landscape from a broader perspective. These indices examine the order, patch density, diversity, and levels of connectivity in the overall structure of the landscape. Patch density (PD) measures the density of patches in the landscape, indicating the level of fragmentation and concentration of the landscape. The aggregation index (AI) analyzes the extent to which similar classes are grouped together in the landscape, providing information on the homogeneity or heterogeneity of the landscape. The contagion index (CONTAG) shows how different classes are connected to each other and how common they are in the landscape. The Shannon diversity index (SHDI) measures biodiversity in the landscape, while the Shannon evenness index (SHEI) analyzes how evenly this diversity is distributed. All of these indices provide a comprehensive analysis to understand the effects of environmental phenomena such as the urban heat island and to track changes in the landscape over time [5,40,53,54]. All these indices were calculated using the Fragstats v4.2.1 software. This comprehensive analysis, covering an extended period (1984–2024), aimed to improve the understanding of the urban heat island (UHI) phenomenon by assessing the impact of urbanization on surface temperatures [5,47,55].

3. Findings

3.1. Spatial and Temporal Changes in Urban Land Use and LST

Figure 1 and Table 4 illustrate the percentage changes in five land-use classes (agricultural land [AL], bare land [BL], built-up land [CL], green land [GL], and water land [WL]) between 1984 and 2024. The spatial distribution of these land uses is shown in Scheme 1a,c,e. Analyses of land use of this nature provide crucial insights into urban development, environmental sustainability, and land-management policies. Addressing these changes concerning global issues such as climate change and rapid urbanization is foundational for developing strategic plans.

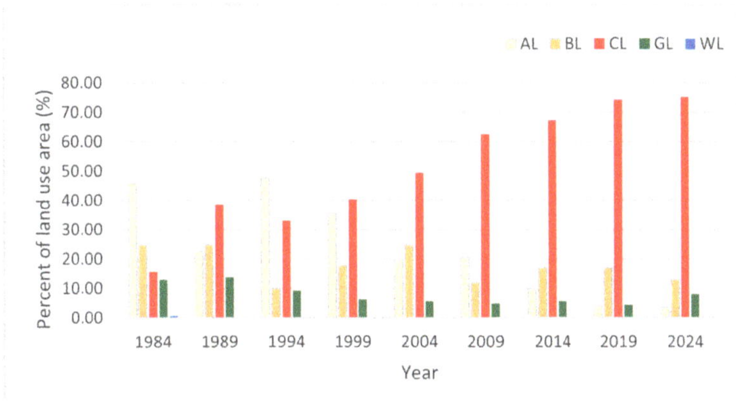

Figure 1. The area percentages of land-use types through the years.

Table 4. The area percentages of land-use types through the years.

Class	1984	1989	1994	1999	2004	2009	2014	2019	2024
AL	45.77	22.64	47.60	35.60	19.89	20.46	10.00	4.13	3.39
BL	24.83	24.78	9.97	17.81	24.90	11.96	17.04	16.96	13.04
CL	15.72	38.51	33.07	40.18	49.46	62.64	67.26	74.39	75.33
GL	12.97	13.78	9.31	6.35	5.71	4.87	5.67	4.48	8.21
WL	0.67	0.25	0.01	0.01	0.01	0.04	0.00	0.00	0.00

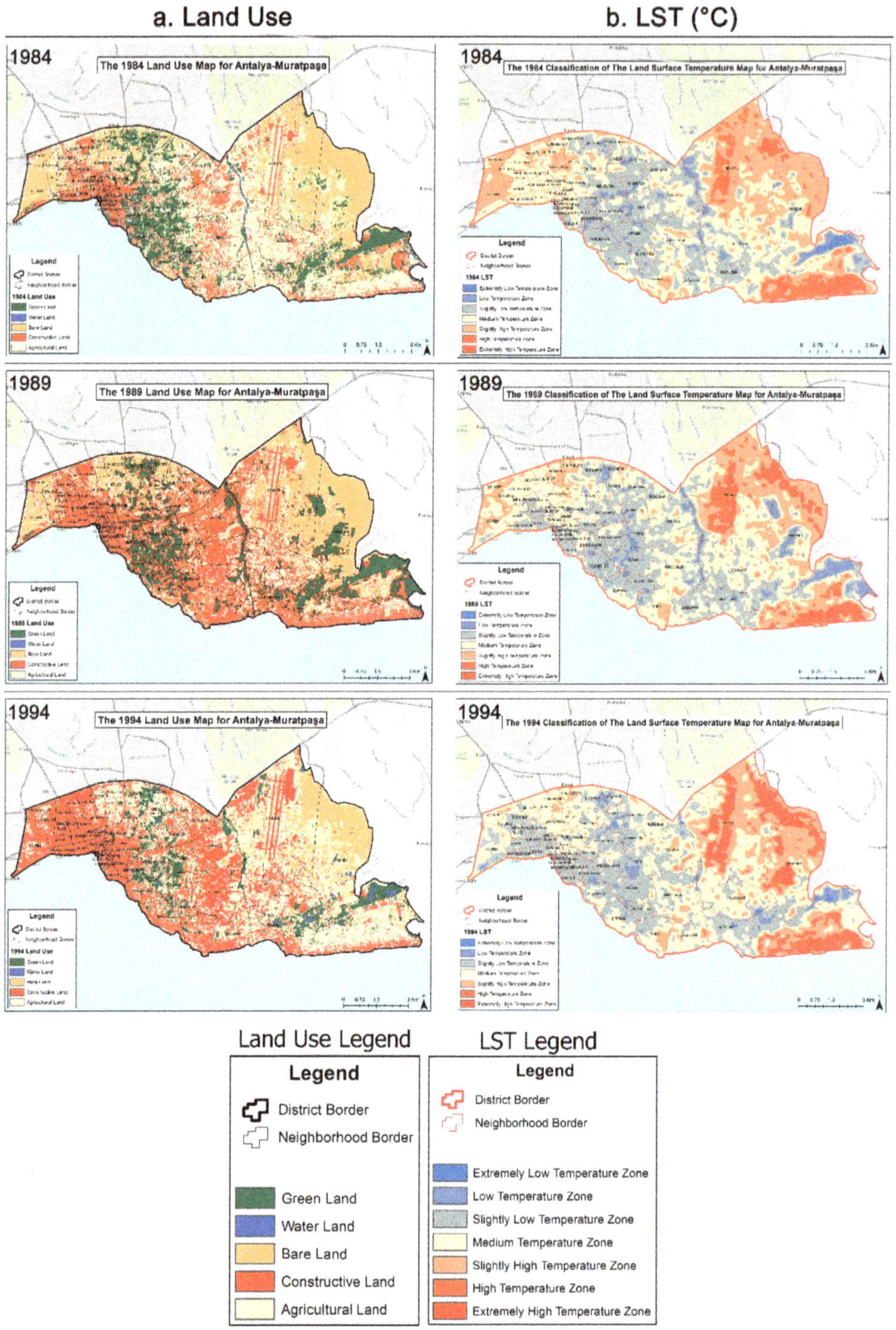

Scheme 1. *Cont.*

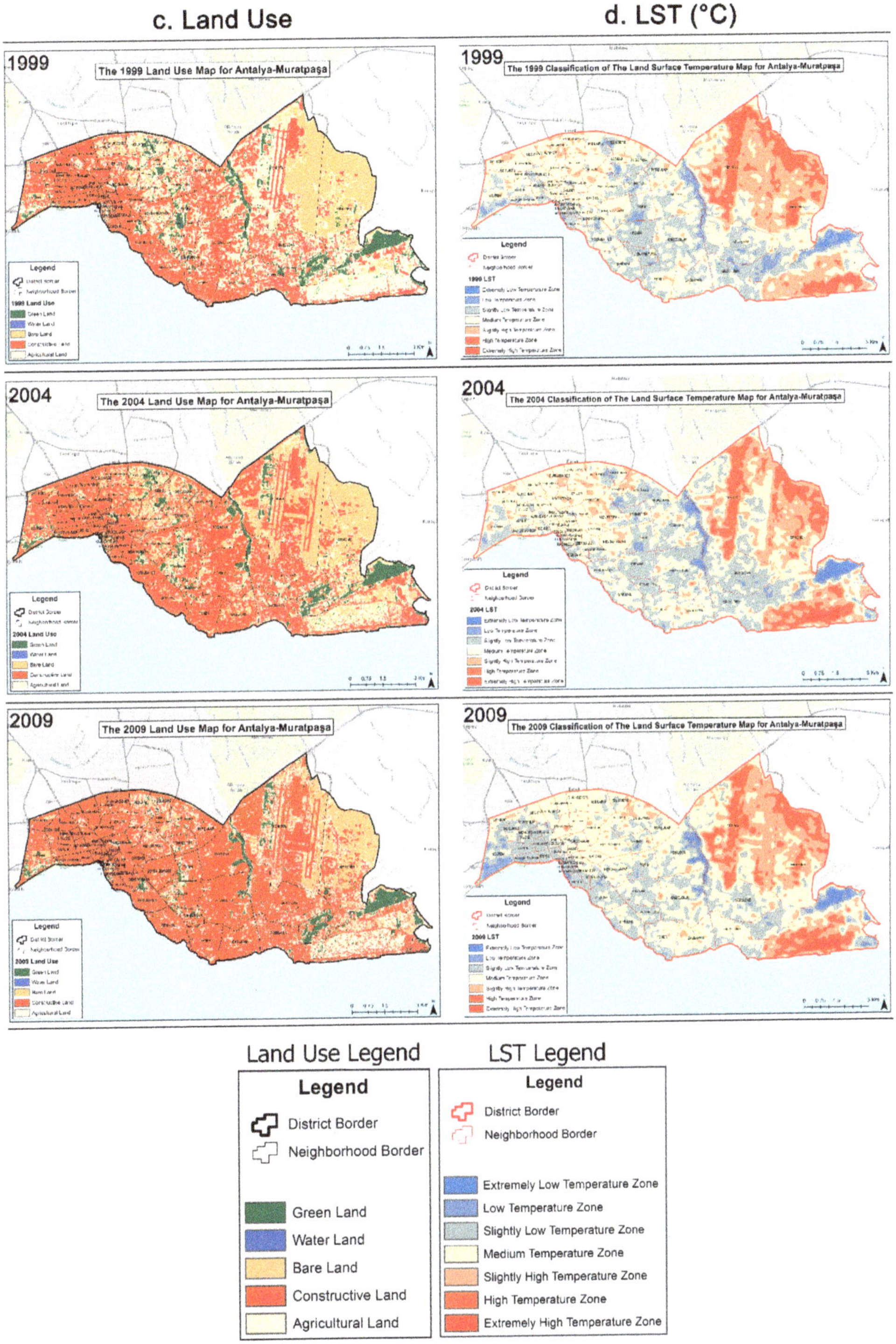

Scheme 1. *Cont.*

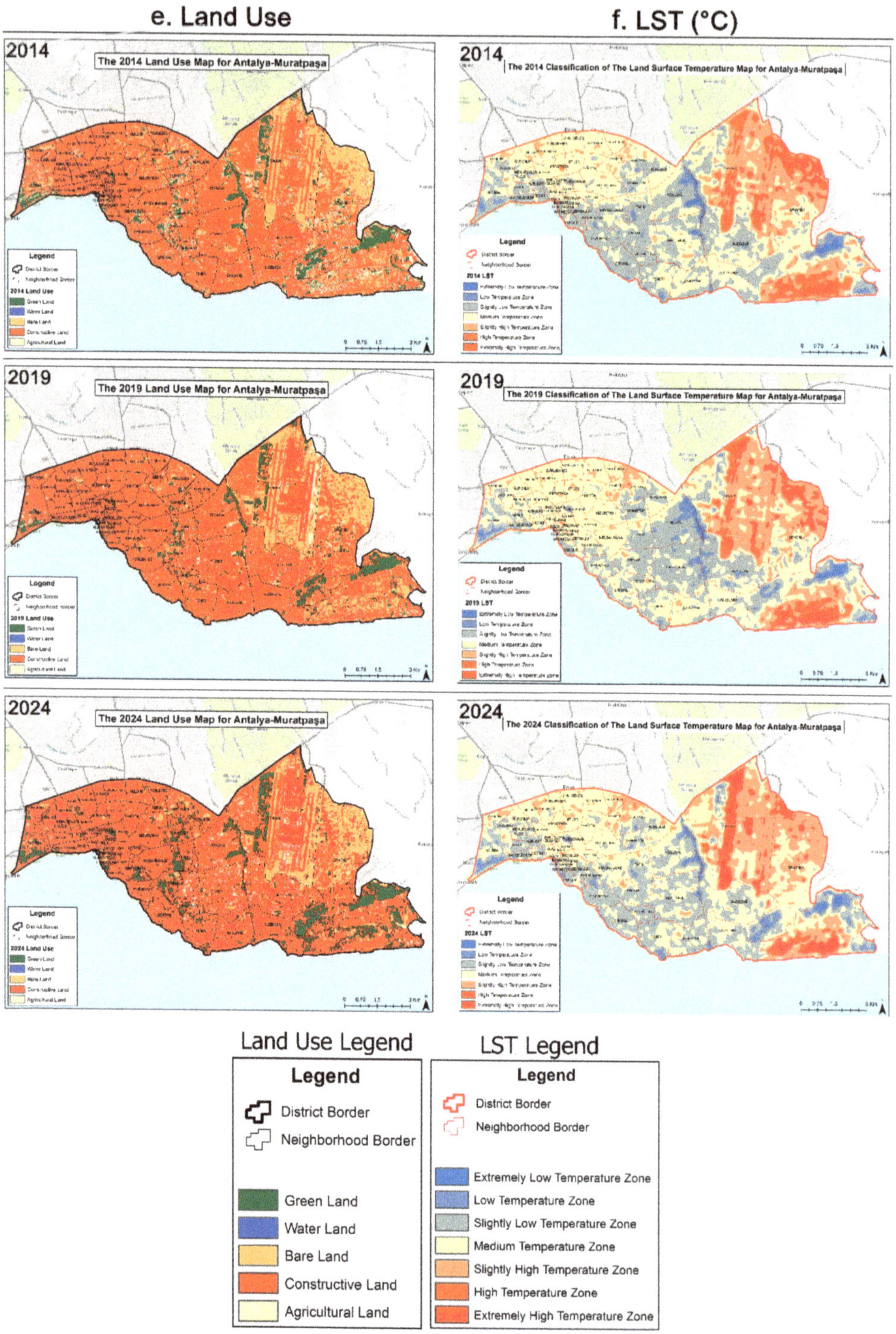

Scheme 1. The land-use maps and LST zones in 1984–2024.

In 1984, agricultural land accounted for the largest share of land use at 45.77%, but it has undergone a significant decline from 1989 to 2024. This decline became especially pronounced after 2004, with the percentage dropping to 19.89% in 2004 and further decreasing to a mere 3.39% in 2024. This reduction can be attributed to rural areas being overtaken by urbanization and the conversion of agricultural lands into industrial and residential areas. This decline raises critical concerns due to its potential adverse impacts on food security and rural economies. The percentages of bare land have fluctuated over time. While bare land had a high share of 24.83% in 1984, it decreased to 9.97% in 1994, rose again to 24.90% in 2004, and fell to 13.04% in 2024. These fluctuations can be explained by land rehabilitation efforts, natural disasters, or industrialization processes. The share of bare land is significant in terms of environmental impacts such as soil erosion and natural hazards. The percentage of built-up land has shown a continuous increase throughout the analyzed period. Starting at 15.72% in 1984, it rose dramatically to 75.33% in 2024. This sharp increase reflects urbanization, population growth, and economic development processes. The post-2004 surge in built-up land highlights the displacement of agricultural and green areas by urban development. This trend has environmental implications, including the loss of ecosystem services and biodiversity and the exacerbation of the urban heat island (UHI) effect. The percentage of green land exhibited a general decline over the study period but showed an increase in 2024. From 12.97% in 1984, it dropped to its lowest point of 6.35% in 1999 before recovering to 8.21% in 2024. This rebound may reflect increased environmental awareness and policies aimed at preserving green spaces. Protecting and increasing green land is critical for urban sustainability and public health. Water land consistently held the lowest share among land-use classes throughout the analyzed period. From 0.67% in 1984, it dropped to 0.00% in 2024. This significant decline is linked to factors such as climate change, excessive water use, and urban expansion. The loss of water bodies raises serious concerns for both ecological balance and sustainable water resource management.

Overall, the data from Table 4 and Figure 1 indicate substantial changes in land use between 1984 and 2024. Notably, the replacement of agricultural and green land with built-up land underscores the environmental consequences of urbanization. In this context, the development and implementation of sustainable land-management policies are vital for preserving ecosystems and combating climate change. Strategies such as increasing green land and protecting agricultural land are pivotal for achieving long-term environmental sustainability goals.

Figure 2, Tables 5 and 6 present the changes in land surface temperature (LST) values for the five land-use classes (AL, BL, CL, GL, and WL) from 1984 to 2024. The spatial distribution of LST is illustrated in Scheme 1b,d,f. LST serves as a key indicator for understanding the thermal properties of land-use types and their interactions with climatic processes.

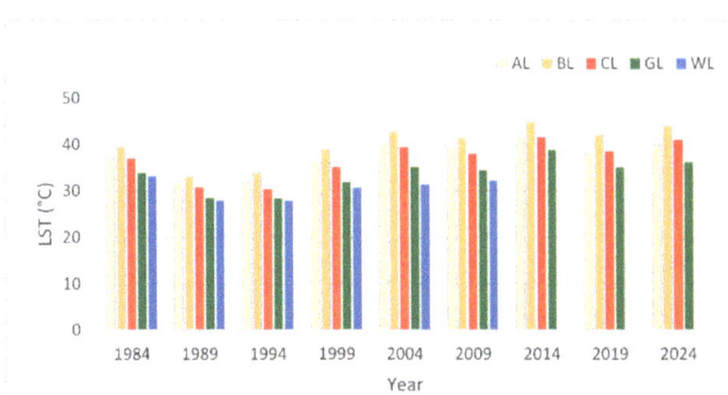

Figure 2. The area LST of each land-use type through the years.

Table 5. The area LST of each land-use type through the years (°C).

Class	1984	1989	1994	1999	2004	2009	2014	2019	2024
AL	37.00	31.94	31.76	36.16	40.13	39.37	41.48	38.07	39.66
BL	39.41	33.01	33.87	38.96	42.72	41.36	44.77	42.00	43.98
CL	36.95	30.80	30.27	35.05	39.37	37.89	41.53	38.50	40.92
GL	33.77	28.44	28.34	31.86	35.11	34.42	38.72	35.06	36.11
WL	33.11	27.97	27.90	30.65	31.32	32.31	0.00	0.00	0.00

Table 6. The area max–min LST of each land-use type through the years (°C).

Class	1984 Max–Min	1989 Max–Min	1994 Max–Min	1999 Max–Min	2004 Max–Min	2009 Max–Min	2014 Max–Min	2019 Max–Min	2024 Max–Min
AL	38.42–35.21	33.65–30.13	33.45–30.02	38.11–34.22	42.56–37.12	41.82–36.91	44.18–39.72	40.11–36.03	42.34–37.81
BL	41.83–37.02	35.76–30.92	36.98–31.24	41.29–36.55	46.12–40.45	45.44–38.71	48.62–41.85	45.92–39.78	47.81–41.62
CL	39.27–34.12	32.98–28.14	32.76–27.52	36.84–32.04	41.73–36.21	40.67–34.51	44.53–38.32	41.63–35.48	43.92–38.21
GL	35.87–31.12	30.82–26.32	30.74–26.11	34.95–29.77	38.91–33.51	37.64–31.34	41.72–36.41	38.65–33.28	40.13–34.67
WL	34.98–30.14	29.76–26.21	29.62–26.08	32.41–28.35	34.78–29.12	35.21–30.48	0.00–0.00	0.00–0.00	0.00–0.00

Agricultural land consistently exhibited high LST values throughout the studied years. The average LST increased from 37.00 °C in 1984 to a peak of 40.12 °C in 2004. In 2014, the LST rose again to 41.48 °C but dropped to 38.06 °C in 2019. By 2024, it had risen slightly to 39.65 °C. These fluctuations can be explained by factors such as the intensity of agricultural activities, vegetation changes, and irrigation practices. Additionally, the decline of agricultural land may correlate with the thermal influence of the built-up land that replaced it. Bare land exhibited the highest LST values among all classes. The LST rose from 39.40 °C in 1984 to 42.72 °C in 2004 and reached its maximum of 43.98 °C in 2024. These high temperatures can be attributed to the low vegetation cover and high solar radiation absorption of bare land. The increasing LST of bare land has significant environmental implications, such as soil moisture loss, higher erosion risk, and environmental degradation. The LST of built-up land showed a general upward trend during the period. From 36.94 °C in 1984, it increased to 39.36 °C in 2004 and again to 40.91 °C in 2024. This increase is directly associated with the UHI effect. The expansion of built-up land, replacing green areas, contributes to rising surface temperatures in urban regions. This phenomenon poses serious challenges to environmental sustainability, energy consumption, and public health in urban areas. Green land had relatively lower LST values compared to other classes, reflecting the thermal regulation capacity of vegetation. The LST increased from 33.76 °C in 1984 to 35.10 °C in 2004 and reached 36.11 °C in 2024, indicating a decline in its cooling effect. This trend may result from the degradation of vegetation due to the reduction of green land and environmental pressures. Preserving and expanding green land is crucial for mitigating urban temperatures and addressing climate-change impacts. Water land, due to its thermal properties, exhibited the lowest LST values among all land-use classes. The LST decreased from 33.11 °C in 1984 to 31.32 °C in 2004, but after 2014, it was recorded as 0 °C. This indicates a significant loss or complete disappearance of water land after 2014. The loss of water bodies can have severe adverse effects on local ecosystems and environmental processes. Protecting water resources is vital for maintaining the water cycle and climate regulation.

The effects of land-use types on LST are directly linked to global challenges such as urban expansion, environmental sustainability, and climate change. Land-use classes such as agricultural land, bare land, and built-up land contribute to environmental degradation and increased energy demands with their high LST values. Conversely, classes like green land and water land play a critical role in maintaining thermal balance with their lower LST values. Overall, in the years when agricultural land (AL) was replaced by built-up land (CL), a significant increase in land surface temperature (LST) values was observed. This trend indicates that the transformation of extensive agricultural areas into settlements and built-up spaces has led to a warmer region and an intensified urban heat island effect. During this period of continuous growth in CL, changes in land use have generally contributed to rising temperatures.

3.2. The Relationship Between LST and Land-Use Percentages

Figure 3 elaborates on the complex relationship between land use and surface temperatures, detailing changes over 40 years from 1984 to 2024. Analyzing the transformations within each temperature range alongside land-use observations provides a deeper understanding of the underlying environmental and urban dynamics driving these changes.

In 1984, green land (GL) accounted for 97.56% of the Extremely Low Temperature Zone (ELTZ), but this share dropped to 86.35% in 2024. This decline highlights the reduction in the cooling effect of green areas, which in turn led to a decrease in Low Temperature Zones. During the same period, the share of constructive land (CL) and agricultural land (AL) within the ELTZ increased from 0.83% and 0.82% in 1984 to 12.21% and 1.24%, respectively, by 2024. This shift reflects the contraction of the ELTZ boundary due to increased urbanization, with green spaces giving way to built-up areas. In 1984, GL covered 54.51% of the Low Temperature Zone (LTZ), but this share increased to 62.66% in 2024. While this suggests that green areas still play a significant role in Low Temperature Zones, the overall cooling effect has diminished. Meanwhile, CL's share within the LTZ rose significantly from 3.91% in 1984 to 34.23% in 2024. This increase indicates that urban influences are becoming more pronounced, even in low temperature areas. The most notable changes in the Slightly Low Temperature Zone (SLTZ) were observed in the CL and AL classes. CL expanded dramatically from 13.87% in 1984 to 83.07% in 2024, underscoring the dominance of urban sprawl within the SLTZ. Conversely, AL's share decreased sharply from 53% in 1984 to 4.81% in 2024, reflecting the significant loss of agricultural land to urbanization. In the Medium Temperature Zone (MTZ), AL accounted for 49.91% in 1984 but plummeted to 3.15% by 2024. This indicates that agricultural lands have become more vulnerable to temperature increases, with urbanization playing a dominant role. In contrast, CL increased its share within the MTZ from 23.94% in 1984 to 87.00% in 2024, signifying a marked expansion of urban areas in this temperature range. Bare land (BL) also exhibited a notable increase, rising from 19% in 1984 to 6.16% in 2024. Within the Slightly High Temperature Zone (SHTZ), CL's share grew from 10.48% in 1984 to 57.15% in 2024. This reflects the increasing dominance of built-up areas in higher temperature zones and highlights the intensification of the urban heat island effect. Meanwhile, AL's share within the SHTZ dropped from 31.09% in 1984 to 3.04% in 2024, indicating a declining presence of agricultural lands in these temperature ranges as they were replaced by urban development. In the High Temperature Zone (HTZ), the most significant change was observed in the CL class, whose share rose from 9.11% in 1984 to 54.76% in 2024. This indicates that the HTZ is now largely dominated by built-up areas. AL, on the other hand, saw its share within the HTZ decrease from 58.92% in 1984 to 1.22% in 2024, suggesting that agricultural lands have almost completely disappeared from high temperature areas. In the Extremely High Temperature Zone (EHTZ), the impact of CL was particularly striking. While CL was completely absent in this zone in 1984 (0%), its share reached 56.06% by 2024. This dramatic change demonstrates the extent to which urban expansion has become influential, even in extreme temperature areas. In contrast, AL, which accounted for 100%

of the EHTZ in 1984, dropped to just 0.42% in 2024, indicating the near-total disappearance of agricultural lands from the hottest zones.

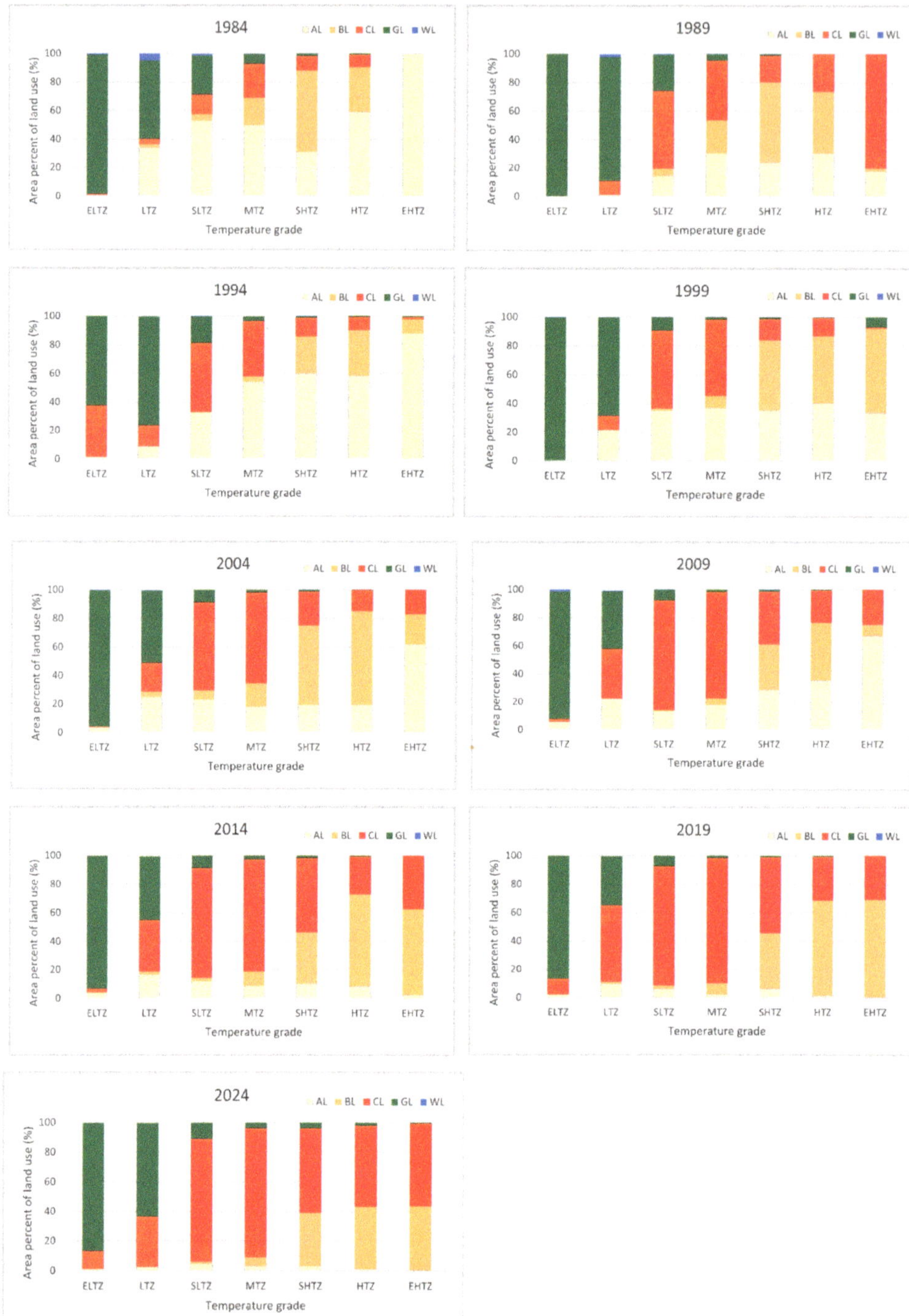

Figure 3. The area percent of each land-use type in different LST zones in 1984–2024.

The changes in land use from 1984 to 2024, as shown in Figure 3, are directly linked to increases in surface temperatures. The CL class exhibited the most significant growth across nearly all temperature ranges, confirming that the expansion of urban areas and built-up land has intensified surface temperatures. In contrast, the substantial percentage decreases in the AL and GL classes highlight the weakening of their temperature-regulating effects and emphasize the role of urbanization in exacerbating warming trends.

3.3. Analysis of LST Correlations Among the Years

The correlation matrix illustrates the linear relationships among the land surface temperatures (LST) of 1984, 1989, 1994, 1999, 2004, 2009, 2014, 2019, and 2024 (Table 7). Correlation coefficients range from −1 to 1, where 1 indicates a perfect positive correlation, −1 signifies a perfect negative correlation, and 0 denotes no correlation. This matrix is a critical tool for understanding how various environmental variables are interconnected over the years.

Table 7. Correlation matrix of LST values between 1984 and 2024.

Years	1984	1989	1994	1999	2004	2009	2014	2019	2024
1984	1	0.77448	0.64581	0.6272	0.59015	0.55147	0.56433	0.53931	0.51412
1989	0.77448	1	0.71567	0.67682	0.62447	0.59491	0.58347	0.55544	0.55729
1994	0.64581	0.71567	1	0.76627	0.69127	0.67531	0.62601	0.57528	0.59327
1999	0.6272	0.67682	0.76627	1	0.80171	0.76522	0.72035	0.70318	0.7135
2004	0.59015	0.62447	0.69127	0.80171	1	0.76769	0.73337	0.71679	0.69547
2009	0.55147	0.59491	0.67531	0.76522	0.76769	1	0.82422	0.77731	0.77667
2014	0.56433	0.58347	0.62601	0.72035	0.73337	0.82422	1	0.8341	0.79066
2019	0.53931	0.55544	0.57528	0.70318	0.71679	0.77731	0.8341	1	0.80401
2024	0.51412	0.55729	0.59327	0.7135	0.69547	0.77667	0.79066	0.80401	1

The correlation coefficients derived from the LST data, which were categorized into seven distinct classes, are positive, indicating that LST exhibited a largely consistent pattern of variation over the years. This consistency reflects the persistent influence of environmental factors on LST. The correlation coefficient between 2014 and 2019, at 0.83410 (the highest value), indicates a very strong positive relationship, signifying that the LST data for these years are highly similar and that the changes observed were largely parallel. This high correlation suggests that environmental conditions between 2014 and 2019 remained relatively stable, resulting in comparable effects on LST. It also implies that a comparative analysis between these two years would yield reliable insights. On the other hand, the correlation coefficient between 1999 and 1984, at 0.62720 (the lowest value), indicates a moderate positive relationship. While there is still a similarity in LST data between these two years, certain differences are evident. This moderate correlation suggests that specific environmental changes, likely influenced by urban development, land-use alterations, or climatic shifts, contributed to some divergence in LST patterns during this period.

Overall, the correlation matrix demonstrates high positive correlations between LST values across the years, with notably stronger correlations observed between consecutive years. This indicates that LST variations followed a consistent trajectory over time. The strongest positive correlation, between 2014 and 2019, highlights the similarity in LST data and stable environmental conditions during these years. In contrast, the lowest correlation, between 1999 and 1984, reflects greater variability in environmental influences on LST.

These findings are invaluable for urban planning, assessing the impacts of climate change, and developing future environmental management strategies.

3.4. Correlation Analysis Between LST and Land-Use Types

To evaluate the impact of land use on land surface temperatures (LST) in Muratpaşa District, correlation analyses were performed using data from the respective years. However, to better visualize and understand the changes, a comparison was made between the first year of the study, 1984, and the last year, 2024. This comparison reveals the transformation in the influence of land use on LST, categorized into seven classes, over time.

In 1984 (Table 8), it was observed that green areas and water bodies in Muratpaşa District exerted a notable cooling effect on surface temperatures. The strong negative correlation between green areas and LST (−0.904) underscores the natural climate-regulating role of these areas, while water bodies demonstrated a similar cooling effect with a negative correlation (−0.788). Conversely, bare areas (r = 0.770) and built-up areas (r = 0.234) exhibited positive correlations with LST, indicating that these land-use types contributed to increased surface temperatures. These findings provide critical insights into how land use influenced LST in 1984, emphasizing the importance of sustainable land-management strategies. This analysis reveals the dynamic relationship between land use and LST and highlights the transformative impact of urbanization, land-cover changes, and environmental management strategies on the region's thermal environment.

Table 8. Correlation matrix between LST and land-use types in 1984.

Model	LST	AL	BL	CL	GL	WL
LST	1	0.281131546	0.770478126	0.234268777	−0.904131177	−0.788383493
AL	0.281131546	1	−0.346097019	0.477740551	−0.346207515	−0.269627897
BL	0.770478126	−0.346097019	1	0.03332565	−0.729321488	−0.63620595
CL	0.234268777	0.477740551	0.03332565	1	−0.565144457	−0.60029086
GL	−0.904131177	−0.346207515	−0.729321488	−0.565144457	1	0.868467106
WL	−0.788383493	−0.269627897	−0.63620595	−0.60029086	0.868467106	1

By 2024 (Table 9), significant changes can be observed in the impacts of land use on LST within the district. The correlations between LST and both bare lands and built-up areas have reached remarkably high values of 0.873 and 0.965, respectively. This indicates that urban sprawl and land-use changes have substantially intensified their effects on surface temperatures. Agricultural areas also exhibit a positive correlation (r = 0.753), revealing that their influence on LST has strengthened over time. Green areas continue to have a strong negative correlation (−0.969), underscoring their role in mitigating surface temperatures. The data regarding water areas point to a decrease or complete loss of water resources in the region.

Overall, an analysis of the correlation data from 1984 and 2024 reveals a significant transformation in the impacts of land use on surface temperatures in the Muratpaşa District. In 1984, green areas (−0.904) and water areas (−0.788) played a substantial role in lowering LST. By 2024, these effects persist, yet the dominant influence of urbanization and bare lands has significantly increased. Built-up areas (r = 0.965) and bare lands (r = 0.873) exhibit strong warming effects on LST, highlighting the growing pressure of urbanization and land-use changes on surface temperatures.

Table 9. Correlation matrix between LST and land-use types in 2024.

Model	LST	AL	BL	CL	GL	WL
LST	1	0.7526	0.8735	0.9651	−0.9686	0
AL	0.7526	1	0.4056	0.8537	−0.8396	0
BL	0.8735	0.4056	1	0.8222	−0.8370	0
CL	0.9651	0.8537	0.8222	1	−0.9996	0
GL	−0.9686	−0.8396	−0.8370	−0.9996	1	0
WL	0	0	0	0	0	1

3.5. Changes in LST and Landscape-Level Indices

The analyses conducted between 1984 and 2024 comprehensively illustrate the substantial changes in land use and their effects on LST, categorized into seven classes, within the Muratpaşa District. During this period, the environmental impacts of urban sprawl, particularly the increase in built-up areas and the loss of natural areas, are evident.

In 1984, agricultural areas (AL) covered the largest surface area at 45.01%, followed by bare lands (BL) at 24.95%, green areas (GL) at 12.99%, and built-up areas (CL) at 16.34%. Water areas comprised only 0.71%. At this time, agricultural and green areas dominated the landscape, while urban areas occupied relatively smaller proportions. The landscape metrics included an LPI of 34.94%, an LSI of 50.42, an SHDI of 1.302, and an SHEI of 0.809. The high SHDI and SHEI values indicate a diverse and heterogeneous landscape.

In 1989, the proportion of agricultural areas (AL) decreased to 23.25%, while built-up areas (CL) increased significantly to 38.06%. This shift reduced the surface areas of other land-use classes, signaling a trend toward urbanization. Water areas remained low at 0.31%. The landscape metrics (LPI = 34.94%, LSI = 50.42, SHDI = 1.302, and SHEI = 0.809) reflected the growing dominance of urban areas and the decline in agricultural and green areas.

In 1994, agricultural areas (AL) expanded to 47.06%, while built-up areas (CL) declined to 33.22%. The proportion of green areas (GL) was 9.42%, and bare lands (BL) accounted for 10.29%. Water areas remained minimal at 0.014%. The metrics (LPI = 32.43%, LSI = 42.85, SHDI = 1.1785, and SHEI = 0.7322) suggest a temporary increase in agricultural areas and a decline in urbanization but also highlight reduced landscape diversity and heterogeneity.

In 1999, the proportions shifted again, with agricultural areas (AL) at 35.64%, built-up areas (CL) at 39.76%, green areas (GL) at 6.46%, and bare lands (BL) at 18.13%. Water areas were nearly negligible (0.0115%). The metrics (LPI = 29.28%, LSI = 44.05, SHDI = 1.222, and SHEI = 0.7593) reflect a rapid increase in urban areas and a loss of green spaces, underscoring the environmental impacts of urban expansion.

In 2004, built-up areas (CL) had risen to 48.49%, while green areas (GL) decreased to 5.73%. Bare lands (BL) accounted for 25.11%, and agricultural areas (AL) dropped to 20.65%. Water areas were nearly absent (0.0135%). The metrics (LPI = 41.60%, LSI = 50.37, SHDI = 1.1889, and SHEI = 0.7387) indicate the dominance of urban areas and the ongoing destruction of natural habitats, with urban areas exerting increasing control over the landscape.

In 2009, built-up areas (CL) rose to 61.61%, while green areas (GL) declined further to 4.90%. Bare lands (BL) decreased to 12.03%, and agricultural areas (AL) slightly increased to 21.41%. Water areas remained at 0.0432%. The metrics (LPI = 56.94%, LSI = 43.55, SHDI = 1.0344, and SHEI = 0.6427) emphasize the unsustainable environmental impacts of urban sprawl and the declining presence of green spaces.

In 2014, built-up areas (CL) reached 65.63%, with green areas (GL) at 5.68%, bare lands (BL) at 17.73%, and agricultural areas (AL) at 10.96%. Water areas were entirely absent. The metrics (LPI = 63.32%, LSI = 43.89, SHDI = 0.9884, and SHEI = 0.713) reveal the intensification of urban development, the continued loss of natural habitats, and the growing dominance of urban areas in the landscape.

In 2019, built-up areas (CL) continued to rise, reaching 72.98%, while green areas (GL) decreased to 4.61%. Bare lands (BL) remained at 17.50%, and agricultural areas (AL) dropped to 4.91%. Water areas were nonexistent. The metrics (LPI = 71.37%, LSI = 33.69, SHDI = 0.8247, and SHEI = 0.5949) highlight the severe environmental consequences of urban sprawl and the ongoing reduction of green spaces.

In 2024, built-up areas (CL) reached their highest proportion at 73.90%, while green areas (GL) increased slightly to 8.69%. Bare lands (BL) accounted for 13.47%, and agricultural areas (AL) dropped to 3.95%. Water areas remained absent. The metrics (LPI = 72.15%, LSI = 32.10, SHDI = 0.8334, and SHEI = 0.6012) reflect the peak dominance of urban areas, the limited but noticeable increase in green spaces, and the ongoing loss of natural habitats in the district.

4. Discussion

This study conducted in Antalya's Muratpaşa District demonstrates how the urban heat island (UHI) effect evolves through strong correlations between land-use changes and land surface temperatures (LST). During a period characterized by rapid urbanization and a decline in natural surfaces, the expansion of impervious surfaces has reduced evapotranspiration, ultimately leading to increased LST. These findings confirm the earlier literature that highlights the strong alignment between built-up areas (CL) and High Temperature Zones (HTZ), as well as the association of green spaces (GL) with Low Temperature Zones (LTZ) [5,22,26,47,49,56–58].

Our findings clearly illustrate the dramatic changes in urban land use and their impact on LST between 1984 and 2024. The percentage of agricultural areas declined from 45.77% to 3.39%, while water areas decreased from 0.67% to complete disappearance. These trends are striking indicators of urban sprawl and land transformation. Conversely, the proportion of built-up areas increased from 15.72% to 75.33%, emphasizing the dominant role of urban development in Muratpaşa. These results are consistent with the trends of urban expansion reported in other studies [5,44,56,58]. Similarly, in other geographic contexts, the loss of agricultural areas and the increase in built-up areas highlight the environmental impacts of urbanization.

The findings of this study reveal that the relationships between land-use classes (CL, AL, BL, GL, and WL) and LST are closely tied not only to the size of these surface areas but also to their distribution and structural characteristics. For instance, large and homogeneous CL areas tend to produce higher LST values, whereas fragmented and irregularly shaped GL and WL areas are more effective in reducing LST. This emphasizes the importance of considering not only the size of surface areas but also their spatial configuration in urban landscape planning, aligning with findings in previous studies [1,28,59]. Extensive and cohesive green areas (GL) with low fragmentation levels demonstrate significant potential to lower LST, while dense and extensive CL areas contribute to elevated LST. This finding, consistent with earlier research, underscores the critical role of landscape composition and configuration in managing the UHI effect. However, further research is necessary to better understand the influence of landscape configurations on LST. In particular, advanced multivariate analyses and long-term datasets are essential to fully comprehend the contributions of landscape pattern metrics to cooling and heating effects, consistent with prior studies [47,51,60].

The former presence of water areas in Muratpaşa, before their complete disappearance, aligns with previous studies emphasizing the critical role of water bodies in mitigating

LST [5,19,46]. This finding highlights the importance of prioritizing the conservation of water resources to ensure environmental sustainability and mitigate climatic impacts.

The increase in built-up areas has shown a clear impact on LST in our findings. The average LST for built-up areas increased from 36.94 °C in 1984 to 40.91 °C in 2024. This rise serves as an indicator of the UHI effect and reflects the environmental consequences of urban expansion. On the other hand, green areas are associated with lower LST values, reaffirming their critical role in maintaining thermal balance. These findings align with the various studies that have highlighted the environmental benefits of green areas [19,26,57].

In summary, the findings of this study underscore the necessity of developing sustainable land-use policies in Antalya's Muratpaşa District. To reduce the UHI effect and balance the local climate, efforts should focus on expanding green infrastructure, minimizing impervious surfaces, and promoting diverse and complex landscape configurations in urban planning. Such strategies not only have positive implications for the local climate but also contribute to enhancing urban quality of life [5,8,13,47].

5. Conclusions

This study comprehensively examined the relationship between land surface temperatures (LST) and land-use classes in Antalya's Muratpaşa District from 1984 to 2024 at five-year intervals. Understanding the impacts of land-use changes on LST is critically important for urban planning and environmental management. In this context, the relationship between LST and land-use classes, especially built-up areas (CL) and green areas, emerged as a key focus of this analysis [5].

The findings revealed that the expansion of urban areas has significantly increased surface temperatures. While the influence of built-up areas (CL) on LST was limited in 1984 (22.67%), this effect markedly intensified by 2024, substantially elevating surface temperatures. This increase highlights a direct correlation between the intensity of urbanization and the urban heat island effect. Notably, the correlation coefficient between built-up areas and LST reached 0.965 by 2024, underscoring the rapid rise in surface temperatures in densely urbanized regions and the pressing influence of built-up areas on the local climate [47,59]. Similar studies have reported parallel findings, demonstrating that the growth of impervious surfaces in urban centers not only contributes to increased LST but also intensifies heat concentration within densely populated zones [26,58]. Moreover, impervious surface expansion has been linked to reduced evapotranspiration, further exacerbating the urban heat island effect [5,47].

The relationship between green areas and LST, on the other hand, exhibited a negative trend over this period. The reduction in green areas from 1984 to 2024 accelerated the increase in surface temperatures. In 1999, when green areas reached their lowest share (6.35%), a significant rise in LST was observed. However, by 2024, the resurgence of green areas (8.21%) played a crucial role in mitigating surface temperature increases. A strong negative correlation coefficient of −0.969 between green areas and LST indicates the potential of green spaces to balance urban surface temperatures and reduce the urban heat island effect [50,57]. Previous research has consistently shown that cohesive and large-scale green areas are highly effective in reducing LST through cooling effects such as shading and evapotranspiration [28,59]. Fragmented or poorly connected green spaces, however, lose their cooling efficiency, emphasizing the need for integrated landscape planning [1,47].

Water areas underwent notable changes during this period. While limited in 1984, some water areas had disappeared entirely by 2024. The reduction in water areas weakened their cooling effect on LST, a trend attributed to factors such as climate change, excessive water consumption, and urban sprawl. The loss of water bodies adversely affected not only surface temperatures but also ecosystem services and biodiversity [5,57]. This aligns with findings in the literature, where water bodies have been identified as key elements in mitigating urban heat due to their thermal inertia and evaporative cooling effects. The disappearance of water areas has been linked to reduced resilience against heat waves and increased urban heat stress in other studies, as well [19,46].

Agricultural areas also played a significant role in influencing surface temperatures. The share of agricultural areas decreased from 45.77% in 1984 to just 3.39% in 2024, largely due to their conversion into built-up areas. This decline posed risks to both local economies and food security while indirectly contributing to LST increases [5]. Similar findings in other regions have demonstrated that the conversion of agricultural areas to urban land not only raises local temperatures but also disrupts regional climate regulation and soil health [5,22]. Furthermore, the loss of agricultural land is often irreversible, making its preservation crucial for sustainable urban development.

In conclusion, this study demonstrates the profound impact of land-use changes on LST in the Muratpaşa region of Antalya. The expansion of built-up areas has rapidly elevated surface temperatures, while the preservation and expansion of green areas have been critical in balancing these temperature increases. In particular, areas such as the Yıldız, Yüksekalan, Gebizli, and Varlık neighborhoods are experiencing significant reductions in green space, which contributes to higher local temperatures. Local authorities should consider developing green-space expansion strategies specifically tailored to these districts, such as converting vacant lots into urban parks and introducing more street-level greenery. With the addition of the necessary equipment for rainwater harvesting from rooftops in the parks to be created, water needs can be met and used to reduce the surface temperature of the land in that area in summer. The decline in water and agricultural areas has raised serious concerns for climate and ecosystem stability in the region. To address this, policies that promote the conservation and restoration of wetland areas around the Tarım, Kırcami, and Yeşilova regions, as well as the preservation of agricultural land near the eastern part of Muratpaşa, should be prioritized. This could include zoning regulations that limit urban encroachment on these vital areas. Developing sustainable urban-planning and land-management policies is essential to controlling LST increases and ensuring ecosystem continuity. To mitigate LST increases, the implementation of green infrastructure, such as green roofs and permeable pavements, should be mandated in new commercial and residential developments, especially in dense urban areas like the Meydankavağı and Varlık districts. Encouraging green infrastructure practices and the conservation of water resources are highlighted as effective strategies to mitigate the urban heat island effect [5]. Consistent with prior studies, the implementation of green roofs, urban forests, and permeable pavements has proven effective in reducing the negative impacts of the urban heat island effect [19,26,57]. These strategies not only contribute to cooling but also enhance biodiversity and water management in urban environments. Specific policies such as tax incentives or subsidies for building owners who install green roofs or adopt permeable paving could be introduced in key areas with high impermeable surfaces, including the Sinan and Zerdalilik neighborhoods. Overall analysis of the findings reveals that urbanization processes in Antalya's Muratpaşa District are in conflict with the expansion and preservation of green spaces. Urban planning fails to align with sustainability goals, prioritizing policies that focus on increasing construction and commercial areas. In the case of rapid development in the Muratpaşa and Kılınçarslan neighborhoods, a more integrated approach is needed, where new developments are planned with sufficient green space allocations. Zoning laws should require that any new building projects incorporate green space elements within their plans. This approach not only leads to the destruction of existing green spaces but also creates significant obstacles to establishing new ones. The insufficient measures taken by local authorities to protect green spaces, the inadequacy of alternative energy policies, and the lack of environmentally friendly infrastructure projects further exacerbate the problem. In other words, actions aimed at increasing green spaces often conflict with economic growth and urban development priorities. It is crucial to integrate landscape ecology principles into urban planning practices, as highlighted in other studies. For instance, in areas with high development pressure, such as the Çağlayan and Güvenlik neighborhoods, the creation of ecological corridors and buffer zones around green spaces could help reduce fragmentation and preserve biodiversity. Strategies such as creating buffer zones, maintaining ecological corridors, and prioritizing mixed-use development have been shown to mitigate the negative impacts of urban sprawl on LST and biodiversity [28,47,51]. These key findings are summarized as follows:

1. Rising Urban Heat Island Effect: The urban heat island effect in Muratpaşa, Antalya, has intensified over the past 40 years. In 1984, with limited built-up areas, heat hotspots were concentrated in the city center. By 2024, the rapid and extensive expansion of built-up areas had spread heat hotspots across the entire city, exacerbating the thermal environment. This highlights the significant pressure rapid urbanization places on surface temperatures.
2. LST Differences Among Land-Use Classes: Notable differences in LST were observed among land-use classes, with built-up areas (CL) exhibiting the highest LST values. According to 2024 data, the LST hierarchy is as follows: CL > agricultural areas (AL) > bare land (BL) > green areas (GL) > water areas (WL). Built-up areas were identified as the largest contributors to the urban heat island. Although green areas provided cooling effects through transpiration and shading, their limited presence in the region constrained their impact. Water areas, with their expansive surfaces, were the most effective in reducing LST, but their decline reduced this cooling effect.
3. Land Use and Patch Indices: Relationships between land-use classes and patch indices revealed the impact of land fragmentation on LST. Larger green areas (GL) with lower fragmentation, denser distribution, and more complex shapes were associated with lower LST, contrasting with built-up areas (CL). The reduction and fragmentation of agricultural areas (AL) contributed to rising LST. These findings underscore the importance of landscape design and land-management strategies in controlling surface temperatures.
4. LST and Landscape Metrics: The relationship between LST and landscape-level indices revealed complex dynamics affecting surface temperatures. LST showed negative correlations with patch density (PD), the landscape shape index (LSI), interspersion and juxtaposition index (IJI), Shannon diversity index (SHEI), and Shannon homogeneity index (SHDI), and positive correlations with the contagion index (CONTAG), largest patch index (LPI), and area index (AI). These findings emphasize that increasing landscape diversity and reducing fragmentation are critical for mitigating surface temperatures. Diverse and less fragmented landscapes proved more effective in balancing surface temperatures compared to homogeneous and expansive landscapes.
5. Recommendations for Mitigation: To reduce surface temperatures, landscape diversity should be increased, and connectivity among different landscape types should be minimized. Expanding green areas and reducing the density of built-up areas are essential strategies for alleviating the urban heat island effect. Urban planning and land-management policies must be revisited in line with sustainability principles. This study provides critical insights into the environmental impacts of urbanization in Muratpaşa, Antalya, and offers a valuable guide to prevent similar challenges in the future [5,61].

Author Contributions: Conceptualization, M.T.Ş., H.H., Ç.Ç. and Ü.Y.; Methodology, M.T.Ş., Ç.Ç. and F.G.; Investigation, F.G.; Resources, H.H., Ü.Y. and F.G.; Data curation, F.G.; Writing—original draft, M.T.Ş., H.H., Ç.Ç., Ü.Y. and F.G.; Writing—review & editing, F.G. All authors have read and agreed to the published version of the manuscript.

Funding: This research received no external funding.

Data Availability Statement: The original contributions presented in the study are included in the article, further inquiries can be directed to the corresponding author.

Conflicts of Interest: The authors declare no conflict of interest.

References

1. Khan, M.S.; Ullah, S.; Chen, L. Comparison on land-use/land-cover indices in explaining land surface temperature variations in the city of Beijing, China. *Land* **2021**, *10*, 1018. [CrossRef]
2. Li, S.; Yang, X.; Cui, P.; Sun, Y.; Song, B. Machine-learning-algorithm-based prediction of land use/land cover and land surface temperature changes to characterize the surface urban heat island phenomena over Harbin, China. *Land* **2024**, *13*, 1164. [CrossRef]
3. Available online: https://www.worldbank.org/en/topic/urbandevelopment/overview/ (accessed on 11 August 2024).
4. Available online: https://data.worldbank.org/indicator/SP.URB.TOTL.IN.ZS?contextual=default/ (accessed on 11 August 2024).
5. Liu, P.; Liu, C.; Liu, Q. Effects of landscape pattern on land surface temperature in Nanchang, China. *Sci. Rep.* **2024**, *14*, 3832. [CrossRef]
6. Mathew, A.; Arunab, K.S.; Sharma, A.K. Revealing the urban heat island: Investigating spatiotemporal surface temperature dynamics, modeling, and interactions with controllable and non-controllable factors. *Remote Sens. Appl. Soc. Environ.* **2024**, *35*, 101219. [CrossRef]
7. Wang, Q.; Wang, H.; Ren, L.; Chen, J.; Wang, X. Hourly impact of urban features on the spatial distribution of land surface temperature: A study across 30 cities. *Sustain. Cities Soc.* **2024**, *113*, 105701. [CrossRef]
8. Abolmaali, S.M.-R.; Tarkesh, M.; Mousavi, S.A.; Karimzadeh, H.; Pourmanafi, S.; Fakheran, S. Impacts of spatio-temporal change of landscape patterns on habitat quality across Zayanderud Dam watershed in central Iran. *Sci. Rep.* **2024**, *14*, 8780. [CrossRef]
9. Li, Z.-L.; Duan, S.-B. Land surface temperature. In *Comprehensive Remote Sensing*; Liang, S., Ed.; Elsevier: Amsterdam, The Netherlands, 2018; Volume 5, pp. 264–283. [CrossRef]
10. Li, Z.L.; Wu, H.; Duan, S.B.; Zhao, W.; Ren, H.; Liu, X.; Leng, P.; Tang, R.; Ye, X.; Zhu, J.; et al. Satellite remote sensing of global land surface temperature: Definition, methods, products, and applications. *Rev. Geophys.* **2023**, *61*, e2022RG000777. [CrossRef]
11. Wang, R.; Li, X.; Li, Z.; Zhu, Y.; Lu, Y.; Zhang, H. Detecting multi-temporal land cover change and land surface temperature in Pearl River Delta by adopting local climate zone. *Urban Clim.* **2019**, *28*, 100455. [CrossRef]
12. Alexander, C. Influence of the proportion, height and proximity of vegetation and buildings on urban land surface temperature. *Int. J. Appl. Earth Obs. Geoinf.* **2021**, *95*, 102265. [CrossRef]
13. Cammalleri, C.; Vogt, J. On the role of land surface temperature as proxy of soil moisture status for drought monitoring in Europe. *Remote Sens.* **2015**, *7*, 16849–16864. [CrossRef]
14. Smith, I.A.; Fabian, M.P.; Hutyra, L.R. Urban green space and albedo impacts on surface temperature across seven United States cities. *Sci. Total Environ.* **2023**, *857*, 159663. [CrossRef] [PubMed]
15. Song, J.; Chen, W.; Zhang, J.; Huang, K.; Hou, B.; Prishchepov, A.V. Effects of building density on land surface temperature in China: Spatial patterns and determinants. *Landsc. Urban Plan.* **2020**, *198*, 103794. [CrossRef]
16. Wang, L.; Wang, G.; Chen, T.; Liu, J. The regulating effect of urban large planar water bodies on residential heat islands: A case study of Meijiang Lake in Tianjin. *Land* **2023**, *12*, 2126. [CrossRef]
17. Zhang, Z.; Luan, W.; Yang, J.; Guo, A.; Su, M.; Tian, C. The influences of 2D/3D urban morphology on land surface temperature at the block scale in Chinese megacities. *Urban Clim.* **2023**, *49*, 101553. [CrossRef]
18. Zhao, K.; Qi, M.; Yan, X.; Li, L.; Huang, X. Dynamic impact of urban built environment on land surface temperature considering spatio-temporal heterogeneity: A perspective of local climate zone. *Land* **2023**, *12*, 2148. [CrossRef]
19. Zhou, W.; Qian, Y.; Li, X.; Li, W.; Han, L. Relationships between land cover and the surface urban heat island: Seasonal variability and effects of spatial and thematic resolution of land cover data on predicting land surface temperatures. *Landsc. Ecol.* **2014**, *29*, 153–167. [CrossRef]
20. Chang, Y.; Xiao, J.; Li, X.; Frolking, S.; Zhou, D.; Schneider, A.; Weng, Q.; Yu, P.; Wang, X.; Li, X.; et al. Exploring diurnal cycles of surface urban heat island intensity in Boston with land surface temperature data derived from GOES-R geostationary satellites. *Sci. Total Environ.* **2021**, *763*, 144224. [CrossRef]

21. Ghaderpour, E.; Mazzanti, P.; Mugnozza, G.S.; Bozzano, F. Coherency and phase delay analyses between land cover and climate across Italy via the least-squares wavelet software. *Int. J. Appl. Earth Obs. Geoinf.* **2023**, *118*, 103241. [CrossRef]
22. Yu, P.; Zhao, T.; Shi, J.; Ran, Y.; Jia, L.; Ji, D.; Xue, H. Global spatiotemporally continuous MODIS land surface temperature dataset. *Sci. Data* **2022**, *9*, 143. [CrossRef]
23. Li, J.; Song, C.; Cao, L.; Zhu, F.; Meng, X.; Wu, J. Impacts of landscape structure on surface urban heat islands: A case study of Shanghai, China. *Remote Sens. Environ.* **2011**, *115*, 3249–3263. [CrossRef]
24. Cafaro, R.; Cardone, B.; D'Ambrosio, V.; Di Martino, F.; Miraglia, V. A new GIS-based framework to detect urban heat islands and its application on the city of Naples (Italy). *Land* **2024**, *13*, 1253. [CrossRef]
25. Intergovernmental Panel on Climate Change (IPCC). *Climate Change 2022—Impacts, Adaptation and Vulnerability: Working Group II Contribution to the Sixth Assessment Report of the Intergovernmental Panel on Climate Change*; Cambridge University Press: Cambridge, UK, 2023.
26. Estoque, R.C.; Murayama, Y.; Myint, S.W. Effects of landscape composition and pattern on land surface temperature: An urban heat island study in the megacities of Southeast Asia. *Sci. Total Environ.* **2017**, *577*, 349–359. [CrossRef] [PubMed]
27. Chirico, F.; Magnavita, N. The significant role of health surveillance in the occupational heat stress assessment. *Int. J. Biometeorol.* **2019**, *63*, 193–194. [CrossRef]
28. Li, H.; Zhou, Y.; Li, X.; Meng, L.; Wang, X.; Wu, S.; Sodoudi, S. A new method to quantify surface urban heat island intensity. *Sci. Total Environ.* **2018**, *624*, 262–272. [CrossRef]
29. Maimaitiyiming, M.; Ghulam, A.; Tiyip, T.; Pla, F.; Latorre-Carmona, P.; Halik, U.; Sawut, M.; Caetano, M. Effects of green space spatial pattern on land surface temperature: Implications for sustainable urban planning and climate change adaptation. *ISPRS J. Photogramm. Remote Sens.* **2014**, *89*, 59–66. [CrossRef]
30. Santamouris, M.; Cartalis, C.; Synnefa, A.; Kolokotsa, D. On the impact of urban heat island and global warming on the power demand and electricity consumption of buildings—A review. *Energy Build.* **2015**, *98*, 119–124. [CrossRef]
31. Sun, R.; Chen, A.; Chen, L.; Lü, Y. Cooling effects of wetlands in an urban region: The case of Beijing. *Ecol. Indic.* **2012**, *20*, 57–64. [CrossRef]
32. Weng, Q. Thermal infrared remote sensing for urban climate and environmental studies: Methods, applications, and trends. *ISPRS J. Photogramm. Remote Sens.* **2009**, *64*, 335–344. [CrossRef]
33. Zhang, Y.; Odeh, I.O.A.; Han, C. Bi-temporal characterization of land surface temperature in relation to impervious surface area, NDVI and NDBI, using a sub-pixel image analysis. *Int. J. Appl. Earth Obs. Geoinf.* **2009**, *11*, 256–264. [CrossRef]
34. Zhou, D.; Zhao, S.; Liu, S.; Zhang, L.; Zhu, C. Surface urban heat island in China's 32 major cities: Spatial patterns and drivers. *Remote Sens. Environ.* **2014**, *152*, 51–61. [CrossRef]
35. Naserikia, M.; Hart, M.A.; Nazarian, N.; Bechtel, B. Background climate modulates the impact of land cover on urban surface temperature. *Sci. Rep.* **2022**, *12*, 15433. [CrossRef]
36. Hasan, M.A.; Mia, M.B.; Khan, M.R.; Alam, M.J.; Chowdury, T.; Al Amin, M.; Ahmed, K.M.U. Temporal changes in land cover, land surface temperature, soil moisture, and evapotranspiration using remote sensing techniques-a case study of Kutupalong Rohingya refugee camp in Bangladesh. *J. Geovis. Spat. Anal.* **2023**, *7*, 11. [CrossRef]
37. Shu, B.; Chen, Y.; Zhang, X.; Dehghanifarsani, L.; Amani-Beni, M. Urban engineering insights: Spatiotemporal analysis of land surface temperature and land use in urban landscape. *Alex. Eng. J.* **2024**, *92*, 273–282. [CrossRef]
38. Roy, B.; Bari, E. Examining the relationship between land surface temperature and landscape features using spectral indices with Google Earth Engine. *Heliyon* **2022**, *8*, e10668. [CrossRef]
39. Javaid, K.; Ghafoor, G.Z.; Sharif, F.; Shahid, M.G.; Shahzad, L.; Ghafoor, N.; Hayyat, M.U.; Farhan, M. Spatio-temporal analysis of land use land cover change and its impact on land surface temperature of Sialkot City, Pakistan. *Sci. Rep.* **2023**, *13*, 22166. [CrossRef] [PubMed]
40. Ramachandra, T.V.; Bharath, S.; Gupta, N. Modeling landscape dynamics with LST in protected areas of Western Ghats, Karnataka. *J. Environ. Manag.* **2018**, *206*, 1253–1262. [CrossRef]
41. Available online: http://www.muratpasa.gov.tr/nufus-durumu (accessed on 1 August 2024).
42. Available online: http://www.muratpasa.gov.tr/muratpasanin-cografi-yapisi (accessed on 1 August 2024).
43. Available online: https://muratpasa-bld.gov.tr/icerik/tanitim (accessed on 1 August 2024).
44. Du, H.; Zhou, F.; Li, C.; Cai, W.; Jiang, H.; Cai, Y. Analysis of the impact of land use on spatiotemporal patterns of surface urban heat island in rapid urbanization: A case study of Shanghai, China. *Sustainability* **2020**, *12*, 1171. [CrossRef]
45. Eneche, P.S.U.; Atun, F.; Zeng, Y.; Pfeffer, K. Robust drivers of urban land surface temperature dynamics across diverse landscape characters: An augmented systematic literature review. *Ecol. Indic.* **2024**, *163*, 112056. [CrossRef]
46. Han, W. Analyzing the scale dependent effect of urban building morphology on land surface temperature using random forest algorithm. *Sci. Rep.* **2023**, *13*, 19312. [CrossRef]
47. Oke, T.R. The energetic basis of the urban heat island. *Q. J. R. Meteorol. Soc.* **1982**, *108*, 1–24. [CrossRef]

48. Quan, J.; Zhan, W.; Ma, T.; Du, Y.; Guo, Z.; Qin, B. An integrated model for generating hourly Landsat-like land surface temperatures over heterogeneous landscapes. *Remote Sens. Environ.* **2018**, *206*, 403–423. [CrossRef]
49. Massaro, E.; Schifanella, R.; Piccardo, M.; Caporaso, L.; Taubenböck, H.; Cescatti, A.; Duveiller, G. Spatially-optimized urban greening for reduction of population exposure to land surface temperature extremes. *Nat. Commun.* **2023**, *14*, 2903. [CrossRef] [PubMed]
50. Shi, Y.; Liu, S.; Yan, W.; Zhao, S.; Ning, Y.; Peng, X.; Chen, W.; Chen, L.; Hu, X.; Fu, B.; et al. Influence of landscape features on urban land surface temperature: Scale and neighborhood effects. *Sci. Total Environ.* **2021**, *771*, 145381. [CrossRef] [PubMed]
51. Guo, A.; Yang, J.; Sun, W.; Xiao, X.; Xiao, X.; Xia, C.J.; Jin, C.; Li, X. Impact of urban morphology and landscape characteristics on spatiotemporal heterogeneity of land surface temperature. *Sustain. Cities Soc.* **2020**, *63*, 102443. [CrossRef]
52. Guo, G.; Wu, Z.; Cao, Z.; Chen, Y.; Yang, Z. A multilevel statistical technique to identify the dominant landscape metrics of greenspace for determining land surface temperature. *Sustain. Cities Soc.* **2020**, *61*, 102263. [CrossRef]
53. Shukla, A.; Jain, K. Analyzing the impact of changing landscape pattern and dynamics on land surface temperature in Lucknow city, India. *Urban For. Urban Green.* **2021**, *58*, 126877. [CrossRef]
54. Tuia, D.; Camps-Valls, G.; Soergel, U.; Starek, M.; Bischof, H.; Morsdorf, F. A survey of active learning algorithms for supervised remote sensing image classification. *IEEE J. Sel. Top. Signal Process.* **2011**, *5*, 606–617. [CrossRef]
55. Yoo, C.; Im, J.; Park, S.; Quackenbush, L.J. Estimation of daily maximum and minimum air temperatures in urban landscapes using MODIS time series satellite data. *ISPRS J. Photogramm. Remote Sens.* **2018**, *137*, 149–162. [CrossRef]
56. Li, W.; Zhang, Y.; Wang, J.; Wu, J. Linking potential heat source and sink to urban heat island: Heterogeneous effects of landscape pattern on land surface temperature. *Sci. Total Environ.* **2017**, *586*, 457–465. [CrossRef]
57. Peng, J.; Zhou, W.; Yu, K.; Zhang, M.; Liu, Y.; Gao, J.; Zuo, W.; Lin, M. Urban thermal environment dynamics and associated landscape pattern factors: A case study in the Beijing metropolitan region. *Remote Sens. Environ.* **2016**, *173*, 145–155. [CrossRef]
58. Xiong, Y.; Peng, F.; Zou, B. Spatiotemporal influences of land use/cover changes on the heat island effect in rapid urbanization area. *Front. Earth Sci.* **2019**, *13*, 614–627. [CrossRef]
59. Jia, X.; Song, P.; Yun, G.; Li, A.; Wang, K.; Zhang, K.; Du, C.; Feng, Y.; Qu, K.; Wu, M.; et al. Effect of landscape structure on land surface temperature in different essential urban land use categories: A case study in Jiaozuo, China. *Land* **2022**, *11*, 1687. [CrossRef]
60. Zhang, Y.; Balzter, H.; Zou, C.; Xu, H.; Tang, F. Characterizing bi-temporal patterns of land surface temperature using landscape metrics based on sub-pixel classifications from Landsat TM/ETM+. *Int. J. Appl. Earth Obs. Geoinf.* **2015**, *42*, 87–96. [CrossRef]
61. Weng, Q.; Yang, S. Managing the adverse thermal effects of urban development in a densely populated Chinese city. *J. Environ. Manag.* **2004**, *70*, 145–156. [CrossRef] [PubMed]

land

MDPI

Article

People–Place Relationships in Regenerative Urban Assemblages: Streetscape Composition and Subjective Well-Being of Older Adults

Takuo Inoue [1,*], Rikutaro Manabe [2], Akito Murayama [1] and Hideki Koizumi [1]

[1] Department of Urban Engineering, Graduate School of Engineering, The University of Tokyo, Tokyo 113-8656, Japan; murayama@up.t.u-tokyo.ac.jp (A.M.); hide@cd.t.u-tokyo.ac.jp (H.K.)
[2] Center for Research and Development of Higher Education, The University of Tokyo, Tokyo 113-0033, Japan; rik@up.t.u-tokyo.ac.jp
* Correspondence: tak@up.t.u-tokyo.ac.jp; Tel.: +81-3-5841-6246

Abstract: Cities are undergoing rapid transformations due to global trends such as population aging, climate change, and increasing social diversity. In order to address these challenges, urban planning must adopt regenerative approaches that enhance subjective well-being by fostering meaningful relationships between people and their surroundings. Streetscapes, which serve as accessible urban landscapes, are important, especially for older adults, who depend on their local environment due to mobility constraints. This study examines the composition of streetscapes and the subjective well-being of older adults in a Japanese municipality. Using streetscape imagery and semantic segmentation, we quantified landscape elements—including vegetation, sky, roads, and buildings—within various walking distances from participants' residences. Subjective well-being was measured using an 11-point Likert scale and analyzed by ordinal logistic regression. The results revealed that specific streetscape elements significantly impacted subjective well-being differently across spatial thresholds, showing that micro-scale urban landscapes are substantially important in promoting well-being among older adults. This study provides evidence-based insights for adaptive, inclusive, and regenerative urban planning strategies that promote the well-being of diverse demographic groups.

Keywords: landscape elements; street-level imagery; semantic segmentation; ordinal logistic regression

1. Introduction

Cities are rapidly transforming in response to global issues such as climate change [1–3], social diversity [4,5], and demographic shifts [6–8]. These challenges highlight the limitations of conventional urban planning, as cities have often been viewed as static entities rather than dynamic systems that are adaptive and resilient [9,10]. In this context, regenerative urbanism has emerged as a conceptual framework that mitigates environmental and social risks, actively adapts to threats, and enhances urban ecosystems [11]. By incorporating nature-based solutions, circular economy, and participatory governance, regenerative approaches seek to establish positive loops throughout the multi-scale urban environment. Such approaches offer a strategic pathway to address the immediate risks associated with the Anthropocene while enhancing urban resilience and livability [12].

In order to promote the design and planning of cities that fully realize their potential for urban renewal, it is essential to recognize that cities are more than just physical infrastructure but are composed of a complex assemblage of diverse and interconnected components, including built form, social relationships, ecological processes, and governance structures [13,14]. Assemblage is the reinterpretation of the concept of "assemblage (agencement)" discussed in the writings of Deleuze and Guattari that has been developed by Manuel DeLanda, offering a dynamic perspective on urban transformation, from macro-scale infrastructure to micro-scale everyday space, highlighting how components of different scales interact and exhibit emergent properties [15]. This perspective is consistent with the multi-scalability that regenerative urbanism emphasizes, which ensures that interventions are not limited to large-scale projects but are across various dimensions. As the most familiar and accessible urban landscape, the streetscape serves as an important site where regenerative urban strategies can directly impact the quality of life. Through their impact on social interactions, environmental experiences, and individual mobility, streetscapes hold great potential for fostering inclusive, resilient, and regenerative urban environments [16,17].

Aging is a pressing issue among the many challenges facing contemporary urban environments, especially in a hyper-aged society such as Japan [18]. Older adults are disproportionately affected by the design and accessibility of their immediate environment, as they often experience mobility constraints [19,20]. In this context, micro-scale urban landscapes, including streetscapes, have been noted to directly influence perceptions of safety, walkability, and social inclusion, which are essential for maintaining quality of life [21–24]. Nevertheless, previous research has focused on limited types of spatial elements and their impact on behavior and cognition, leaving room for further investigation of experiential and place-based aspects of well-being.

When examining the relationship between streetscapes and individual well-being, subjective well-being emerges as a significant concept. This term encompasses an individual's self-evaluation of happiness or satisfaction, attracting enormous attention from diverse academic and practical fields [25]. Many studies have shown a strong connection between subjective well-being and physical health, suggesting that a positive evaluation of well-being is linked to better health outcomes [26,27]. In light of its extensive implications, subjective well-being is a crucial indicator for assessing older adults' health and quality of life.

Studies have been conducted on the effects of urban environments and landscapes on subjective health and well-being, with natural elements as the primary focus [28]. However, the results regarding natural elements vary. For example, one study showed that exposure to the natural outdoor environment within 300 m of their residence improved subjective health [29]. Another study found that exposure to vegetation and waterscapes within 100 and 150 m has been shown to decrease depression [30]. On the other hand, one study has concluded that the amount of greenery within 400 m, 800 m, and 1200 m of one's home has no significant effect on subjective health [31]. These studies use the Normalized Difference Vegetation Index (NDVI) to indicate the amount of greenery [32]. In addition to natural landscapes, studies on the urban environment have also shown the impact of gentrification [33], public safety [34], and walkability [35] on subjective health perceptions.

Based on these arguments, this research aims to investigate the relationship between streetscapes and subjective well-being, which seems essential for older adults, who depend on the local environment for their daily activities. Research that has focused on various landscape elements and analyzed them at the micro-scale of the living area, with the street as the unit, which is a spatial scale directly related to the living experience of older adults, remains scarce. This study examined the relationship between the configuration of streetscapes and the subjective well-being of older adults in a municipality in Japan. Specifically, we quantified how streetscape at various spatial thresholds affects the sense of well-being by semantically subdividing the street-level images and evaluating the configuration of important landscape elements. The introduction and comparison of multiple spatial thresholds are based on the defining characteristic of assemblages, in which individual components express their characteristics as an aggregate at any scale [13], and is an attempt to incorporate the idea of "city as assemblage" as a dimension of urban planning and spatial design. The findings contribute to developing evidence-based urban design strategies that enhance livability through regenerative and inclusive planning methodologies.

The structure of the paper is as follows: Section 2 introduces the methodology, including data collection, image analysis, and statistical modeling. Section 3 presents the results, emphasizing significant relationships between streetscape elements and subjective well-being. Section 4 discusses the implications of these findings for regenerative urban design and planning. Finally, Section 5 concludes with recommendations for future research and policy applications.

2. Materials and Methods

2.1. Study Area

This study was conducted in Kunitachi, Japan. The country provides a suitable context for this study, given that individuals aged 65 and older constitute 26.6% of the population, according to the 2015 national census [36]. Kunitachi, situated approximately 30 km west of central Tokyo, was chosen for its blend of urban and suburban environments that characterize Japanese metropolitan suburbs. From a statistical perspective, Kunitachi is a typical situation in the Japanese metropolitan area. It is ranked as nearly average among municipalities in Tokyo in terms of the municipal aging rate (23.93% as of January 2023, ranked 21st/49th in Tokyo), population density (9346 persons per square kilometer as of January 2023, ranked 31st/49th in Tokyo), per capita income per taxpayer (4,807,000 yen as of 2022, ranked 16th/49th in Tokyo), and average residential land prices (356,500 yen per square meter as of July 2023, ranked 21st/49th in Tokyo) [37,38].

The Fujimidai neighborhood is located in the central part of Kunitachi. The area includes a variety of urban structures, from densely populated, walkable areas and boulevards designed for automobile traffic to agricultural lands. This mixture establishes an optimal setting for investigating the relationship between streetscape composition and subjective well-being. The area has an approximate population of 18,000 and is approximately 2.5 km from east to west and 1.0 km from north to south.

2.2. Data Collection

2.2.1. Street-Level Imagery

Street-level imagery was obtained from Mapillary [39], an online street imagery platform. Images taken in bad weather or at night were excluded from the dataset to ensure the quality of the analysis (see Figure 1 for examples).

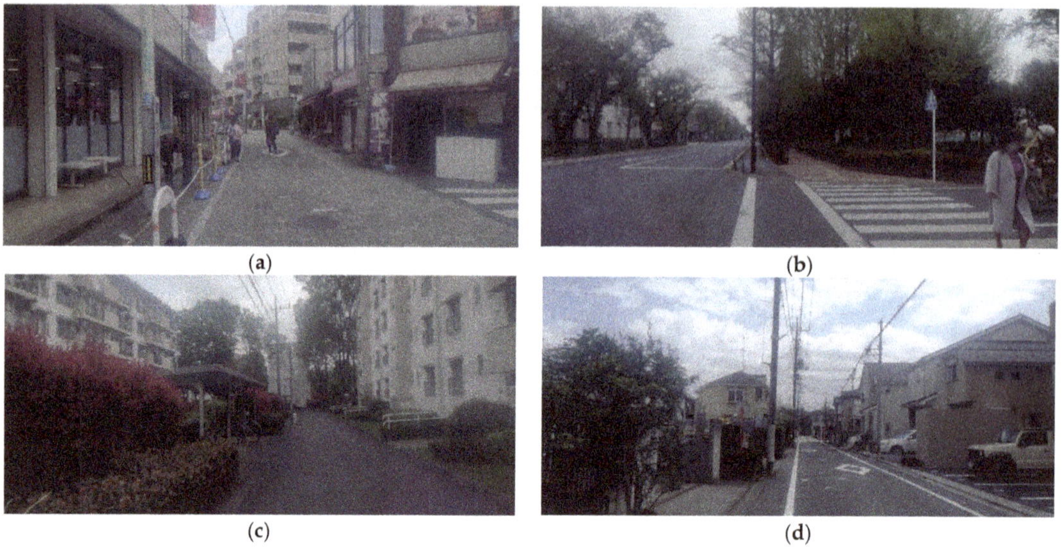

Figure 1. Typical examples of streetscape images: (a) a shopping street near a train station; (b) a lush green boulevard; (c) a public housing complex; and (d) a single-family residential area.

2.2.2. Subjective Well-Being and Other Individual Data

Data on subjective well-being and other personal information were collected from a municipal questionnaire survey, the "Care Prevention and Daily Living Area Needs Survey", in 2016. This survey, mandated for municipalities following the Long-Term Care Insurance Law, includes questions on perceived physical and mental health, social interactions, and financial circumstances. It is aimed at residents aged 75 years and older who have not been certified by the local government as requiring long-term care.

Subjective well-being was assessed by asking participants, "How happy do you feel now?" Responses were recorded on an 11-point Likert scale, ranging from 0 (not at all happy) to 10 (extremely happy). The survey also collected personal information, including age, sex (binary variable), and family structure (categorized as living alone, living with a spouse aged 65 or over, living with a spouse aged under 65, living with children, or other). Housing type was classified into seven categories: owned single-family detached house, owned unit in a housing complex, public rental housing, private rental detached housing, private rental unit in a housing complex, company housing, and others. Additionally, subjective financial status was assessed on a five-point scale (stable, somewhat stable, neither stable nor unstable, somewhat unstable, unstable). The survey also inquired about the frequency of going out, willingness to participate in community activities, and willingness to organize community activities. One thousand one hundred forty-six responses were collected from people living in the Fujimidai neighborhood. Those responses were geocoded to enable spatial analysis. The residential addresses were included in the basic data on individual residents, which was provided under a joint research agreement concluded between Kunitachi City and the authors. To avoid the leakage of personal information through the use of the geocoder, an anonymized ID was assigned to the basic data and questionnaire response data for use in this study, and a file containing only this ID and text indicating their addresses was used to convert addresses to latitude and longitude using the geocoder publicly available online, after which the latitude and longitude were reflected in the questionnaire response data using the ID. We obtained permission to use address information in our analysis, provided that we carried out the above operations and that we would not disclose individual results.

Table 1 presents the sociodemographic and personal characteristics of the survey respondents. The sample consisted of 915 older adults residing in the Fujimidai neighborhood, with a mean age of 80.66 years (SD = 4.09). Among the respondents, 42.2% were male, and 57.8% were female. Regarding family structure, 32.6% lived alone, 43.6% resided with their spouse aged 65 or over, 1.0% lived with their spouse under 65, 10.7% resided with their children, and 10.7% lived with other family members or acquaintances. Housing type varied, with 25.0% owning a single-family detached house, 18.7% living in an owned unit in a housing complex, 44.4% living in public rental housing, and the other respondents residing in private rental, company-provided housing, and others. Subjective financial status, measured on a five-point scale, showed that 5.5% of participants perceived their financial situation as stable and 34.9% as unstable, whereas 59.7% considered it neither stable nor unstable. Regarding daily activity and community engagement, the frequency of going out varied across individuals, with 42.1% reporting daily outings (five or more times a week), 45.2% reporting occasional outings (two to four times a week), 9.2% going out once a week, and 3.5% rarely going out. 60.9% of respondents expressed a willingness to participate in community activities eagerly or somewhat eagerly, while 32.2% showed interest in organizing such initiatives.

Table 1. Sociodemographic and personal characteristics of the survey respondents.

Category	Item	Number of Respondents	Proportion (%)
Age	75–79	312	34.1
	80–84	411	44.9
	85–89	148	16.2
	90–94	37	4.0
	95 or over	4	0.4
Sex	Male	386	42.2
	Female	529	57.8
Family structure	Living alone	298	32.6
	Living with a spouse aged 65 or over	399	43.6
	Living with a spouse aged under 65	9	1.0
	Living with children	111	12.1
	Other	98	10.7
Housing type	Owned single-family detached house	229	25.0
	Owned a unit in a housing complex	171	18.7
	Rental public housing	406	44.4
	Rental private detached house	3	0.3
	Rental private unit in a housing complex	92	10.1
	Company-provided housing	7	0.8
	Other	7	0.8
Subjective financial status	Stable	7	0.8
	Somewhat stable	43	4.7
	Neither stable nor unstable	546	59.7
	Somewhat unstable	253	27.7
	Unstable	66	7.2
Frequency of going out	Five or more times a week	385	42.1
	Two to four times a week	414	45.2
	Once a week	84	9.2
	Rarely	32	3.5

Table 1. *Cont.*

Category	Item	Number of Respondents	Proportion (%)
Willingness to participate in community activities	Would love to participate	70	7.7
	May participate	487	53.2
	Do not want to participate	358	39.1
Willingness to organize community activities	Would love to organize	24	2.6
	May organize	271	29.6
	Do not want to organize	620	67.8

2.2.3. Additional Data

Other datasets, including land use maps and road network data, were integrated for spatial analysis. Land use and building information were extracted from the Tokyo Metropolitan Government's Geographic Information System data, and "street units" were established, comprising segments of streets delimited from intersection to intersection [40]. Street network data were obtained from OpenStreetMap.

2.3. *Image Processing and Feature Extraction*

Streetscape elements were analyzed using semantic segmentation, a deep learning technique that classifies image pixels into predefined categories. This study employed DeepLabv3+ [41] trained on a Mapillary Vistas Dataset [42] with urban landscape elements. Key elements extracted included the following:

- Object: pole, utility pole, traffic sign front, traffic sign back, traffic sign frame, streetlight, billboard, traffic light, trash can, junction box, car, bus, truck, motorcycle, bicycle;
- Construction: building, wall, fence, rail track, road, sidewalk, curb, guard rail, other barrier;
- Human: person;
- Nature: sky, vegetation, terrain, mountain, water;
- Marking: crosswalk zebra, general;

2.4. *Spatial Analysis*

Network-based distance measures were used instead of conventional circular buffers to better reflect the everyday walking experiences of older adults. Street network distances were calculated using GIS-based routing algorithms, providing a more accurate representation of walkable access to urban features. The distances between the center points of all street unit pairs within and around the Fujimidai neighborhood were computed using the QNEAT3 network analysis tool, a QGIS plug-in. Spatial thresholds of 50 m, 100 m, 150 m, 200 m, 250 m, and 300 m were applied to calculate the average percentage composition of each landscape element within all these scales from each street unit.

2.5. *Modeling*

We employed hierarchical ordinal logistic regression to examine the relationship between streetscape composition and subjective well-being. This method was chosen due to the ordinal nature of the dependent variable, which represents subjective well-being on a graded scale. The hierarchical approach was adopted to assess the incremental explanatory power of landscape composition variables beyond demographic and personal factors. By structuring the analysis in stages, we aimed to isolate the contribution of streetscape characteristics while controlling for individual-level covariates.

The variables were hierarchically incorporated into the model as follows:

1. Baseline Model: The first model included only demographic and personal variables, such as age, gender, housing type, subjective socioeconomic status, and self-reported everyday behaviors. This model served as a reference to assess the effect of individual characteristics on subjective well-being.
2. Streetscape Composition Model: The second model introduced streetscape composition variables, calculated as the proportion of different landscape elements within predefined network distance thresholds. These elements, extracted through semantic segmentation of street-level images, included vegetation, sky, road, and building, as listed in the previous section.

As mentioned above, the independent variables for landscape composition were aggregated based on network distances rather than circular buffers, ensuring a more accurate representation of physical accessibility in older adults' daily experiences.

To assess model performance, we examined key goodness-of-fit measures, including the corrected Akaike Information Criterion (AICc) and Nagelkerke R-squared. We used those statistical indicators to compare those models and determine whether including streetscape variables significantly improved explanatory power. By employing this hierarchical modeling approach, we aimed to disentangle the relative contributions of individual elements and streetscape composition to subjective well-being. This approach provides a robust framework for evidence-based urban design interventions.

3. Results

This section presents the results of the analysis based on survey responses and spatial analysis, examining the relationships between subjective well-being among older adults and both demographic/personal factors and the proportions of streetscape elements at multiple spatial thresholds from their residences. The findings indicate statistically significant relationships between certain streetscape elements and subjective well-being, highlighting the importance of micro-scale urban environments. The following subsections offer a detailed account of the descriptive statistics, the relationships between variables and subjective well-being, and the development of hierarchical ordinal logistic regression models.

3.1. Subjective Well-Being

The average subjective well-being score was 7.10 (SD = 2.18), assessed on an 11-point Likert scale. The distribution of subjective well-being (Figure 2) shows three peaks; therefore, we decided to label them as low (marked 0–6), mid (7–8), and high (9–10).

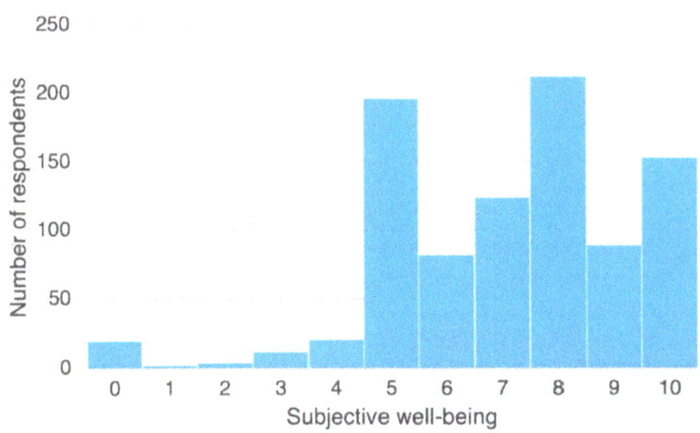

Figure 2. The distribution of subjective well-being in the questionnaire survey.

3.2. Streetscape Configurations

To identify potential impacts of correlations among landscape elements on the analysis, we first calculated the proportion and distribution of each landscape element in the streetscape configuration. Semantic segmentation was performed on 2374 street units in and around the Fujimidai neighborhood, specifying the latitude and longitude of their center points and selecting the street-level image of the nearest point. As a result, the percentage of each landscape element introduced in Section 2.3 was calculated in the image. Figure 3 shows the distribution in terms of the number of street units of the compositional proportions of the four typical elements: road, sky, building, and vegetation (these elements combined, on average, account for about 69% of the total area of the image). Roads accounted for an average of 13.5% of the image (SD = 0.059), and when divided by two percentage points, 10–12% of the image was the largest number of street units (316). Similarly, sky averaged 7.8% (SD = 0.047), buildings averaged 28.6% (SD = 0.116), and vegetation averaged 20.0% (SD = 0.119).

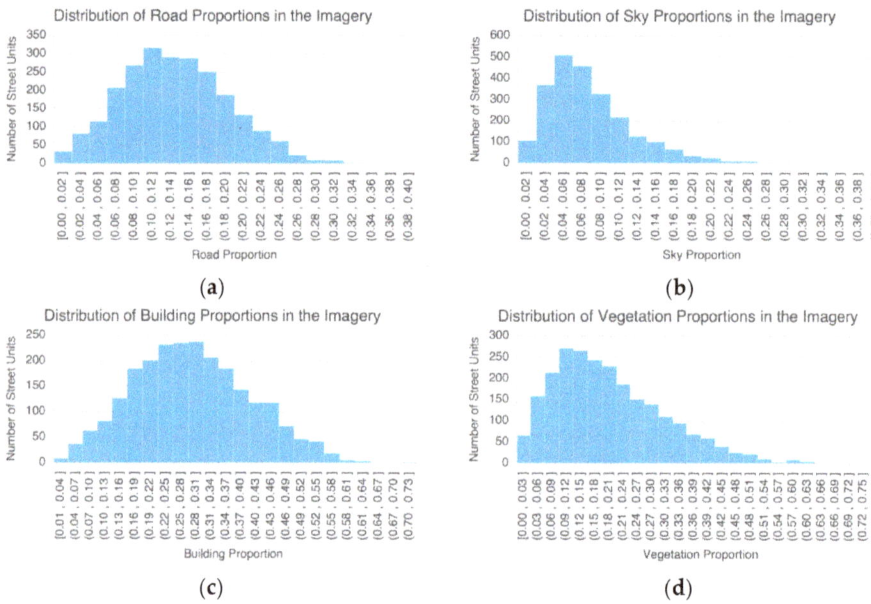

Figure 3. The distribution of the compositions of landscape elements in the imagery among street units: (**a**) road; (**b**) sky; (**c**) building; and (**d**) vegetation.

3.3. Bivariate Analysis

Bivariate analysis of sociodemographic and personal variables and subjective well-being is conducted as a prerequisite for creating a model to identify factors that influence subjective well-being. Since subjective well-being is a three-level ordinal scale of "low", "medium", and "high", Pearson's chi-square test was used to assess their significance. In all cases, the significance level was set at $p < 0.05$. Table 2 shows the percentage of subjective well-being levels and p-value for each ordinal and nominal scale. Women had significantly higher levels of subjective well-being than men, and significant differences were found for housing type, subjective financial status, frequency of going out, willingness to participate in community activities, and willingness to organize community activities. On the other hand, no significant differences were found regarding family structure.

Table 2. Difference between sociodemographic and personal characteristics.

Category	Item	Subjective Well-Being (%)			*p*-Value
		Low	Mid	High	
Sex	Male	41.7	38.3	20.0	0.0004
	Female	33.3	35.5	31.2	
Family structure	Living alone	37.6	39.6	22.8	0.6732
	Living with a spouse aged 65 or over	37.3	34.3	28.3	
	Living with a spouse aged under 65	22.2	33.3	44.4	
	Living with children	36.0	35.1	28.8	
	Other	34.7	39.8	25.5	
Housing type	Owned single-family detached house	27.5	37.6	34.9	0.0005
	Owned a unit in a housing complex	30.4	39.8	29.8	
	Rental public housing	44.3	35.7	20.0	
	Rental private detached house	33.3	0.0	66.7	
	Rental private unit in a housing complex	37.0	33.7	29.4	
	Company-provided housing	57.1	28.6	14.3	
	Other	42.9	57.1	0.0	
Subjective financial status	Stable	14.3	42.9	42.9	<0.0001
	Somewhat stable	9.3	41.9	48.8	
	Neither stable nor unstable	30.4	38.1	31.5	
	Somewhat unstable	49.4	36.4	14.2	
	Unstable	62.1	22.7	15.2	
Frequency of going out	Five or more times a week	31.2	36.4	32.5	<0.0001
	Two to four times a week	36.5	38.9	24.6	
	Once a week	52.4	33.3	14.3	
	Rarely	68.8	21.9	9.4	
Willingness to participate in community activities	Would love to participate	22.9	38.6	38.6	0.0006
	May participate	33.3	39.8	26.9	
	Do not want to participate	44.4	32.1	23.5	
Willingness to organize community activities	Would love to organize	20.8	36.7	26.5	0.0056
	May organize	29.9	40.2	29.9	
	Do not want to organize	40.5	35.3	24.2	

p-values < 0.05 were considered statistically significant.

3.4. Ordinal Logistic Regression

Hierarchical ordinal logistic regression analysis was then conducted. As a first step, a baseline model was created that did not include streetscape configurations but only sociodemographic and personal variables. Based on the results of the previous section, age, sex, housing type, subjective financial status, frequency of going out, willingness to participate in community activities, and willingness to organize community activities were entered as variables. The statistical analysis software JMP Pro (Version 18.0.1) was used for the following analyses.

A baseline model was created by selecting variables using the maximum likelihood method. The model had a negative log-likelihood (NLL) of 921.51, an AICc of 1886.06, and a Nagelkerke R-square (NR2) of 0.1676. The Area Under the Curve (AUC) was fair, with 0.6998 for "medium" and 0.7082 for "high" subjective well-being. Table 3 shows the results of the Wald test for variable effects. These results indicate that the baseline model should include age, sex, housing type, frequency of going out, and subjective financial status.

Table 3. The statistical profile of a baseline model.

Characteristics	Degree of Freedom	Wald Chi-Square	*p*-Value
Age	1	3.8975	0.0484
Sex	1	17.7344	<0.0001
Housing type	6	13.0190	0.0427
Subjective financial status	3	26.7374	<0.0001
Frequency of going out	4	41.7977	<0.0001
Willingness to participate in community activities	2	2.4772	0.2898
Willingness to organize community activities	2	3.1912	0.2028

p-values < 0.05 were considered statistically significant.

The second step is to create a streetscape composition model with the percentage of streetscape elements within a specific road network distance from the respondent's residence as a variable. As mentioned above, the distance thresholds are 50 m, 100 m, 150 m, 200 m, 250 m, and 300 m. In selecting the distances, we are based on the methodology in a previous study [29], where the analysis was conducted in the range of 50 m to 250 m in a radius from the place of residence. For each respondent, we take the center point of the street unit that the geocoded residence faces as the starting point, calculate the mean value of the percentage composition of streetscape elements in all street units closer than a specific distance threshold from it, and use these as the values of the landscape variables linked to the individual. In each distance threshold, the correlations among the streetscape elements were checked. It turned out that, for example, in the 50 m model, the VIF exceeds 15 for the four elements that occupy a large area in the image (road, sky, building, and vegetation). Therefore, a principal component analysis of the four component proportions was performed, and a new VIF was calculated by adding the first and second principal components (PC1, PC2) instead of the four components, resulting in a maximum VIF of 8.275 for the first principal component, which is sufficiently small. The same procedure was followed for the models from 100 m to 300 m.

For each distance threshold from 50 m to 300 m, parameters were again estimated by the maximum likelihood method to create a streetscape composition model. Table 4 presents the statistical details of the model for each distance threshold and the log values and Wald test results for each variable. NLL, AICc, NR2, and AUC were calculated for each model. The 50 m model showed a significant change in NLL compared to the baseline model, improving the model's explanatory power. In comparison, while NLL and AICc remained similar to the baseline model for all models except the 250 m model, they increased AUC by over 0.01. No change in any statistical indicators was found on the 250 m model, and no significant streetscape elements were detected that significantly impacted it. The results indicate that, in addition to sociodemographic and personal variables, different streetscape elements affect the subjective well-being of older adults within different spatial boundaries.

Table 5 shows the range of odds ratios for the significant streetscape elements in each model, expressing the odds ratio of subjective well-being between the largest and smallest value of the relevant streetscape element tied to the individual who responded to the survey.

Table 4. The statistical details and the effects of variables in streetscape composition models.

Variable	Model_50 m		Model_100 m		Model_150 m		Model_200 m		Model_250 m		Model_300 m	
	Log Value	Wald Chi-Square	Log Value	Wald Chi-Square	Log Value	Wald Chi-Square	Log Value	Wald Chi-Square	Log Value	Wald Chi-Square	Log Value	Wald Chi-Square
Age	0.483	1.013	1.091	3.196	1.190	3.530	1.104	3.193	1.160	3.417	0.950	2.618
Sex	3.674	13.722 **	4.247	16.052 **	4.776	18.255 **	4.663	17.848 **	4.893	18.771 **	4.906	18.789 **
Housing type	0.878	7.618	0.850	9.772	0.512	7.389	0.903	10.309	0.568	7.908	1.237	12.552
Frequency of going out	2.867	15.390 **	7.237	35.141 **	6.437	31.452 **	6.503	31.720 **	6.007	29.328 **	6.106	29.909 **
Subjective financial status	5.687	31.081 **	8.311	42.930 **	9.276	46.818 **	9.096	46.122 **	8.417	43.020 **	7.836	40.375 **
PC1	0.201	0.236	0.123	0.103	0.749	1.913	0.755	1.905	0.044	0.015	1.863	6.075 *
PC2	1.136	3.316	0.197	0.225	1.001	2.728	0.106	0.074	0.313	0.500	0.009	0.001
Pole	0.356	0.598	0.540	1.124	0.301	0.465	0.618	1.370	0.034	0.009	1.178	3.438
Utility pole	2.355	7.950 **	0.052	0.020	0.050	0.019	0.469	0.924	0.199	0.230	0.130	0.113
Traffic sign (front)	0.299	0.451	0.067	0.033	0.287	0.430	0.928	2.452	0.092	0.058	0.022	0.004
Traffic sign (back)	0.166	0.175	0.576	1.228	0.731	1.720	0.510	1.022	0.555	1.178	0.162	0.163
Traffic sign (frame)	1.234	3.706	1.982	6.741 **	1.258	3.709	1.626	5.232 *	0.070	0.036	0.491	0.981
Streetlight	1.299	3.562	2.078	6.921 **	2.019	6.841 **	1.642	5.206 *	0.647	1.464	0.058	0.026
Billboard	0.079	0.042	1.462	4.320 *	1.822	5.851 *	1.544	4.801 *	0.984	2.691	0.667	1.593
Traffic light	0.316	0.504	0.231	0.305	0.145	0.135	0.844	2.164	0.654	1.526	0.207	0.257
Trash can	0.104	0.074	0.109	0.082	0.287	0.416	0.003	0.000	0.151	0.143	0.335	0.541
Junction box	2.596	8.826 **	1.070	3.057	0.575	1.282	0.828	2.122	0.347	0.583	0.141	0.127
Car	0.089	0.056	0.060	0.026	0.264	0.356	0.468	0.882	0.318	0.489	0.149	0.142
Bus	1.274	3.763	1.068	3.029	1.437	4.310 *	2.351	8.039 **	0.323	0.511	0.845	2.211

Table 4. *Cont.*

Variable	Model_50 m		Model_100 m		Model_150 m		Model_200 m		Model_250 m		Model_300 m	
	Log Value	Wald Chi-Square	Log Value	Wald Chi-Square	Log Value	Wald Chi-Square	Log Value	Wald Chi-Square	Log Value	Wald Chi-Square	Log Value	Wald Chi-Square
Truck	0.714	1.956	0.128	0.102	0.275	0.401	0.190	0.219	0.189	0.217	0.494	1.013
Motorcycle	1.309	3.824	0.885	2.380	0.451	0.866	0.779	1.888	0.046	0.015	1.404	2.990
Bicycle	1.179	3.484	0.080	0.048	0.268	0.392	0.220	0.272	0.133	0.117	0.072	0.037
Wall	1.264	3.545	0.431	0.796	0.362	0.632	1.892	6.284 *	0.377	0.678	0.052	0.021
Fence	0.334	0.545	0.401	0.722	0.138	0.121	0.311	0.476	0.336	0.556	1.738	5.513 *
Rail track	1.291	3.833	1.338	4.196 *	0.713	1.789	0.176	0.188	0.331	0.568	0.126	0.101
Sidewalk	1.811	5.678 *	0.265	0.367	0.150	0.140	0.938	2.394	1.307	3.788	0.214	0.261
Curb	1.452	4.375 *	0.098	0.063	0.296	0.453	0.244	0.317	0.147	0.138	0.774	1.915
Guard rail	0.262	0.364	0.323	0.527	0.136	0.124	2.291	7.938 **	0.142	0.136	0.221	0.275
Other barriers	0.729	1.674	1.030	2.771	2.176	7.312 **	1.544	4.990 *	0.492	0.999	0.904	2.496
Person	0.053	0.021	0.011	0.001	0.182	0.200	0.297	0.447	0.910	2.403	0.214	0.268
Terrain	1.694	5.126 *	0.756	1.838	0.555	1.187	0.096	0.063	0.322	0.507	0.178	0.189
Mountain	0.307	0.446	0.071	0.037	0.098	0.064	0.024	0.005	0.055	0.023	0.526	1.080
Water	1.394	3.457	0.059	0.024	0.032	0.008	0.807	1.997	0.485	0.944	0.233	0.305
Crosswalk	0.195	0.218	0.605	1.348	1.765	5.868 *	3.021	10.893 **	0.312	0.499	1.569	5.003 *
Marking (general)	1.804	5.745 *	0.436	0.804	1.071	2.967	1.165	3.287	0.943	2.510	0.233	0.296
Statistical details												
NLL	648.81		874.19		905.24		902.53		914.97		911.47	
AICc	1398.92		1847.79		1909.69		1904.26		1929.14		1922.15	
NR2	0.2227		0.1975		0.2011		0.2066		0.1812		0.1884	
AUC (mid)	0.7442		0.7219		0.7211		0.7288		0.7099		0.7149	
AUC (high)	0.7161		0.7197		0.7254		0.7230		0.7116		0.7159	

** $p < 0.01$, * $p < 0.05$.

Table 5. The odds ratio of statistically significant streetscape elements.

	Model_50 m	Model_100 m	Model_150 m	Model_200 m	Model_250 m	Model_300 m
PC1						0.003
PC2						
Pole						
Utility pole	9.728					
Traffic sign (front)						
Traffic sign (back)						
Traffic sign (frame)		0.075		0.007		
Streetlight		8.370	12.613	9.202		
Billboard		0.087	0.098	0.088		
Traffic light						
Trash can						
Junction box	0.079					
Car						
Bus			4.080	10.591		
Truck						
Motorcycle						
Bicycle						
Wall				41.619		
Fence						0.072
Rail track		4.929				
Sidewalk	5.375					
Curb	0.121					
Guard rail				282.808		
Other barriers			0.094	0.113		
Person						
Terrain	6.598					
Mountain						
Water						
Crosswalk			0.212	0.056		24.751
Marking (general)	5.185					

4. Discussion

The hierarchical ordinal logistic regression analysis conducted in this study reveals that certain streetscape elements significantly affect subjective well-being at different spatial scales around the residence of older adults in relatively good health. Notably, these effects are more eminent at smaller spatial scales. It may be due to the diminishing differences between the places of residence as the distance increases. For example, terrain, which showed a statistically significant impact in the 50 m model, represents small-scale greenery and soil, and the distribution per street unit is shown in Figure 4a. In contrast, vegetation represents relatively large greenery, and the distribution per street unit is shown in Figure 4b. While the standard deviation of terrain is greater, street units with large proportions are scattered inside the city blocks, making them accessible to many individuals even in shorter distance thresholds. On the other hand, since large-scale vegetation is concentrated and distributed on some blocks and along arterial roads, many individuals can access them only when the spatial range extends to several hundred meters. Comparing the mean and standard deviation of the composition of terrain and vegetation for each individual from 50 m to 300 m, we see a gradual decrease in individual differences (Figure 5).

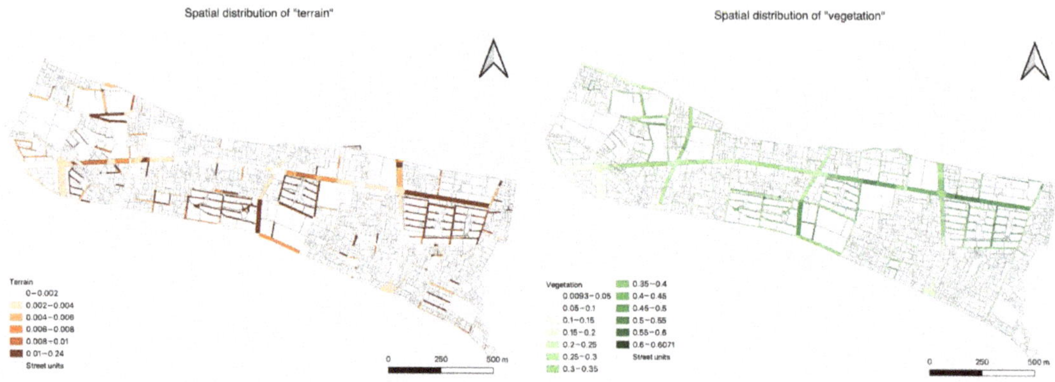

(a) (b)

Figure 4. The geographic distributions: (a) terrain; (b) vegetation.

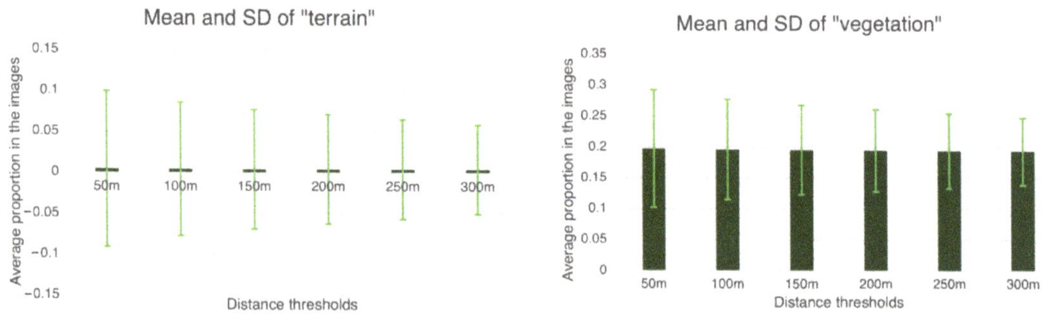

(a) (b)

Figure 5. The mean and standard deviation: (a) terrain; (b) vegetation.

The revealed effects of streetscape elements on subjective well-being vary across spatial scales. In the 50 m model, the smallest spatial scales addressed in this study, the impact of streetscape elements, distinguished from sociodemographic and personal variables, was most eminent, with elements like sidewalks and terrain having a more positive impact. This finding suggests that enhancing a well-designed walking environment is important in the very vicinity of the residence of older adults. At 100 m to 200 m from the residence, streetlights, buses, walls, and guard rails contributed positively, while billboards and crosswalks had a negative effect. These results suggest that primary roads nearby with moderate traffic and pedestrian infrastructure play a role in well-being, while boulevards having commercial billboards and crosswalks may be too large to be present near their homes. Significant streetscape elements were not detected in the 250 m model. Finally, the positive impact of a lower PC1 value in the 300 m model was found.

To better interpret PC1 in the 300 m model, we will now investigate the principal component analysis of the composition ratio of the four main elements of the streetscape: road, sky, building, and vegetation. Figure 6 displays the component vectors of the four elements with PC1 on the horizontal axis and PC2 on the vertical axis for each model from 50 m to 300 m. Across all distance thresholds, PC1 has positive contributions from roads, sky, and buildings and negative contributions from vegetation. In PC2, road and sky are positive, buildings are negative, with vegetation remaining neutral. The range odds ratio of 0.003 for PC1 in the 300 m model suggests that a higher proportion of vegetation at this scale enhances subjective well-being. This finding aligns with previous research findings that access to large green spaces at the neighborhood scale is desirable for improving subjective health and well-being and reducing the risk of certain mental illnesses among older adults.

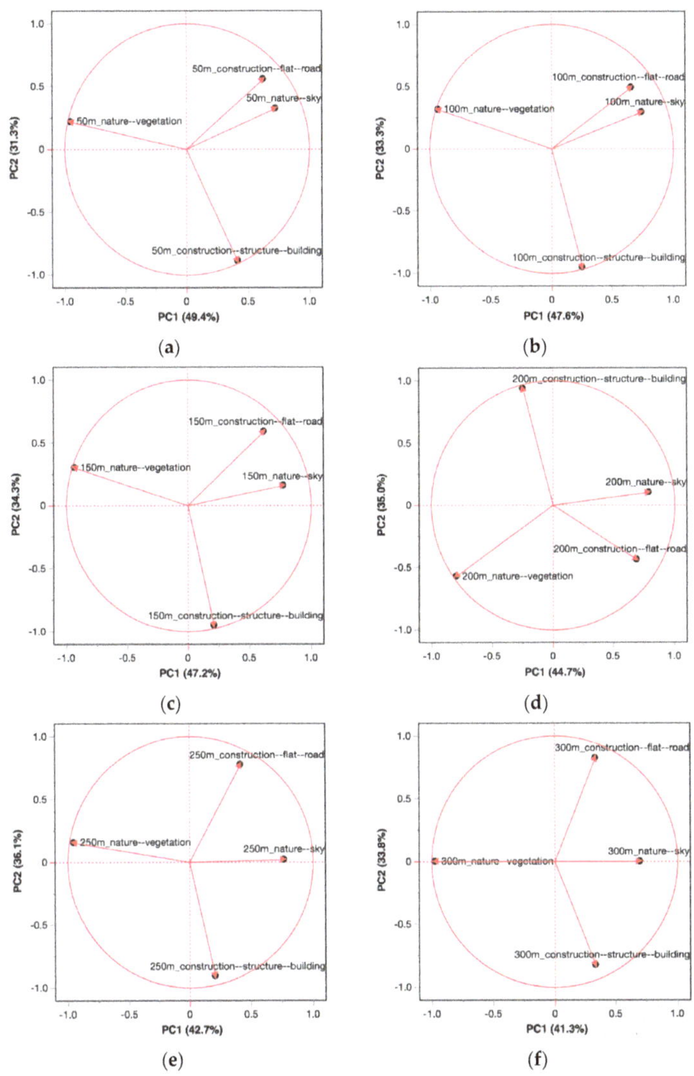

Figure 6. The component vectors of the principal component analysis of four streetscape elements—vegetation, sky, roads, and buildings—in different distance thresholds: (**a**) 50 m model; (**b**) 100 m model; (**c**) 150 m model; (**d**) 200 m model; (**e**) 250 m model; (**f**) 300 m model.

The results of this study contribute to neighborhood-scale planning and design through the lenses of regenerative urbanism and assemblage thinking. The insignificance of PC1 and PC2 below 250 m, coupled with the statistical significance of smaller-scale elements such as terrain and ground-level facilities, suggests a strong connection to the principles of regeneration and resilience. Unlike large-scale infrastructure such as buildings and roads, small-sized elements such as sidewalks, terrain, and guard rails can be modified or experimentally implemented in the short term, opening up the possibility of intervention through tactical efforts in the micro-scale urban environment. Additionally, the difference in significant streetscape elements at different spatial scales suggests that individuals may perceive a certain "semantic cohesion" at a sub-neighborhood scale. This empirical evidence supports the assemblage thinking, in which various components of a city dynamically assemble and disassemble across scales to express their emergent properties.

In discussing the findings obtained in this study, it is also necessary to consider the effect of the urban planning regulations. In the study area, the Kunitachi City Community Development Ordinance stipulates procedures for making development projects meet certain criteria, such as the submission of prior consultation documents, holding meetings to provide explanations to neighborhood residents, and confirmation of conformance with approval criteria. The ordinance sets standards, such as minimum lot size, green space, building height restrictions, road widths, and sidewalk installations [43]. The possibility that these standards influence the high percentage of sidewalks and vegetation, which can be extracted from street-level images, and subsequently contribute to improving the subjective well-being of older adults provides a rationale for the legitimacy of regulations and guidelines related to building and spatial design that have been introduced in various regions.

5. Conclusions

This study examined the relationship between streetscape elements and subjective well-being at different spatial scales and revealed that significant environmental factors vary depending on distance. Specifically, the desirable direction of streetscape design to enhance the subjective well-being of older adults includes the development of a safe and pleasant walking environment with terrains and sidewalks in the very vicinity of their homes and having a living primary road of moderate size located a short distance away from their homes. The study also revealed that large-scale greenery is beneficial if located within a neighborhood scale. Implications of these findings for urban design include the fact that multiple spatial scales should be considered simultaneously rather than relying on a single scale. This can be seen as an indication that assemblage thinking applies to urban design. In addition, given the fact that the streetscape elements for which standards are set in the Kunitachi City Community Development Ordinance, such as the provision of green space and sidewalks, were shown to contribute to the subjective well-being of older adults, it is expected that the findings on streetscape and subjective well-being presented in this study can be used in the formulation of such ordinances and guidelines.

Future research can contribute to a more comprehensive urban design and planning approach by exploring three directions while recognizing the limitations of this study. The first is to incorporate short-term temporal variability. Although open-source street-level imagery was used in this study, the images at each location were only taken at a single moment, and the study could not account for transient factors, such as time of day and weather conditions. This limitation in data availability could be solved by the emergence of new methodologies for capturing and sharing streetscapes. For example, mobility-mounted devices could facilitate more dynamic analyses. Data sources other than movies and images, such as audio, may also be combined to provide a more detailed understanding of the situation. Second, a long-term temporal variation could be added, as subjective well-being and the streetscape are expected to evolve. A longitudinal study tracking physical and perceptual changes in the streetscape could contribute to regenerative and assemblage-based urbanism theories. Finally, although this study uses a typical urban neighborhood in a Japanese metropolitan suburb as its case study, the relationship between streetscape and subjective well-being may differ in different urban contexts. Conducting comparative analyses across a broader range of people and cultures is essential to developing theories that account for cultural differences between countries and regions. This will ultimately lead to practices depending on those contexts.

Author Contributions: Conceptualization, T.I.; methodology, T.I.; software, T.I. and R.M.; validation, T.I. and R.M.; formal analysis, T.I.; investigation, T.I.; resources, R.M. and H.K.; data curation, T.I.; writing—original draft preparation, T.I.; writing—review and editing, R.M., A.M. and H.K.; visualization, T.I.; supervision, A.M. and H.K.; project administration, H.K. All authors have read and agreed to the published version of the manuscript.

Funding: This research received no external funding.

Data Availability Statement: The data presented in this study are available on request from the corresponding author due to privacy reasons.

Acknowledgments: This research was conducted with the cooperation of the city government of Kunitachi, Tokyo, Japan.

Conflicts of Interest: The authors declare no conflicts of interest.

References

1. Betsill, M.M.; Bulkeley, H. Cities and the multilevel governance of global climate change. In *Understanding Global Cooperation: Twenty-Five Years of Research on Global Governance*; Mills, K., Stiles, K., Eds.; Brill: Leiden, The Netherlands, 2021; pp. 219–236. [CrossRef]
2. Hölscher, K.; Frantzeskaki, N. Perspectives on urban transformation research: Transformations in, of, and by cities. *Urban Transform.* **2021**, *3*, 2. [CrossRef]
3. Seto, K.C.; Churkina, G.; Hsu, A.; Keller, M.; Newman, P.W.; Qin, B.; Ramaswami, A. From low-to net-zero carbon cities: The next global agenda. *Annu. Rev. Environ. Resour.* **2021**, *46*, 377–415. [CrossRef]
4. Hamnett, C. The changing social structure of global cities: Professionalisation, proletarianisation or polarisation. *Urban Stud.* **2021**, *58*, 1050–1066. [CrossRef]
5. Vertovec, S. *Superdiversity: Migration and Social Complexity*; Taylor & Francis: London, UK, 2023. [CrossRef]
6. Döringer, S.; Uchiyama, Y.; Penker, M.; Kohsaka, R. A meta-analysis of shrinking cities in Europe and Japan. Towards an integrative research agenda. *Eur. Plan. Stud.* **2020**, *28*, 1693–1712. [CrossRef]
7. Lesthaeghe, R. The second demographic transition, 1986–2020: Sub-replacement fertility and rising cohabitation—A global update. *Genus* **2020**, *76*, 10. [CrossRef]
8. Wolff, M.; Wiechmann, T. Urban growth and decline: Europe's shrinking cities in a comparative perspective 1990–2010. *Eur. Urban Reg. Stud.* **2018**, *25*, 122–139. [CrossRef]
9. Datola, G. Implementing urban resilience in urban planning: A comprehensive framework for urban resilience evaluation. *Sustain. Cities Soc.* **2023**, *98*, 104821. [CrossRef]
10. Nel, D.; du Plessis, C.; Landman, K. Planning for dynamic cities: Introducing a framework to understand urban change from a complex adaptive systems approach. *Int. Plan. Stud.* **2018**, *23*, 250–263. [CrossRef]
11. Thomson, G.; Newman, P. Cities and the Anthropocene: Urban governance for the new era of regenerative cities. *Urban Stud.* **2020**, *57*, 1502–1519. [CrossRef]
12. Thomson, G.; Newman, P. Urban fabrics and urban metabolism–from sustainable to regenerative cities. *Resour. Conserv. Recycl.* **2018**, *132*, 218–229. [CrossRef]
13. DeLanda, M. *Assemblage Theory*; Edinburgh University Press: Edinburgh, UK, 2016.
14. Dovey, K.; Rao, F.; Pafka, E. Agglomeration and assemblage: Deterritorialising urban theory. *Urban Stud.* **2018**, *55*, 263–273. [CrossRef]

15. Durose, C.; van Ostaijen, M.; van Hulst, M.; Escobar, O.; Agger, A. Working the urban assemblage: A transnational study of transforming practices. *Urban Stud.* **2022**, *59*, 2129–2146. [CrossRef]
16. Lara-Hernandez, J.A.; Melis, A.; Lehmann, S. Temporary appropriation of public space as an emergence assemblage for the future urban landscape: The case of Mexico City. *Future Cities Environ.* **2019**, *5*, 1–22. [CrossRef]
17. Thwaites, K.; Simpson, J.; Simkins, I. Transitional edges: A conceptual framework for socio-spatial understanding of urban street edges. *Urban Des. Int.* **2020**, *25*, 295–309. [CrossRef]
18. Arai, H.; Ouchi, Y.; Toba, K.; Endo, T.; Shimokado, K.; Tsubota, K.; Matsuo, S.; Mori, H.; Yumura, W.; Yokode, M.; et al. Japan as the front-runner of super-aged societies: Perspectives from medicine and medical care in Japan. *Geriatr. Gerontol. Int.* **2015**, *15*, 673–687. [CrossRef]
19. Levasseur, M.; Généreux, M.; Bruneau, J.F.; Vanasse, A.; Chabot, É.; Beaulac, C.; Bédard, M.M. Importance of proximity to resources, social support, transportation and neighborhood security for mobility and social participation in older adults: Results from a scoping study. *BMC Public Health* **2015**, *15*, 503. [CrossRef]
20. Maresova, P.; Krejcar, O.; Maskuriy, R.; Bakar, N.A.A.; Selamat, A.; Truhlarova, Z.; Horak, J.; Joukl, M.; Vítkova, L. Challenges and opportunity in mobility among older adults–key determinant identification. *BMC Geriatr.* **2023**, *23*, 447. [CrossRef]
21. Chen, Y.; Huang, X.; White, M. A study on street walkability for older adults with different mobility abilities combining street view image recognition and deep learning- The case of Chengxianjie Community in Nanjing (China). *Comput. Environ. Urban Syst.* **2024**, *112*, 102151. [CrossRef]
22. Moran, M.R.; Werner, P.; Doron, I.; HaGani, N.; Benvenisti, Y.; King, A.C.; Winter, S.J.; Sheats, J.L.; Garber, R.; Motro, H.; et al. Exploring the objective and perceived environmental attributes of older adults' neighborhood walking routes: A mixed methods analysis. *J. Aging Phys. Act.* **2017**, *25*, 420–431. [CrossRef]
23. Yang, L.; Ao, Y.; Ke, J.; Lu, Y.; Liang, Y. To walk or not to walk? Examining non-linear effects of streetscape greenery on walking propensity of older adults. *J. Transp. Geogr.* **2021**, *94*, 103099. [CrossRef]
24. Yap, W.; Chang, J.H.; Biljecki, F. Incorporating networks in semantic understanding of streetscapes: Contextualising active mobility decisions. *Environ. Plan. B Urban Anal. City Sci.* **2023**, *50*, 1416–1437.
25. Diener, E.; Oishi, S.; Tay, L. Advances in subjective well-being research. *Nat. Hum. Behav.* **2018**, *2*, 253–260.
26. Diener, E.; Pressman, S.D.; Hunter, J.; Delgadillo-Chase, D. If, why, and when subjective well-being influences health, and future needed research. *Appl. Psychol. Health Well Being* **2017**, *9*, 133–167. [CrossRef]
27. Steptoe, A.; Deaton, A.; Stone, A.A. Subjective wellbeing, health, and ageing. *Lancet* **2015**, *385*, 640–648. [PubMed]
28. Smardon, R.C. Perception and aesthetics of the urban environment: Review of the role of vegetation. *Landsc. Urban Plan.* **1988**, *15*, 85–106. [CrossRef]
29. Triguero-Mas, M.; Dadvand, P.; Cirach, M.; Martínez, D.; Medina, A.; Mompart, A.; Basagaña, X.; Gražulevičienė, R.; Nieuwenhuijsen, M.J. Natural outdoor environments and mental and physical health: Relationships and mechanisms. *Environ. Int.* **2015**, *77*, 35–41.
30. Zhifeng, W.; Yin, R. The influence of greenspace characteristics and building configuration on depression in the elderly. *Build. Environ.* **2021**, *188*, 107477.
31. Kim, Y. Impacts of the perception of physical environments and the actual physical environments on self-rated health. *Int. J. Urban Sci.* **2016**, *20*, 73–87.
32. Pettorelli, N. *The Normalized Difference Vegetation Index*; Oxford University Press: Oxford, UK, 2013.
33. Izenberg, J.M.; Mujahid, M.S.; Yen, I.H. Health in changing neighborhoods: A study of the relationship between gentrification and self-rated health in the state of California. *Health Place* **2018**, *52*, 188–195.
34. Lamarca, G.A.; Leal, M.D.C.; Sheiham, A.; Vettore, M.V. The association of neighbourhood and individual social capital with consistent self-rated health: A longitudinal study in Brazilian pregnant and postpartum women. *BMC Pregnancy Childbirth* **2013**, *13*, 1. [CrossRef]
35. Soares, J.; Simões, E.J.; Ramos, L.R.; Pratt, M.; Brownson, R.C. Cross-sectional associations of health-related quality of life measures with selected factors: A population-based sample in Recife, Brazil. *J. Phys. Act. Health* **2010**, *7*, S229–S241. [CrossRef] [PubMed]
36. Statistics Bureau of Japan. Available online: https://www.stat.go.jp/english/data/kokusei/2015/final_en/final_en.html (accessed on 9 February 2025).
37. Statistics Bureau of Japan. Available online: https://www.e-stat.go.jp/stat-search/files?page=1&toukei=00200241&tstat=000001039591 (accessed on 17 March 2025).
38. OCN Real Estate. Available online: https://house.ocn.ne.jp/toshi/detail/13215/ (accessed on 17 March 2025).
39. Mapillary. Available online: https://www.mapillary.com/ (accessed on 9 February 2025).
40. Inoue, T.; Manabe, R.; Murayama, A.; Koizumi, H. Aggregation of Geospatial Data on "Street Units": The Smallest Geographical Unit of Urban Places. In *Urban Informatics and Future Cities*; Geertman, S.C.M., Pettit, C., Goodspeed, R., Staffans, A., Eds.; Springer: Cham, Switzerland, 2021; pp. 501–526.

41. Chen, L.C.; Zhu, Y.; Papandreou, G.; Schroff, F.; Adam, H. Encoder-decoder with atrous separable convolution for semantic image segmentation. In Proceedings of the European Conference on Computer Vision (ECCV), Munich, Germany, 8–14 September 2018.
42. Neuhold, G.; Ollmann, T.; Rota Bulo, S.; Kontschieder, P. The mapillary vistas dataset for semantic understanding of street scenes. In Proceedings of the IEEE International Conference on Computer Vision, Venice, Italy, 24–27 October 2017.
43. Kunitachi City. Available online: https://www.city.kunitachi.tokyo.jp/machi/keikaku/9355.html (accessed on 17 March 2025).

www.ingramcontent.com/pod-product-compliance
Lightning Source LLC
Chambersburg PA
CBHW040751250926
51850CB00007BB/1468
* 9 7 8 3 7 2 5 8 4 6 1 9 1 *

MDPI AG
Grosspeteranlage 5
4052 Basel
Switzerland
Tel.: +41 61 683 77 34

Land Editorial Office
E-mail: land@mdpi.com
www.mdpi.com/journal/land